THE MEDIEVAL ENGLISH SHERIFF

TO 1300

BY

WILLIAM ALFRED MORRIS, Ph.D.

Professor of English History in the University of California

MANCHESTER UNIVERSITY PRESS
BARNES & NOBLE INC., NEW YORK

© 1927 MANCHESTER UNIVERSITY PRESS
Published by the University of Manchester at
THE UNIVERSITY PRESS
316–324 Oxford Road, Manchester 13

U.S.A.
BARNES & NOBLE, INC.
105 Fifth Avenue, New York, N.Y. 10003

First published 1927
Reprinted 1968

G.B. SBN 7190 0342 3

Printed in Great Britain by Lowe & Brydone (Printers) Ltd.,
London

Mills, he is indebted for an insight into the exchequer procedure of the thirteenth century which he could not otherwise have attained. The results of the most diligent and careful work of Miss Mills have been placed at his disposal before their publication, in one instance while in proof. Through the kindness of Dr. Curtis H. Walker of Rice Institute, Houston, Texas, the writer was also permitted to see the proof of his article dating the terms of the sheriffs of Henry I.

Finally, obligation to the printed works of several scholars is so great as to deserve special acknowledgement. First among these are the various studies of Dr. J. Horace Round. Without their assistance and inspiration this work could hardly have been undertaken. Professor Sidney K. Mitchell's book on taxation in the thirteenth century presents useful material which has been of very decided assistance. The same is true of Miss Helen M. Cam's *Studies in the Hundred Rolls.* Moreover, the careful work of Professor H. W. C. Davis on the documents of the first two Norman reigns and that of Mr. William Farrer on those of Henry I., have afforded much chronological data for the earlier chapters of the book, and have performed a service of great value to the writer as to all students of the history of Norman England.

W. A. M.

BERKELEY, CALIFORNIA,
    31st *December* 1925.

The writer is grateful to Professor James Tait, Chairman of the Press Committee of Manchester University Press, for very helpful criticism while the work was in progress.

THE PRINCIPAL ABBREVIATIONS OF TITLES OF BOOKS, MANUSCRIPTS AND PERIODICALS USED IN THE FOOTNOTES OF THIS VOLUME

A.H.R. . . .	*The American Historical Review*, New York, etc., 1895, etc.
Abingdon Chron. .	*Chronicon Monasterii de Abingdon.* Ed. Joseph Stephenson. Rolls Series. London, 1858.
Adams, *English Constitution*	*The Origin of the English Constitution.* By George Burton Adams. New Haven and London, 1912.
Add. MS. . . .	Additional Manuscript, British Museum.
Ann. Monast. . .	*Annales Monastici.* Ed. H. R. Luard. London, 1864–1869.
A.-S. Chron. . .	*The Anglo-Saxon Chronicle.* Ed. Benjamin Thorpe. London, 1861.
Benedict . . .	*Gesta Regis Henrici Secundi Benedicti Abbatis.* Ed. William Stubbs. Rolls Series. London, 1867.
Bigelow, *Placita Anglo-Norm.*	*Placita Anglo-Normannica : Law Cases from William I. to Richard I. preserved in Historical Records.* Ed. M. M. Bigelow. Boston, 1879.
Birch, *Chart.* . .	*Cartularium Saxonicum : a Collection of Charters relating to Anglo-Saxon History.* Ed. W. de Gray Birch. London, 1885–1893.
Britton . . .	Britton : the French text carefully revised with an English translation. Ed. F. M. Nichols. Oxford, 1865.
Brit. Mus. . . .	British Museum.
Cam, *Studies* .	*Studies in the Hundred Rolls : Some Aspects of Thirteenth Century Administration.* By Helen M. Cam. *Oxford Studies in Social and Legal History.* Vol. vi. no. xi. Oxford, 1921.
Chadwick, *Studies* .	*Studies on Anglo-Saxon Institutions.* By H. M. Chadwick. Cambridge, 1905.
C. Ch. R. . . .	*Calendar of the Charter Rolls* [1226–1272]. Rolls Series. London, 1903, etc.
C. Cl. R. . . .	*Calendar of Close Rolls* [1272–1307]. Rolls Series. London, 1900–1908.
Cal. Inq. Misc. . .	*Calendar of Inquisitions Miscellaneous (Chancery).* Rolls Series. London, 1916.
Chronicles of Stephen .	*Chronicles of the Reigns of Stephen, Henry II. and Richard I.* Ed. Richard Howlett. Rolls Series. London, 1886–1889.
Chron. Mon. de Bello .	*Chronicon Monasterii de Bello.* [Ed. J. S. Brewer.] Anglia Christiana Society. London, 1846.

Chron. Ramsey . . Chronicon Abbatiae Ramsiensis. Ed. W. D. Macray. Rolls Series. London, 1886.

Cl. Rolls . . . Close Rolls of the Reign of Henry III. [1227–1251]. Rolls Series. London, 1902–1922.

C.P.R. . . . Calendar of the Patent Rolls [1232–1307]. Rolls Series. London, 1893–1913.

D.B. . . . Domesday Book seu Liber censualis Wilhelmi Primi Regis Angliae. [Ed. Abraham Farley. London, 1783.] Vols. iii. iv. [Ed. Henry Ellis.] Record Commission. [London], 1816.

Davis, England under Normans   England under the Normans and Angevins, 1066–1272. By H. W. C. Davis. London, [1905].

Davis, Regesta . . Regesta regum Anglo-Normannorum, 1066–1154. Ed. H. W. C. Davis. Oxford, 1913.

Dialogus . . . De necessariis observantiis scaccarii dialogus, commonly called Dialogus de scaccario. Ed. Arthur Hughes, C. G. Crump and C. Johnson. Oxford, 1902.

D.N.B. . . . Dictionary of National Biography. Ed. Leslie Stephen and Sidney Lee. London, 1885–1901.

Dignity of a Peer . Reports from the Lords Committees appointed to Search the Journals of the House, Rolls of Parliament and other Records for all Matters touching the Dignity of a Peer. London, 1820–1829.

E.H.R. . . . English Historical Review. London, etc., 1886, etc.

Exch. Misc. Bks. . Exchequer Miscellaneous Books, Public Record Office.

Farrer, Itinerary . An Outline Itinerary of Henry I. By William Farrer. London, 1920.

Florence of Worcester . Chronicon ex Chronicis by Florence of Worcester. Ed. Benjamin Thorpe. English Historical Society. London, 1848–1849.

Freeman, Norm. Conq. History of the Norman Conquest. By E. A. Freeman. Oxford, 1869–1879.

Gesta Regis . . See Benedict.

Gesta Stephani . . Gesta Stephani Regis Anglorum. Ed. Richard Howlett in Chronicles and Memorials of Stephen, Henry II. and Richard I., iii. 3-136. Rolls Series. London, 1886.

Glanville . . . Tractus de legibus et consuetudinibus regni Angliae. By Ranulph de Glanville. Ed. John Rayner. London, 1780.

Hist. Mon. S. Augustini Historia Monasterii S. Augustini Cantuariensis. By Thomas of Elmham. Ed. Charles Hardwick. Rolls Series. London, 1858.

Hist. Monast. Selebiensis   Historia Selebiensis Monasterii. In Coucher Book of Selby. Ed. J. F. Fowler, i. 1-54. Yorkshire Archaeological and Topographical Association. Record Series, vol. x. [Durham], 1893.

Inq. Misc. . . See Cal. Inq. Misc.

Kemble, Cod. Dipl. . Codex diplomaticus aevi Saxonici. By J. M. Kemble. English Historical Society. London, 1839.

L.T.R. Mem. Roll . Lord Treasurer's Remembrancer's Memoranda Roll. Manuscript in Public Record Office [1216–1307].

Lapsley, Durham . The County Palatine of Durham : a Study in Constitutional History. By G. T. Lapsley. New York, etc., 1900.

Leges Edw. Conf. . Leges Edwardi Confessoris. In Liebermann F. Gesetze der Angelsachsen.

Liber de Legibus . . De antiquis legibus liber : Cronica majorum et vicecomitum Londoniarium, 1188–1274. By Arnold fitz Thedmar. Ed. Thomas Stapleton. Camden Soc. London, 1846.

Liebermann, Gesetze . Die Gesetze der Angelsachsen, herausgegeben im Auftrage der Savigny-Stiftung. By Felix Liebermann. Halle, 1898–1916.

Liebermann, Nat. Assembly . The National Assembly in the Anglo-Saxon Period. By Felix Liebermann. Halle, 1913.

Lit. Claus . . . See Rot. Lit. Claus.

Lit. Pat. . . . See Rot. Pat.

Madox, Exchequer . The History and Antiquities of the Exchequer. By Thomas Madox. Second ed. London, 1769.

Maitland, Domesday Book . Domesday Book and Beyond : Three Essays in the Early History of England. By F. W. Maitland. Cambridge, 1897.

Maitland, Pleas for Gloucester . Pleas of the Crown for the County of Gloucester before the Justices Itinerant, 1221. Ed. F. W. Maitland. London, 1884.

Matthew Paris . . Chronica Majora. By Matthew Paris. Ed. H. R. Luard. Rolls Series. London, 1872–1883.

Mitchell . . . Studies in Taxation under John and Henry III. By S. K. Mitchell. Yale Historical Studies, vol. 2. New Haven, 1914.

Monasticon . . Monasticon Anglicanum. Ed. John Caley, Henry Ellis and Bulkeley Bandinel. London, 1817–1830.

Northumberland, A.R. . Three Early Assize Rolls of Northumberland. [Ed. William Page.] Surtees Society. London, etc., 1891.

Ordericus Vitalis . Historia ecclesiastica by Ordericus Vitalis. Ed. Auguste le Prevost. Societé de l'Histoire de France. Paris, 1838–1855.

Parow, Compotus . Compotus vicecomitis. By [Walter] Parow. Berlin, 1906.

P.R. . . . Pipe Roll.

Pat. Rolls . . . Patent Rolls of the Reign of Henry III. [1216–1232]. Rolls Series. London, 1901–1903.

Plac. de Quo War. . Placita de Quo Warranto, Edward I.-Edward III. [Ed. Wilham Illingworth.] Record Commission. [London], 1818.

Plac. Abbrev. . . Placitorum abbreviatio, Richard I.-Edward II. Record Commission. London, 1811.

Pollock and Maitland . The History of English Law before the Time of Edward I. By Frederick Pollock and F. W. Maitland. Second ed. Cambridge, 1898.

Poole, *Exchequer* . *The Exchequer in the Twelfth Century.* By R. L. Poole. Oxford, 1912.

P.R.O. . . . Public Record Office.

Ramsey *Chartul.* . *Cartularium Monasterii de Rameseia.* Ed. W. H. Hart and P. A. Lyons. Rolls Series. London, 1884–1893.

Ramsey *Chron.* . . *Chronicon Abbatiae Ramesiensis.* Ed. W. D. Macray. Rolls Series. London, 1886.

Red Book . . . *The Red Book of the Exchequer.* Ed. Hubert Hall. Rolls Series. London, 1896.

Rot. *Chart.* . . *Rotuli chartarum in Turri Londoniensi asservati.* Ed. Thomas Duffus Hardy. Record Commission. London, 1837.

Rot. *Curiae Regis* . *Rotuli Curiae Regis.* Ed. Francis Palgrave. Record Commission. [London], 1835.

Rot. *Hund.* . . *Rotuli hundredorum, temp. Henry III. and Edward I.* Record Commission. [London], 1812–1818.

Rot. *Lit. Claus.* . . *Rotuli litterarum clausarum* [1204–1227]. Ed. T. D. Hardy. Record Commission. [London], 1833–1834.

Rot. *Parl.* . . *Rotuli parliamentorum ; ut et petitiones et placita in parliamento, n.p., n.d.—Index,* 1832.

Rot. *Pat.* . . *Rotuli litterarum patentium,* 1201–1216. Ed. T. D. Hardy. Record Commission. [London], 1835.

R.S. . . . Rolls Series.

Rot. *Selecti* . . *Rotuli selecti ad res Anglicas et Hibernicas spectantes.* Ed. Joseph Hunter. Record Commission. [London], 1834.

Round, *Calendar* . *Calendar of Documents preserved in France illustrative of the History of Great Britain and Ireland.* Ed. J. H. Round. Rolls Series. London, 1899.

Round, *Commune* . *The Commune of London and other Studies.* By J. H. Round. Westminster, 1899.

Royal Letters . . *Royal and other Historical Letters illustrative of the Reign of Henry III., from the Originals in the Public Records Office.* Ed. W. W. Shirley. Rolls Series. London, 1862–1866.

Salt Arch. Soc. . . *Publications of the William Salt Archæological Society. Collections for a History of Staffordshire.* Birmingham, etc. [1888, etc.].

Sel. *Coroners' Rolls* . *Select Cases from the Coroners' Rolls,* A.D. *1265–1413, with a Brief Account of the History of the Office of Coroner.* Ed. Charles Gross. Selden Society. London, 1896.

Select Pleas . . *Select Pleas of the Crown.* Ed. F. W. Maitland. Selden Society. London, 1888.

State *Trials of Edward I.* *State Trials of the Reign of Edward I., 1289–1293.* Ed. T. F. Tout and Hilda Johnstone. Royal Historical Society [Camden Society, Third Series]. London, 1906.

Stats. *of Realm* . . *Statutes of the Realm* [1235–1713. Ed. A. Luders, T. E. Tomlins, J. Raithby and others]. Record Commission. [London], 1810–1828.

Stubbs, *Const. Hist.* . *The Constitutional History of England in its Origin and Development.* By William Stubbs. Vol. i., 6th ed., Oxford, 1903 ; vol. ii., 4th ed., Oxford, 1906.

ABBREVIATIONS XV

Stubbs, *Sel. Charters* . *Select Charters and other Illustrations of English Constitutional History.* Ed. William Stubbs. Eighth ed. Oxford, 1900.

Thorpe, *Diplomatarium* *Diplomatarium Anglicum aevi Saxonici.* Ed. Benjamin Thorpe. London, 1865.

Trans. R.H.S. . . *Transactions of the Royal Historical Society.* London, 1872, etc.

V.C.H. . . . *Victoria History of the Counties of England.* Ed. H. A. Doubleday, William Page and others. Westminster, etc., 1900, etc.

CONTENTS

PAGE

PREFACE vii

CHAPTER I

INTRODUCTION : THE KING'S REEVE 1

CHAPTER II

THE OFFICE OF SHERIFF IN THE ANGLO-SAXON PERIOD . . 17

CHAPTER III

THE BARONIAL SHRIEVALTY, 1066–1100 41

CHAPTER IV

THE PROBLEM OF CENTRAL CONTROL UNDER HENRY I. AND STEPHEN,
1100–1154 75

CHAPTER V

THE ADMINISTRATIVE SHRIEVALTY OF THE FIRST TWO ANGEVIN
REIGNS, 1154–1199 111

CHAPTER VI

THE OFFICE AT THE CULMINATION OF ANGEVIN ABSOLUTISM, 1199–1216   143

CHAPTER VII

THE APPOINTMENT, STAFF, AND JUDICIAL DUTIES OF THE SHERIFF,
1216–1307 167

B

CHAPTER VIII

PAGE

THE EXECUTIVE AND PEACE FUNCTIONS OF THE THIRTEENTH CENTURY
SHERIFF 207

CHAPTER IX

THE FISCAL FUNCTION 241

CHAPTER X

THE REWARDS AND ABUSES OF OFFICE ABOUT 1300 . . . 275

INDEX 287

I

INTRODUCTION : THE KING'S REEVE

THE office of sheriff is one of the most familiar and most useful to be found in the history of English institutions. With the single exception of kingship, no secular dignity now known to English-speaking people is older. The functions, status and powers of the office, like those of kingship itself, have undergone change, but for over nine centuries it has maintained a continuous existence and preserved its distinguishing features. Throughout this very long career the sheriff has performed many duties. He has always been the agent of monarchy, and his constitutional importance is well recognized. The faithfulness of his stewardship is attested by its duration. Until the fourteenth century the activities of his office largely recapitulate not only local government but also the royal administration of the county.

The title *scirgerefa* or shire-reeve, from which is derived that of sheriff, occurs first in a document which belongs to the early years of the eleventh century. The view held by Kemble and Stubbs,[1] that the office is of very early origin, has recently been assailed by Larson, Chadwick and Liebermann.[2] The shire to which the sheriff was attached, and from which he took his title, may be regarded as a primitive Anglo-Saxon institution only through acceptance of the mark theory, now practically abandoned as an explanation of English institutional origins. The theory that the shire is an early kingdom mediatized[3] is applicable in but comparatively few instances. Although the earliest shire divisions may rest upon a

---

[1] Kemble, *Saxons in England*, ii. 158 ; Stubbs, *Const. Hist.* i. 126 ; and preface to *Benedict of Peterborough*, ii. p. lix.

[2] Larson, *The King's Household in England before the Norman Conquest* (Bulletin, University of Wisconsin, History Series, i. no. 2), pp. 105-6 ; Chadwick, *Studies*, 229-31 ; Liebermann, *Gesetze*, ii. 480 ; also *The National Assembly in the Anglo-Saxon Period*, p. 36.

[3] Henry Adams in *Essays in Anglo-Saxon Law*, 20-21.

basis originally tribal, their appearance is under monarchial rather than democratic auspices. So far as may be seen they are not older kingdoms, but territorial divisions of one kingdom. There has survived no evidence of their existence in Northumbria and Mercia.[4] But the *subreguli*, mentioned in Wessex toward the end of the seventh century, have been identified with a high degree of probability [5] as aldermen who were heads of shires.

The West Saxon alderman of King Ine's time indeed bears a dignity befitting such a *subregulus*. His social rank is the same as that of the bishop.[6] Aldermen not only sit among the *witan* of the kingdom, but have the chief responsibility as judicial officials. The purpose assigned by Ine in the enactment of his code is " the establishment of just and kingly dooms which none of the aldermen hereafter shall pervert ". Their punishment for failure to hold criminals to account was different from that of minor officials and upon a different plane : they lost their shire.[7] Doubt has been expressed as to whether the word bears its familiar meaning in this passage.[7a] It may mean merely office or functions, although elsewhere in the same code *scir* is clearly a territorial designation.[8]

[4] As to Northumberland, see Larson, *The King's Household in England before the Norman Conquest*, 106 ; as to Mercia, Brownhill, ' The Tribal Hidage ", *E.H.R.* xxvii. 632. There are *subreguli* in the outlying regions subject to Mercia, but they rule earlier kingdoms such as the Hwicci, Kent, Sussex, Surrey and, it would also appear, Essex. Kemble, *Codex Diplomaticus*, nos. 113, 117, 999 ; Birch, *Chartularium Saxonicum*, nos. 132, 202, 328. The twenty-three shires over which Offa is said to have ruled (Wilkins, *Concilia Magnae Britanniae*, i. 156) are, of course, reckoned upon a basis of the shires established in Mercian territory by West Saxon kings subsequent to the reconquest of the Danelaw in the tenth century. " The ancient system of the civil government of the Mercians is as clean gone as are the satrapies of the Persians " (C. S. Taylor in *Proc. Bristol and Gloucestershire Archæol. Soc.*, 1898, p. 36).

[5] Chadwick, *Studies*, 282-90. Cf. Bede, *Hist. Eccl.* iv. 12 ; Larson (*King's Household*, 105) seems to give too little weight to the existence of under-kings in Wessex even earlier than the reign of Ine. Chadwick holds that these are members of the royal family. The *Victoria County History*, Somerset, ii. 179, follows the more traditional view.

[6] Cam, *Francia and England*, p. 45, n. 5.

[7] Ine, 36, 36. 1 ; Liebermann, *Gesetze*, i. 104-5. For similar offences lesser officials paid wergeld.

[7a] *Ibid.* ii. 480, art. " Grafschaft ", 3.

[8] Ine, 39. For the early use of the word *scir* in the sense of office or functions, see Larson, *King's Household*, 105. Cf. *groefscire* in the citation from the *Lindisfarne* gospels in Stubbs, *Const. Hist.* i. 122 ; the *tollscire* held by the Apostle Matthew in Elfric, *Homilies*, ii. 468 ; and the *scire* of the manorial reeve in *Rectitudines Singularum Personarum*, 4, 6, and *Gerefa*, 2 (Liebermann, *Gesetze*, i. 448, 453). An example of the use of the word in the sense of territory occurs in an Anglo-Saxon version of a grant of privileges by Pope Sergius in 701 : *Gesett ond araerd on Angel sexena scire* (Birch, *Chartularium Saxonicum*, no. 106). In

The *scirman*, mentioned here as a judge, Stubbs assumed to be a
sheriff,[9] because the sheriff was sometimes called a *scirman* three
centuries later. This is the only tangible evidence that has been
adduced to prove the existence of the sheriff earlier than the tenth
century. But since a *scirman* is one in charge of a *scir*, in Ine's
day, so it has been shown, he corresponds to the alderman ;[10]
hence the use of the word does not establish the sheriff's existence.
The alderman's headship of the shire, it may be added, is convincingly demonstrated somewhat later by the story told in the
Chronicle of civil commotion in the middle of the eighth century
and of the Mercian and Danish wars of the ninth. Here the shires
of the West Saxon kingdom bear the names which they still retain,[11]
and in time of war the men of each fight under the leadership of
their alderman.

The precursor of the sheriff is to be sought, then, not among the
early heads of shires but among the king's reeves, a class of officials
whose administrative rank is inferior to that of the alderman. It is
impossible to show definitely that any of these are associated with
shires until towards the end of the tenth century. The king's
gerefa or reeve of the West Saxon laws of Ine, as well as the Kentish
praefectus of the seventh century, appears to be a dignitary of the
royal household,[12] The earliest definite mention of public reeves
who are purely administrative officials occurs not far from the year
800. The *Chronicle* tells a well-known story of a *gerefa*, described by
Ethelward at a later time as an *exactor*,[13] who in the reign of Beortric
of Wessex lost his life because he mistook the character of the first
band of Danes he had ever seen and attempted to drive them into
the king's *tun*, probably to pay toll.[14] A Kentish charter of the
Mercian king Cenwulf in the time of Egbert mentions public *tributa*
and royal dues both in property and labour as rendered to *principes*,

---

the *Chronicle* for the year 894 Alfred is said to advance against the Danes with the
scire (division) *the mid him fierdedon*. The more familiar use of the word as a
territorial district occurs in Alfred's Orosius (E.E. Text Soc. 19) : *sio scir hatte
Halgoland.*

[9] Stubbs, *Const. Hist.* i. 126.
[10] Ine, 8. See Chadwick, *Studies*, 231.
[11] Cf. Cam, *Francia and England*, 41 ; Liebermann, ii. 480.
[12] Larson, *King's Household*, 104-13. Even the *wicgerefa* of London seems to
have had duties which he performed at the king's hall (Hlothaire and Eadric,
7, 16). Cf. Cam, *Francia and England*, 38. This may be true also of the *gerefa* in
Wihtred, 22, who corrects the king's *esne* for his misdeeds.
[13] *Monumenta Historiae Britanniae*, ed. Petrie and Sharpe, 509.
[14] *Anno* 787. See Plummer's edition, ii. 59.

duces or *procuratores*.[15]   The last named are elsewhere called
exactores.[16] They seem to be the officials who are mentioned in the
same period as the *subditi* or *juniores* of the alderman.[17] A Mercian
grant to Abingdon Abbey, confirmed in 835 by Egbert of Wessex,
forbids either *princeps* or *graphio* from interfering with the abbey in
the enjoyment of privileges which include exemption from the king's
feorm.[18] This *graphio* by function, position and etymology [19]
may be identified as a reeve, just as the *princeps* represents the
alderman.

When charters in the vernacular begin to appear in the ninth
century, the title *gerefa* [20] corresponds to that of *exactor* or *procurator*.
That this king's reeve was by no means a person of humble rank like
the ordinary *tungerefa* or *villicus* of Bede [21] has been well demonstrated.[22] In Kent, however, his activity centres about a royal
tun.[23] In the Mercia of the early ninth century labour dues were
rendered at royal *villulae*; here the king's *tun* was also the centre of
a criminal jurisdiction; [24] and individuals had to reckon in their
dealings over public dues and services with an *exactorum conflictio*.[25]
Furthermore the *gerefa* of Alfred's law, like the one mentioned in the

[15] To the Archbishop of Canterbury, 811, Birch, *Chart.* no. 355; Kemble, *Cod. Dipl.* no. 196.
[16] Egbert to the church of Rochester, 828, Birch, no. 395; Kemble, no. 223; also note 25 below. The title is used in charters of Offa which Kemble (nos. 161, 162) stars because of defects in chronology.
[17] *Pascua regis et principis vel subditorum eorum*: Cenwulf to the Bishop of Worcester, 816 (Birch, no. 357; Kemble, no. 210). Quittance *ab omnibus servitutibus et tributalibus rebus, magnis vel modicis regis vel principis aut juniorum illorum* occurs in a Mercian charter to Worcester in 855 (Birch, no. 487; Kemble, no. 277). Cf. Heming, *Chartulary*, ed. Hearne, i. 47, 63, 69, 70, 101.
[18] Birch, *Chart.* nos. 366, 413; Kemble, *Cod. Dipl.* nos. 214, 236.
[19] Cf. *greve* (*Leges Edw. Conf.* 32), *tungravius* (7 Ethelred, 2. 5) and *tungravio* (4 Ethelred, 3).
[20] *Naefre ne kyninge, ne aethelingc, ne biscop, ne ealdorman, ne thegen, ne gerefa*: Ethelwulf's doubtful confirmation of the privileges of Winchester Cathedral for land in Hampshire, 856 (Birch, no. 493; Kemble, no. 1057). A somewhat similar formula is used in Birch, no. 551; Kemble, no. 313.
[21] *Hist. Eccles.* in Petrie's *Monumenta*, 238. According to Ine, 63, the *gerefa* is an important personage in the household of the nobleman.
[22] Larson, *King's Household*, 111-12. Concerning the king's *praefectus, ibid.* 111, and Liebermann, *Nat. Assembly*, 36, n. 1.
[23] Chadwick, *Studies*, 249-53.
[24] Cenwulf to Bishop Denebert, Kemble, no. 210; Birch, no. 357. Labour dues at royal *villulae* are mentioned in Offa's confirmation of privileges to the Kentish churches in 892 (Birch, no. 848). A charter of Wiglaf of Mercia seems to show that a royal tun is a *burh* where wall work is performed: *liberabo a pastu regis et principum et ab omni constructione regalis ville* (Birch, no. 416).
[25] Birch, no. 351; Kemble, no. 206. In Kemble, no. 196, these Mercian officials are called *procuratores*.

time of his predecessor, Beortric, performs duties at the king's *tun*.[26]

It seems well established that several of these officials existed in a shire at the same time.[27] Ethelnoth, to whom Cuthred, sub-king of Kent, brother of the Mercian king Cenwulf, refers as his *praefectus fidelissimus in provincia Cantiae*, was clearly not a shire-reeve, for he is better known under his English title *gerefa* at Eastry.[28] The functions of the office assume clearer outline, and its importance apparently increases in the reigns of Alfred and Edward the Elder. Asser represents Alfred as exercising a close supervision over the work of his *gerefan*, so close that when they need reproof the king gives it by word of mouth.[29] The reeve is mentioned as president of a *gemot* and also as a steward who feeds the prisoners at the king's *tun*.[30] The death of the *gerefa* Lucumon in naval battle, recorded in the *Chronicle* for the year 897, suggests either that the king is appointing a special high reeve to serve as a military leader or that the king's ordinary reeve, like the later sheriff, is assisting the alderman in the discharge of his military duties.[31]

The *gerefa* now resembles the later sheriff in that he exercises judicial functions in a capacity subordinate to that of the alderman. The penalty for fighting before the alderman in a folkmote is four times that for fighting before his *gingra* (*junior*) or before a king's priest.[32] The *gerefa* (*praepositus*) is mentioned by Asser [33] as a lay official who presided in these assemblies. It is interesting to note in this relation that the translator of the Anglo-Saxon Bede regarded an alderman as the superior of a *tungerefa* and apparently the official above him.[34] What territorial unit the men assembled in these *gemots* represented is a difficult and crucial question. There seems to be no mention of a shiremote prior to this period other than the *tota scira*, the members of which,[35] according to a unique and dubious entry, witnessed a charter of King Offa. But it is likely that the alderman already presides over such an assembly. The *contiones*

---

[26] Alfred, 1. 2-3 ; also Chadwick, *Studies*, 257-8.
[27] For the Kentish evidence of the early ninth century see *ibid.* 251-2.
[28] Birch, *Cod. Dipl.* no. 318 ; Kemble, *Chart.* no. 191.
[29] Stevenson's ed., sects. 91, 106.
[30] Alfred, 1. 2, 1. 3 ; Liebermann, *Gesetze*, i. 48-9.
[31] Cf. Liebermann, *Nat. Assembly*, 36, 43.
[32] Alfred, 38, 38. 1, 38. 2.   [33] Stevenson's ed., sect. 106.
[34] . . . *tha eoden hio in sumes tungerefan giastern and hine baedon thaette hio onsende to thaem aldormen the ofer hine waes* (*Bede*, E.E. Text Soc., v. 10).
[35] Apparently in Kent or Sussex (Birch, *Chart.* no. 208).

comitum et praepositum found in Asser [36] are apparently different bodies, for reeves as well as aldermen hold them and pronounce judgement in them. According to the laws of Alfred certain routine matters came regularly before the *gerefa* in the folkmote.[37] These are entirely different from the known items of business which are said to come before the alderman,[38] presumably also in his *folkmote*.[39] The *gerefa* apparently was not presiding over a *shiremote*, for he and the alderman do not seem to be present at the same assembly. The evidence concerning the beginnings of the hundred as set forth by Dr. Liebermann [40] forbids the assumption that his was a *gemot* held for such an area. Only the conclusion of Mr. Chadwick [40a] seems reasonable. The *folkmote* held by the king's *gerefa* of Alfred's reign was for a subdivision of a shire laid out originally about a royal *tun*. The evident importance of these centres during the Danish wars,[41] and the mention by Asser of *burhs* and royal *tuns* reconstructed or newly built,[42] seem to account for the fact that a half-century later the *burh* appears as the important administrative centre.

This administrative unit of the earlier decades of the tenth century seems to bear a relation to the territorial divisions consisting of multiple hundreds of hides of land which are found in the so-called burghal hidage.[43] It has been supposed that a development and strengthening of this organization by burghal areas was due to the stress of the military situation. This apparently reduced the relative importance of the alderman in ordinary administration.[44]

---

[36] Asser, ed. Stevenson, sect. 106.

[37] Accusation of guilt is made before him (Alfred, 22) ; those who accompany merchants are led before him to be placed in suretyship (*ibid*. 34). Kemble (*Saxons in England*, ii. 169) simply assumes that this reeve is a sheriff.

[38] His witness is required before a man goes to seek a new lord (*ibid*. 38).

[39] The man who went to seek a new lord without the alderman's witness incurred a heavy *wite*, half of which was paid in the *scir* he left, half in the *scir* to which he went (*ibid*. 37, 37. 1). It may be held with reason that the word *scir* in this passage probably does not signify a shire in the later sense, but only an administrative district. But this is none the less the *scir* of an alderman. Cf. the aldermen of Kent, Essex, Hants, Wilts or Devon in *A.-S. Chron.*, years 897, 898, 911. The *boldgetael*, another name for this same administrative unit, Liebermann believes is the county (*Gesetze*, ii. 480, 1 b).

[40] Liebermann, *Gesetze*, ii. art. " hundred ".   [40a] *Studies*, 258.

[41] Cf. Asser, ed. Stevenson, sects. 27, 52.

[42] *Ibid*. sects. 55, 91. *Urbs* as well as *arx* is a ninth-century word for *burh*.

[43] Maitland, *Domesday Book*, 502-6 ; Chadwick, *Studies*, 204-8, 219-27. Cf. Oman, *England before the Norman Conquest*, 469-70.

[44] The view that the burghal districts came "in place of the counties" (Chadwick, *Studies*, 224-5) is unnecessarily radical. Cf. Cam, *Francia and England*, 51, 63.

From Edward the Elder to Canute the constant care of king and *witan*, as attested by their dooms, is not the judicial work of the alderman but that of the reeve.   In the laws of the former king the king's reeves appear as the important judicial officials.   They enforce folkright, they see that each plea has a term, they give judgement according to the testimony of the witnesses produced,[45] and they deal not alone with criminal [46] but also with civil cases, cases in which is involved right to both bookland and folkland.[47]   Each reeve is enjoined to have a *gemot* once in four weeks,[48] a regulation which tends to substantiate the evidence from Alfred's time that this is not a shiremote.[49]

The laws of Athelstan give indication that a process of reorganizing administrative divisions is under way.   In the earlier laws of the reign the *burh* is seen to hold the older place of the king's *tun* as a centre of administration.   The king directs the *gerefan* in each *burh* that they render tithes of the king's property, both of the young and of the fruits of the earth.[50]   The folkmote reeve is clearly distinguished from the portreeve,[51] yet those who are enjoined to ride and enforce ordinary court process and also to execute judgement on the thief are the most distinguished men who belong to the borough.[52]   It seems clear that administrative and judicial areas are laid out about boroughs.   The reeve is associated with a definite region, his *folgothe* [53] or *manung*, both of these names suggesting specific duties which he performed.[54]   The reeve's *manung* is, moreover, a district for which are named trustworthy men to be compurgators in judicial cases.[55]   A peculiar code of this reign, the so-called *Judicia Civitatis Lundonie*, enacted by the bishops and reeves belonging to the borough of London, some of whose acts are said to be in conformity with certain demands of the king and *witan* extend-

---

[45] 1 Edward, prol. ; 2 Edward, 2.   [46] See Alfred, 22.
[47] 1 Edward, 2.   [48] 2 Edward, 8.
[49] The convening of the shire of this period at frequent intervals seems hardly possible, as Chadwick points out (*Studies*, p. 235).   Moreover, *each gerefa* holds monthly sessions.
[50] 1 Athelstan, prol.
[51] 2 Athelstan, 12.
[52] 2 Athelstan, 20. 1-4.   Maitland (*Domesday Book*, 185) inclined to the view that this was a shire *gemot* held in a borough.
[53] 2 Athelstan, 25. 1.
[54] The *folgothe* denoted originally his following (Schmid, *Gesetze*, 578), the *manung* his warning or summons to them (Bosworth-Toller, *Dictionary*, art. " manung ").
[55] 5 Athelstan, 1. 5.

ing to the kingdom in general,[56] introduces complications. Not
only does it provide for the organization of police units consisting of
a hundred men [57] each, but it reveals for the first time a reeve's
district, which is called a *scir*. Within this he takes pledges for the
preservation of the peace ; [58] he must lead its men in pursuit of the
thief ; [59] and when called upon he must aid the *gerefa* in the next
scir.[60] The character of this district obviously determines whether
or not a shire-reeve or sheriff of the familiar sort has appeared.

Within some twenty years after the death of Athelstan the
organization and duties of the hundred as well as the sessions of
the shiremote twice a year are prescribed in the laws, although the
existence of the sheriff is not apparent until about the end of Edgar's
reign. It is evident that territorial organization did not everywhere
at the same time take final form.[61] Although with its development
appear reeves attached to hundreds, wapentakes and shires,[62] old
English law mentions only king's reeves without territorial designation.[63] The *tungerefa* [64] and *portgerefa* [65] are named, but they stand
apart from the others.[66] The high reeve of Ethelred, known through
the *Chronicle*, is a household rather than a territorial official.[67]
The attempt to differentiate the sheriff from the other folkmote
reeves is best deferred until an examination is made of the position
and functions of the king's reeve as described during the final

[56] 6 Athelstan, 10, 11, 12. 1, 12. 3.
[57] *Ibid.* 3.
[58] *Ibid.* 10.   [59] *Ibid.* 8. 2, 8. 4.
[60] *Ibid.* 8. 4. If he fail to do this, the two reeveships deal with the matter
in common. The penalty is the king's *oferhyrnesse*.
[61] As late as the reign of Ethelred there may have been but one alderman and
one reeve over the whole region of the five boroughs (3 Ethelred, 1. 1). Liebermann
(*Gesetze*, ii. 480, art. " Grafschaft ", 4 a) holds that the county boundaries were not
final about the year 1000. It has been shown that the creation of the Mercian
shires belongs to the period between 1000 and 1016 (C. S. Taylor, *Transactions
Bristol and Gloucestershire Archæol. Soc.*, 1898, pp. 32 ff.).
[62] Chadwick, *Studies*, 234, and p. 25, below.
[63] The reeve of the wapentake is not so called, but is distinguished through his
presence at its *gemot* (3 Ethelred, 3. 1).
[64] 3 Athelstan, 7. 1 ; 4 Edgar, 13, 13. 1 ; 7 Ethelred, 2. 5 ; 7 a Ethelred, 2. 3.
In 4 Ethelred, 3, he is an official who collects toll.
[65] 1 Edward, 1; Kemble, *Cod. Dipl.* no. 1289; 4 Ethelred, 7. 3; 2 Canute, 8-8. 2.
[66] Above, p. 4, and 4 Ethelred, 3. They are distinguished also by function.
As to the portreeve, see Haddan and Stubbs, *Councils*, i. 688-90, and 2 Canute,
8. 1, 8. 2.
[67] Concerning him as well as the earlier high reeve of Northumbria, who was
probably a territorial official, see Larson, *King's Household*, 113-16 ; Chadwick,
Studies, 231-2 ; Liebermann, *Gesetze*, ii. 498-9, 694, art. " sheriff ", 1 b. The writer
of the *Blicking Homilies* (171, 177) uses the title to describe a Jewish ruler, or
member of the sanhedrin.

century of Anglo-Saxon legislation, extending to and including the reign of Canute.

The responsibility of the king's *gerefa* to the king is clear. According to Asser, King Alfred tells his aldermen and reeves who deliver unsatisfactory judgements either to devote themselves to study or surrender their offices.[68] In the tenth century the penalty for neglecting the directions given in the dooms was the king's *oferhyrnesse*, or contempt fine, of a hundred and twenty shillings.[69] According to the law of Athelstan the *gerefa* who will not enforce the prohibition against trafficking on Sunday incurs the same penalty as the offender, and the king will find another in his place who is willing to enforce the enactment.[70] For failure to guard the peace as the king and *witan* required, as well as for neglect of duty in other respects, the *gerefa* lost his *folgothe* and the friendship of the king, making amends to the latter with a hundred and twenty shillings.[71] Canute, likewise, commanded his *gerefan* that they faithfully do their duty as they valued the king's friendship.[72] By these strong kings *reeves* were evidently both appointed and removed.

The king's reeve, at least when acting in a judicial capacity, had at his disposal minor officials to carry out the directions of the court. An under-gerefa who acted as the *gingra* of the judge was a well-known figure about the year 1000.[72a] Elsewhere it appears who this officer was. Dr. Liebermann shows from the Anglo-Saxon Scriptures that the judge's officer who imprisons the condemned is a beadle, and finds that he is also called *oefgroefa*.[73] This subordination of beadle to *dema* appears more than once in Elfric's homilies.[74]

Highly important duties in regard to police and criminals in Athelstan's day appertained to the office of king's reeve. According to the London *frithgild* laws, he was not only the leader of the men of the *scir* in the pursuit of thieves but was also an authority on police questions. Any of the reeves of the district, like the king himself, had the right to suggest additions to these *frithgild* regulations.[75] The reeve, moreover, as well as the ealdorman was a

[68] Asser, ed. Stevenson, sect. 106.   [69] 5 Athelstan, 1. 2 ; Edward, 2.

[70] 2 Athelstan, 25.   [71] 6 Athelstan, 11.

[72] Canute's letter of 1027, sect. 12, in Liebermann, *Gesetze*, i. 277.

[72a] Elfric, *Lives of the Saints*, E.E. Text. Soc., pp. 108, 182.

[73] Liebermann, *Gesetze*, ii. 722 (19).

[74] " As the day-star goes before the sun, as the *bydel* before the *deman* " (*Homilies*, ed. Thorpe, i. 354). Again, "men in orders are God's messengers (*bydelas*), and who shall proclaim the Judge to come if the beadle is silent ? "

[75] 6 Athelstan, 8.

guardian of the peace. The frithgild, when need arose, sent to the
gerefan on both sides for the aid of the men who were necessary to
vindicate justice against a *maegth*, or kindred group, and kill the
thief defended by it.[76] It was the duty of these officials promptly to
execute justice, even on the thief who took asylum.[77] He who had
been convicted of theft at the ordeal and redeemed by his lord or his
kindred, and who afterwards stole, became again justiciable to the
gerefa and was to be killed.[78] One of the dooms of Ethelred shows
the alderman or the king's reeve proclaiming the *grith* in the *thing* or
gemot of the five boroughs.[79] The fact that the king's reeve placed
men under official pledge for the observance of the peace,[80] designates him as the one who maintained the remarkable *borh* or suretyship system [81] of the Anglo-Saxon period. In the latter dooms
especial directions are given for bringing under suretyship persons
of bad reputation, *tyhtbysig* men, a class who from the time of
Edmund were all required to be included.[82] The king's *gerefa* was
to go and bring under *borh* the *tyhtbysig* man ; if no *borh* held him
he was to be slain and buried with the damned.[83] Neglect in
enforcing requirements regarding *borh*,[84] or failure to do justice on
the thief,[85] no less than neglect of duty in other directions, made the
reeve liable to the heavy penalty of a hundred and twenty shillings.
Moreover, some of the Anglo-Saxon laws provide penalties to be laid
upon *gerefan* for committing actual crime. For consenting to a
thief they were held similarly guilty ; [86] for committing robbery they
made amends by a sum double what others paid ; [87] for permitting
falsification of the coinage they suffered the loss of a hand like the
ordinary counterfeiter.[88] Their reputation for honesty and efficiency
was obviously not always above reproach.

Witnessing officially to various matters constituted another of
the *reeve's* duties. In the reign of Athelstan he was a necessary
witness to transactions who vouched for all purchases not made in
port, that is to say in market. No cattle were to be bought without
the witness of the *gerefa*, priest, *hordere* or some other reliable man,

[76] 6 Athelstan, 8. 2-3.
[78] 6 Athelstan, 1, 4.
[80] Above, p. 8.
[81] See the writer's *Frankpledge System*, 15-19.
[82] 3 Edmund, 7. 1, 7. 2.
[83] 1 Ethelred, 4, 4. 1 ; 2 Canute, 33, 33. 1.
[84] 3 Edmund, 7. 2.
[86] 2 Athelstan, 3. 2.
[88] 2 Canute, 8. 1, 8. 2.

[77] 4 Athelstan, 6, 6. 1, 6. 2, 7.
[79] 3 Ethelred, 1. 1.

[85] 4 Athelstan, 6. 1, 6. 2, 7.
[87] 7 Ethelred, 6. 3.

and no buying above twenty pence done out of port. Bargaining
was to be done either on the testimony of the portreeve or of some
other credible man, or on that of the *gerefan* in the folkmote.[89] By the
reign of Edgar these officers seem to be relieved of this function.[90]
Other duties grew out of their regular attendance at the *folkmote*.
Thus in Alfred's time persons travelling with merchants had to
appear before the *reeve* in this assembly.[91] The significance of the
reeve's witnessing transactions lies of course in his ability to vouch
in the folkmote for or against the lawful title to goods found in any
man's possession. A similar idea may be traced in the appearance
of the *gerefa* as a witness to manumissions.[92] In Ethelred's time it
was the rule that no one should pay the composition for an offence
except on witness of the king's reeve.[93] At about the same period
the king's *gerefa* appears in a familiar rôle as official witness of
proceedings in the shiremote.[94]

The manner in which the reeve performed his judicial duties
in the *folkmote* evidently caused great solicitude to the old English
legislator of the tenth and the earlier part of the eleventh century.
For perverting judgement through the acceptance of a bribe the
reeve paid the king's *oferhyrnesse* and forfeited the royal favour.[95]
Both Edgar and Canute imposed a monetary penalty of the same
amount upon the *dema* who gave false judgement, unless he could
show that he knew no better,[96] and the enactment of the last-named
king specifies hatred or bribery as motives of this offence. In a
special charge to the *gerefan* of the realm the same ruler forbids them
to do unjust violence either to rich or poor, and bids them preserve
to both classes the right to have just law,[97] deviating from this rule
neither on account of the royal favour nor that of any powerful
person. Supervision over the judicial work of the *reeve* between the
time of Athelstan and that of Canute is known to have fallen to the
bishop,[98] whose duty it was to collect the payment imposed upon
this official when he broke the laws.[99] This arrangement is first
mentioned in Athelstan's dooms. Here the bishop takes charge of

[89] 2 Athelstan, 10, 12.   [90] Below, p. 20.   [91] Alfred, 34.
[92] Birch, *Chart.* 639. Cf. Haddan and Stubbs, *Councils*, i. 688-90.
[93] 1 Ethelred, 1. 14.
[94] Kemble, *Cod. Dipl.* no. 693, *anno* 990-94.
[95] 5 Athelstan, 1. 3.
[96] 3 Edgar, 3 ; 2 Canute, 15. 1.
[97] Canute's letter of 1027, sect. 12, in Liebermann, i. 277.
[98] Zinkeisen in *Pol. Science Quart.* x. 140.
[99] 3 Edgar, 3.

the *bote*, or fine, levied upon the *gerefa* who neglects to enforce Sabbath observance.[100]   According to a homilist of Canute's time, the bishop shall in accusations direct proceedings so that no man wrong another, either in oath or ordeal, and shall consent to no injustice nor wrong measure nor false weight.[101]   In the *gemot* he is to remember his sacred calling and promulgate divine lore.[102] Canute, moreover, commands all his reeves that they everywhere hold his people rightly and deem right dooms by the shire bishop's witness.   Clearer still is the evidence of the prelate's oversight in the precaution taken to check the judicial severity of the reeves. They are to show such mercy in their judgements as the shire bishop thinks right.[103]

The possibility that the president of a *folkmote* should work so much injustice has occasioned comment.[104]   In the time of Alfred members of the assemblies often denounced the judgement declared by the alderman or the reeve.[105]   Even if he fairly stated it, the *gerefa* was in a position powerfully to influence this decision.   He was official witness concerning various matters.   Moreover, the all-important question which determined criminal judgement in a tribunal of this sort was apparently that of the previous character of the defendant.[105a]   To place *tyhtbysig* men in *borh* the *gerefa* of necessity had information on this point.   Athelstan's law commanded that in each reeve's *manung* law-worthy men be named who were to be *in testimonio* in each case, without selection of the parties.[106]   A familiar law of Ethelred's reign shows a direct attempt to control the reeve in a similar matter, and incidentally reveals him as the leading figure at the *gemot* of the *wapentake*.   The twelve senior thegns are to go out, the reeve with them, and swear on the relic that they will accuse no innocent man.[107]   Whether this is special procedure to determine if an accused person be *tyhtbysig* or merely community accusation to give common report, the important place which the reeve fills is evident.   His large powers of initiative seem to justify the demand made by the homilists of the later tenth

[100] 2 Athelstan, 25. 1.
[101] Institutes of Polity, 7, in Thorpe, *Ancient Laws*, 426.
[102] *Ibid.* chap. 8, p. 427.
[103] Canute's charter of 1020, sect. 11, in Liebermann, *Gesetze*, i. 274.
[104] Oman, *England before the Norman Conquest*, 473-4.
[105] Asser, ed. Stevenson, sect. 106.
[105a] This decided whether he escaped the ordeal or not, and whether he went to the single or triple ordeal.
[106] 5 Athelstan, 1. 5.   [107] 3 Ethelred, 3. 1.

century that the *dema* not only refrain from positive injustice and wrong but that he also severely punish doers of wickedness.[108]

The reeve's opportunities for bribe-taking were manifold, for he was also a fiscal official. The laws of Ethelred mention the *praepositi* or reeves as toll-collectors.[109] In Elfric's homilies the publicans of the New Testament are called *gerefan*.[110] Priests and deacons are to undertake no mongering nor reeveship because they are chosen from worldly men to God's labours.[111] This prohibition is also found in a late and spurious version [112] of Egbert's Penitential,[113] where a loophole is left through the proviso that churchmen may be especially nominated, evidently by ecclesiastical authority, for such secular service. The rule is of especial interest, since the first sheriff whose name is known was a priest.[114]

Closely connected with the collection of dues was the enforcement of services and other royal rights. Elfric represents Shadrach, Meshach and Abed-nego as Nebuchadnezzar's work-reeves.[115] King Ethelred bids all his *gerefan* spare the abbots in every secular necessity as best they may.[116] The king's *feorm*, or food rent, was in existence as early as the days of Mercian supremacy,[117] but a statement of the value of a two-nights' *feorm* in the time of Athelstan has been preserved through the direction given by that king to the reeves that they distribute its amount in alms.[118] Their payment of tithes and feeding of the prisoners at the king's *tun* illustrate the same class of duties. Canute told his reeves to collect only what was rightly the king's and serve him with this. The provision which is added, that no one be compelled to give any *feorm fultum* except

---

[108] *Blicking Homilies*, p. 63. Cf. Institutes of Polity, 11, in Thorpe, *Ancient Laws*, 429.

[109] 4 Ethelred, 3.

[110] Elfric, *Homilies*, Thorpe's ed. i. 339.

[111] *Ibid.* ii. 94-5. The inhibition upon men in sacred orders from giving judgement involving the shedding of blood is the reason assigned for this rule by a twelfth-century pope (*Benedict of Peterborough*, i. 85).

[112] Haddan and Stubbs, *Councils*, iii. 413-16.

[113] iii. 8, in Thorpe, *Ancient Laws*, 374.

[114] Birch, *Chart.* no. 1097; Kemble, *Cod. Dipl.* no. 1288,—earlier than 988. Authority is against Thorpe's rendering of the passage in *Diplomatarium*, 273. See Stubbs, *Const. Hist.* i. 126, n. Kemble suggests that this rule may apply to private rather than public business (*Saxons in England*, ii. 168, n. 1).

[115] Elfric, *Homilies*, Thorpe's ed. ii. 68.

[116] 8 Ethelred, 32.

[117] Above, note 17.

[118] Athelstan, *Alms*, prologue, Liebermann, i. 148-9. In Alfred's laws occurs mention of *mansiones* of churches belonging to the king's *feorm* (Alfred, 2).

voluntarily,[119] appears to be the first reference to purveyance. The *gerefa* not only attended to the collection of court compositions,[120] but also took possession of felons' chattels. In the reign of Edward the Elder the reeve Eanulf Penearding seized the property of a thief adjudged to the king, apparently on the ground that the offender was a king's man.[121] When in Edgar's reign the king received land which had been fraudulently withheld from him, the *gerefa* Wulfstan made the necessary arrangements to take it into the king's hands.[122]

In the exercise of both the reeve's judicial and his fiscal functions avaricious motives were often attributed to him. His likeness to the publican did not end with his title and his toll-collecting. Common reproach is echoed by the ecclesiastical regulation contained in Elfric's Pastoral Letter : " no priest may rightfully be a rapacious monger, nor a public spoiler in reeveship ".[123] The politic Canute, sending to his English subjects a pleasing message from Rome, included a command to the reeves of the realm. As they valued his friendship and their own safety they were to inflict injury upon none in collecting *pecunia*, inasmuch as there was no need for collecting for the king by unjust exaction.[124]

The duties of the king's reeve were increased and his powers strengthened through the overlapping of the functions of church and state. Through his dispensing the king's alms under the bishop's oversight comes the clearest glimpse of the reeve as disbursing officer. The exercise of mercy in his judgements as well as the prevention of buying and selling on Sunday are under the same supervision. As the guardian of morality and religion the church showed a deep interest in the question whether judges should punish severely not only thieves but manswearers, adulterers and those who practised divination. " To correct God's people ", so the homilist held, was one of their most important duties.[125] Moreover the direction of ordeals and oaths, part of the ordinary criminal procedure, belonged to the church and were subject to its regulation.[126] But the king, on the other hand, employed his reeves to aid the church. The "law of God as well as the secular law" was enforced in the shire court.[127] Wulfstan says that the *scirbiscop* and *scirwitan* judge

[119] 2 Canute, 69. 1.   [120] Ine, 73 ; 1 Ethelred, 1. 14.
[121] Kemble, *Cod. Dipl.* no. 328 ; Birch, *Chart.* no. 591.
[122] *Ibid.* no. 1258.   [123] Thorpe, *Ancient Laws*, 462.
[124] Canute's letter of 1027, sect. 12, in Liebermann, i. 277.
[125] *Blicking Homilies*, p. 63.   [126] Wulfstan, *Homilies*, 117.
[127] 3 Edgar, 5. 2.

those who break fasts or withhold rightful alms.[128] In Athelstan's reign the reeves aided the bishops to collect the tithe on the decollation of St. John the Baptist.[129] The same king directed his reeves to see that church-scot and soul-scot [130] no less than alms and tithes [131] were duly rendered. The laws of Edgar, Ethelred and Canute provide that unless this is done the king's *gerefa* shall aid in levying execution, co-operating with the priest and the bishop's and the lord's reeve to enforce collection [132] both of the tithe itself and the penalty incurred by its non-payment.

The homilies of the later tenth and the earlier eleventh century throw some interesting light on reeves just at the time when it is becoming possible in a few instances to distinguish the sheriff from the other *gerefan* of the king. They describe a *gerefa* who is one of the most important of public functionaries. He represents the possibility of much that is extortionate and unscrupulous, and is withal a person to be dreaded. His judicial iniquities and severities call for the condemnation of the pulpit. The homilist of Edgar's time denounces evil *gerefan* who are unjust judges, perverters of right judgements, and reminds his hearers that Judas is now in eternal torment because he sold Christ for a bribe.[133] Elfric condemns the unrighteous *deman* of his day who are blinded by bribes, who offer justice for sale, and who sell themselves for money.[134] The *Iudex* [135] complains that it is the custom of evil *gerefan* to take whatever they can and leave scarcely anything to the poor. Wulfstan inveighs against the perverse judgements of *reeves* and other officials, but is still more inclined to denounce their pride and haughtiness. He exhorts earls and *heretogan*, judges and *gerefan*, that they become poor in spirit, that they diligently abandon injustice and love justice, that they never neglect wisdom through unjust judgement for reward nor for friendship so that wrong give

---

[128] Wulfstan, *Homilies* (Napier), 173-4.
[129] Athelstan, 1.
[130] 1 Athelstan, 4, Liebermann, i. 146-7.
[131] 1 Athelstan, prol. Reeves were to pay alms of their own as well as on the king's property.
[132] 2 Edgar, 3. 1 ; 8 Ethelred, 7-8 ; 1 Canute, 8. 2.
[133] *Blicking Homilies*, 61, 63. The writer uses especially strong language against judges. He contends for the necessity of mercy in their judging the poor (*ibid.* 95). They have the names of judges and the deeds of thieves, for they are ravenous wolves when for the sake of bribes they condemn the innocent poor. Yet some are much readier to correct God's people than to rob the poor (*ibid.* 63).
[134] Elfric, *Lives of the Saints*, E.E. Text Soc., i. 430.
[135] *Iudex*, 13, Liebermann, *Gesetze*, i. 476.

c

way to right, nor decree unjust judgement to the misery of the poor.[136] There is coming a time when there shall be an end of the kingdoms of kings and the injustice, plundering, subtlety, perverse judgements and trickery of reeves.[137] In another passage reeves are among the rulers who stand as fleeting types of worldly glory. " Where are the aldermen that were proud and the reeves (*rican gerefan*) that gave law (*laga*) and commandment ? Where are the judgement-seats of the judges ? Where is their ambition and wantonness and pride but covered in earth and turned to misery ? " [138] If the homilies are severe in their estimates of these officials, so are the royal dooms of the same period.

Functionaries who are entrusted with the detail of the king's administration, who hold folkmotes, enforce law and maintain order, administer royal estates, collect tolls and dues, apprehend and punish criminals and possibly perform military duties, are rightly classed among the rulers of the realm. Their title is a symbol of earthly power. They are not manorial bailiffs of humble rank. These king's *gerefan* presumably include officials with the functions of sheriff, though they include also administrators of subdivisions of the shire. But for a very few passages the dooms would give the impression that even in the days of Ethelred and Canute there is only one grade among the folkmote *gerefan*. The type portrait which has been drawn with the aid of the laws will be misleading if all that is said of king's reeves is applied to any one of them or at any single time. But it is not difficult to trace the broad outlines of a very important variety of administrative reeveship and to recognize in it a public office indispensable to the state which for two centuries preserves in law and usage its chief characteristics.

[136] Wulfstan, *Homilies*, ed. Napier, 267-8. Cf. Institutes of Polity, 11, in Thorpe, *Ancient Laws*, 429.

[137] *Ibid.* 245.

[138] *Ibid.* 148.

II

THE OFFICE OF SHERIFF IN THE ANGLO-SAXON PERIOD

THE administrative and constitutional importance of the early shrievalty are incontestable, and it throws light on the question whether the elements which combined to form the English state of the twelfth century were in origin native or Norman. Yet no systematic attempt has hitherto been made to trace this development. Kemble's account, which is based mainly on the charters, is brief and contains some untested surmise. Stubbs made use of other materials as well, but the needs of his work called only for an outline of the subject, and this he borrowed largely from Kemble. The great work of Dr. Liebermann necessarily looks first to the laws, a class of sources which for this particular period contains suggestion rather than definite information. Although this has deterred Liebermann from making a critical study of the office of sheriff in the Anglo-Saxon period, it has not prevented him from adopting the newer views of its origin. The older conception of an elective sheriff ruling over a primitive community [1] is sustained by nothing more substantial than general ideas dependent upon the mark theory. The title of the *scirman* of Ine's time, upon which Stubbs further based the antiquity of the sheriff, may not to-day be regarded as proof.[2] Furthermore, the fact that, so far as record goes, the more important type of king's reeve until about the middle of the tenth century occurs in association with a king's *tun* or a burghal district rather than a shire destroys confidence in the familiar theory.[3] The fact calls for some modification of older views concerning the sheriff's functions. It also requires that the bearing

---

[1] Above, p. 1.   [2] Above, p. 3.
[3] Above, p. 4. "The office of sheriff did not branch off from the rest of the reeves until the tenth century" (Liebermann, *Nat. Assembly*, p. 36).

of the evidence be critically examined. The king's reeve of the
earlier laws may be a prototype of the sheriff, but not the sheriff
himself. " Shire-reeve " must not be understood, as has so often
been the case, where " king's reeve " is mentioned, and the laws
speak only of the latter. The laws, therefore, are of less value for
the present purpose than the charters and the scanty monastic
records, annals and miscellaneous works bearing on the subject ;
and most detailed of all are the hitherto somewhat neglected Domesday entries concerning the sheriffs of Edward the Confessor. These
facts seem to justify a fresh study of a familiar subject.

The starting-point of this investigation is the enactment of the
reign of Athelstan, known as the *Iudicia Civitatis Londonie*, and
made in what appears to be a local *witan*. This designates as a
scir the district over which the *gerefa* is placed. The name of the
territory and some of the functions of its officials are older than this
document, for in part it repeats more general legislation of the same
reign.[4] For this district the reeve has apparently been holding a
folkmote ;[5] within it he takes pledges for the observance of the
peace,[6] and in the London region its men are to be led by him in
pursuit of the thief.[7] Some have hesitated to identify this shire
with the county,[8] and the fact that the word *scir* was used in several
senses [9] justifies caution. But Professor H. W. C. Davis has clearly
shown that the body responsible for the London enactment contained
dignitaries from more than one county.[10] We may be dealing
with a shire in the usual sense of the term, and the reeve of this
scir in any event may have been called a shire-reeve or sheriff.[11]

When the king's reeve began to administer a shire instead of
some other territory and thus, from the later point of view, became
a sheriff, it would not appear that he assumed an essentially new
character. This probably explains why the laws never distinguish

[4] 6 Athelstan, 10.
[5] 5 Athelstan, 1. 5.
[6] 6 Athelstan, 10.
[7] 6 Athelstan, 8. 4.
[8] Chadwick, *Studies*, p. 231, n. 2 ; Liebermann, *Gesetze*, ii. 649, art. "Sheriff",
9 a, citing 6 Athelstan, 10.
[9] Land or territory, Birch, *Chartularium Saxonicum*, no. 106 ; a division of a
county in Cornwall and in the north of England, Stubbs, *Const. Hist.* i. 111-12 ;
the charge of the manorial reeve, *Rectitudines Singularum Personarum*, 4, 6, and
Gerefa, 2 (Liebermann, *Gesetze*, i. 448, 453).
[10] *E.H.R.* xxviii. 429.
[11] Dr. Liebermann (*Gesetze*, i. 179 ; cf. "Sheriffsämter" in 6 Athelstan, 8. 4)
identifies the official as a sheriff and his territory as a "grafschaft", though in his
second volume he appears doubtful as to whether the shire in question is a county.
See above, note 8.

between the *scirgerefa* and any other king's *gerefa*, and why the former title is not to be found in any extant record earlier than the reign of Canute. A supposed charter of Edgar which mentions the shire-reeve dates in its present form only from about the twelfth century.[12] In the absence of direct evidence it is therefore a difficult matter to determine when the king's reeve of the laws first became placed regularly over a shire. But there are some indications which serve roughly to mark the period. One may agree with Mr. Chadwick[13] that the king's reeve in the time of Edward the Elder could hardly have presided over a shire. The facts brought forward to prove the point certainly corroborate the earlier evidence for Alfred's reign. It should be added, moreover, that, while in the reign of Alfred the centre of financial administration mentioned is the king's *tun* and in the reign of Athelstan the burghal district, the law of Edgar speaks of the rights of his kingship, which his father had also possessed, both in borough and in shire.[14] This implies that in the reign of Edgar, if not in that of Edmund, the shire was a fiscal district and that a reeve was probably attached to it. The earliest recorded case of the holding of a shiremote by an official other than an alderman, a shireman who must be regarded as a reeve,[15] seems to occur in a document dating between the years 964 and 988.[16] It

[12] *A.-S. Chron.*, ed. Plummer, i. 116 ; ii. 156 ; Birch, *Chart.* no. 1281. The supposed charter of Wulfhere of Mercia (Birch, *Chart.* no. 22 ; Kemble, *Cod. Dipl.* no. 984) which mentions sheriff's aid is a rank forgery not older than the latter part of the twelfth century. The Anglo-Saxon account of the Council of Bapchild (Birch, *Chart.* no. 94) dates from the eleventh century. (See note 45.) The version of Offa's charter in Heming, *Chartulary*, i. 96, is apparently that of the eleventh century. The supposed charter of Wiglaf of Mercia to Croyland Abbey (Kemble, *Cod. Dipl.* no. 233 ; Birch, *Chart.* no. 409) names sheriffs who held office in the Norman period. The foundation charter of Ramsey Abbey (Kemble, no. 581 ; Birch, nos. 1310, 1311) and the charter of Edred to Croyland Abbey (Kemble, no. 420 ; Birch, no. 872) are further examples of spurious documents which assume the existence of the shrievalty before the year 975.

[13] *Studies*, p. 235. In the laws of Edward's reign the king's reeves appear as his most important judicial officials. They enforce folkright, they see that each plea has a term, they give judgement according to the testimony of the witnesses produced (see 1 Edward, prol. ; also 2 Edward, 2), and they deal not alone with criminal (see Alfred, 22) but also with civil cases, cases in which is involved right to both bookland and folkland (1 Edward, 2). But each reeve is enjoined to have a *gemot* once in four weeks (2 Edward, 8), a term altogether too short for the regular meeting of the men of the shire at this date.

[14] 4 Edgar, 2 a.   [15] Below, pp. 22-3.

[16] Kemble, *Cod. Dipl.* no. 1288. The words *coram scyre hominibus vel aliis iudicibus*, evidently intended by the compiler of the appendix to Alfred and Guthrum's *Peace* (chap. 1 ; Liebermann, *Gesetze*, i. 394) as a rendering of Ine, 8, show unfamiliarity with the title *scirman* a little earlier than 960. Dr. Liebermann suggests 940–56 as the date of the document.

may therefore be inferred that the office of sheriff originated in the half-century between the enactment of the laws of Edward the Elder and the death of Edgar.   This is the period during which in the south of England administration by burghal areas gives way in legal enactment to that by shires and hundreds.   The *gerefa* of Athelstan's *scir* seems actually to mark the transition from the reeve of the burghal district to the sheriff.

The appearance of the office of sheriff seems to coincide with the rise of the police and judicial powers of the hundred.   Leaving aside the question of the fiscal significance of the old usage which estimated territories in hundreds of hides of land, it is clear that the break-up of units of police and judicial administration intermediate in size between the hundred and the shire would tend to throw a heavy burden upon the sheriff.   The same enactment which first mentions a *gerefa* of a *scir* shows a number of these officials with the bishops about London ordering the formation of groups of a hundred men each for the exercise of police functions.[17]   Had the hundred as organized according to the ordinance ascribed to Edgar been already in working order, it would be difficult to discern the need for such legislation.   The hundred of Edgar and Canute, moreover, discharged duties, such as the pursuit and trial of thieves,[18] which belonged to the *scir* of Athelstan.   Again, the taking of *wedds* or pledges for the observance of the peace was in Athelstan's time the work of the reeve of the *scir* ; [19] but by Canute's time the maintenance of *borh* or suretyship for keeping the peace rested with the hundred court,[20] and in the twelfth century it was the sheriff who supervised the matter in the same assembly.   Moreover, the witnessing of transactions, which was the business of the *gerefa* in the folkmote of the older administrative area,[21] belonged by the reign of Edgar to the representatives of the hundred and the wapentake.[22]   Apparently the sheriff acquired the functions of the reeve of the burghal district and, to lighten the burden, somehow shifted part of it to the hundreds.   The fact fits well into the newer theory of the sheriff's origin, and it explains the control which that official in later times exercised over the hundred.

[17] 6 Athelstan, 3, 8. 1.
[18] Cf. 6 Athelstan, 1. 4, with 3 Edmund, 2, and 1 Edgar, 2 and 5.
[19] 6 Athelstan, 10.
[20] See the writer's *Frankpledge System*, p. 26, n. 1.
[21] Presumably the burghal district (2 Athelstan, 10, 12).
[22] 4 Edgar, 5, 6.

If these views of institutional development be correct, the rise of the sheriff is one phase of an extensive movement in the tenth century making for the centralization of local government. The reign of Athelstan, when the administrative strain incident to rapid territorial expansion seems to have been greatest, is a natural period for such a process. It has been attributed to the grouping of shires in the tenth century under the rule of individual aldermen, a plan presumably necessitated by the ascertained diminution in their number.[23] Indeed, Mr. Chadwick has shown that the sheriff is never mentioned so long as each shire has its own alderman.[24] It became impossible for the alderman to attend in person to the duties of his office in several shires each of which retained its identity. He required a deputy in the shire, and the officer in the shire next in rank had long been a reeve. But there were other changes. The powers of the alderman were judicial and military, yet the possession of these alone would not make the reeve a sheriff. As the burghal areas, which were largely military in character,[25] tended to disappear in Wessex after the cessation of the Danish wars in the south of England, the fiscal and police powers of their *gerefan* seem to have been centralized in the hands of the sheriff. At the same time the sheriff's duties were increased by the alderman's frequent absence from the shire. The process of transfer is thus accelerated.

The clearly recorded history of the sheriff begins, as already shown, about the time of Edgar. General provisions in the laws concerning king's reeves at an earlier time may refer to the reeve of a burghal area or a group of royal estates, and at a later time to the reeve of the hundred or wapentake.[26] For the known period of the sheriff's existence, then, it is evident that what the laws say of *gerefan* in general may often apply to him, though not certainly to him alone. Thus the inference may be drawn that he was appointed and was removable by the king,[27] that in some places he proclaimed the *grith*[28] and was a guardian of the peace[29] with summary powers of action against offenders and suspects,[30] that he

[23] Chadwick, p. 231 ; for the number of aldermen see pp. 172-90.
[24] *Ibid.* p. 230.   [25] Maitland, *Domesday Book*, p. 504.
[26] Below, notes 66, 67.
[27] 2 Athelstan, 25 ; 6 Athelstan, 11 ; Canute's letter of 1027, sect. 12, in Liebermann, *Gesetze*, i. 277.
[28] 3 Ethelred, 1. 1.   [29] 6 Athelstan, 1. 4 ; 8. 2.
[30] 1 Ethelred, 4, 4. 1 ; 2 Canute, 33, 33. 1.

took charge of certain court compositions for offences,[31] that he was a presiding judicial officer who required admonition to do justice and show mercy,[32] and that as the king's fiscal officer [33] he needed instructions to collect only what was rightly the king's and serve him with this.[34] It cannot be affirmed positively that any *gerefa* mentioned in the laws of the tenth or eleventh century, with the possible exception of the reeve mentioned in Athelstan's London enactment, is a sheriff. But it is extremely likely that such were the king's reeve who led the men of the shire in Ethelred's wars,[35] and the reeve who sat in judgement with the bishop and thus apparently took the place of the alderman.[36] The reeves whom Canute bids give judgement by the shire bishop's witness and show such mercy as the latter thinks right [37] at first sight seem to be none but sheriffs ; yet the famous writ of William the Conqueror, which represents the bishop as having in the past held pleas in the hundred,[38] may warn us against presuming an identification which may not hold for all cases.

With the person who in the reign of Ethelred, if not earlier, appears in the alderman's absence as the leading lay official in the shiremote begins the recorded history of the sheriff as differentiated from that of the king's reeve. About the last decade of the tenth century this official is termed a *gerefa* and also a *scirman*.[39] The shireman is mentioned in Kent, and the natural assumption that he is a reeve in charge of the shire in the alderman's absence [40] seems to be confirmed by the identity of *scirman* and *scirgerefa* as estab-

---

[31] 1 Ethelred, 1. 14.

[32] Canute's charter of 1020, sect. 11 (Liebermann, *Gesetze*, i. 274).

[33] 8 Ethelred, 32.   [34] 2 Canute, 69. 1.

[35] *A.-S. Chron.* a. 1001 (Hampshire), a. 1011 (Kent).   [36] See 3 Edgar, 5. 1, 2.

[37] Canute's charter of 1020, sect. 11 (Liebermann, *Gesetze*, i. 274).

[38] Chap. 2, Liebermann, *Gesetze*, i. 485 ; Stubbs, *Sel. Charters*, 85.

[39] Kemble, *Cod. Dipl.* nos. 929, 1288. The mention of a priest who served as *scirman* (no. 1288) does not weaken this conclusion. The ecclesiastical ban upon the holding of reeveships by priests seems to belong to the age of Elfric (Elfric, *Homilies*, Thorpe's edition, i. 339 ; Pastoral Letter, sect. 49, in Thorpe, *Ancient Laws*, p. 462). The version of Archbishop Egbert's Penitential in Thorpe (iii. 8, p. 374) is not so old as Egbert's time, but may date from the ninth century (see Haddan and Stubbs, *Councils*, iii. 413-16). It provides that ecclesiastics may be especially nominated for such secular duty. The one mentioned in the instance given above acted as *scirman* in the presence of Archbishop Dunstan, so his conduct may be regarded as regular.

[40] The whole usage of the Anglo-Saxon period places a reeve in the judicial position next to the alderman or earl. See Asser, sect. 106 ; 6 Athelstan, 11 ; 4 Ethelred, 8 ; Polity, 11, in Thorpe, *Ancient Laws*, p. 429 ; Florence of Worcester, a. 1066.

lished by Kentish documents dating from the reign of Canute.[41] In the eleventh century his title was rendered by *iudex comitatus*.[42] By Canute's time there is evidence of the sheriff's activity in several directions. The *gerefa* of the shire of Kent is the fiscal official whom the king forbids to infringe the archbishop's temporalities.[43] In Heming's narrative the sheriff of this period is mentioned as receiving money in payment of Danegeld, and Ævic, the sheriff of Staffordshire, is accused of having occupied, during the struggle between Canute and Edmund Ironside, lands of the church of Worcester which long remained in the hands of his successors in office.[44]

Materials to illustrate in detail the general position and activities of the pre-Conquest shrievalty exist only for the reign of Edward the Confessor; hence the account which follows must deal almost exclusively with this limited period. The statement that the king appointed the shire-reeves, which has been believed to be of ancient date, is in fact the work of an eleventh-century forger,[45] and may well be taken as descriptive of a practice with which he was familiar. That the tenure of office was not for life was a well-recognized rule;[46] but holding during-pleasure might then, as later, mean that it was for many years.[47] A number of sheriffs of King Edward remained in office in the reign of Harold and even in that of William the Conqueror.[48] As early as the reign of the Confessor a sheriff was

---

[41] Kemble, *Cod. Dipl.* nos. 731, 732. A *scirman* for Middlesex is named (Kemble, *ibid.* no. 972), but in a document which is of questionable authenticity, as some of the witnesses belong to the earlier part of the reign of Ethelred and others to the reign of Canute. Edric, shireman of Norfolk (Kemble, *ibid.* no. 785), seems to have been remembered as a contemporary of Alderman Ethelwine, who died in 992.

[42] See Birch, *Chart. Sax.* no. 1098.   [43] Kemble, *Cod. Dipl.* no. 1323.

[44] Heming, *Chartulary*, i. 277, 278.

[45] Haddan and Stubbs, *Councils*, iii. 241, 245; Kemble, *Saxons in England*, ii. 165, n. 1; Liebermann, *Gesetze*, ii. 649. See also Birch, *Chart. Sax.* no. 94.

[46] *E. habuit ipsa [sic] dimidiam hidam quam Godricus vicecomes ei concessit quamdiu vicecomes esset (D.B.* i. 149).

[47] Toli, who seems to have been sheriff of Suffolk as late as about 1065 *(D.B.* ii. 140, 334), is mentioned as sheriff in a writ along with Bishop Grimketel (Kemble, *Cod. Dipl.* no. 1342), who died in 1047, and who was recognized as bishop in East Anglia only for a short time after he was appointed to that see in 1038 (Florence of Worcester, a. 1038, 1047).

[48] Those who continue under William are Marloswein (Kemble, *Cod. Dipl.* no. 806; *D.B.* i. 376; Freeman, *Norm. Conq.* iii. 421), Robert fitz Wymarc (below, note, p. 43; *D.B.* ii. 98), Norman (below, note 148; *D.B.* ii. 438; Round, *Feudal England*, pp. 427-8), Touid or Tofig (Kemble, *Cod. Dipl.* nos. 837, 839; Davis, *Regesta*, i. nos. 7, 23), Edric (*D.B.* i. 72; Round, *Feudal England*, p. 422), and apparently Elfwine (Alwin, *D.B.* i. 238 b, 242 b) and Edwin (*ibid.* 157 b, 238 b, 241). For Harold's reign we should add the name of Godric (below, p. 42) and in all probability that of Ezi (*D.B.* i. 43).

sometimes placed over two counties. Thus Norfolk and Suffolk formed the bailiwick of Toli, though each county had its own shire-mote.[49] Godric was sheriff of Berkshire and Buckinghamshire.[50] The same custom which gave *æfgrefan* or *bydelas* to *deman* and *gingran* to the more important administrative officials [51] doubtless attached deputies also to sheriffs. The perquisites of the office, apart from grants of land and other advantages due to the personal favour of the king, included the privilege of farming his estates and some other sources of his income. It is probable that when the sheriff was a tenant on the royal demesne his land was often exempt from the obligations of the king's *feorm*; [52] but, as will appear later, Kemble was wrong in supposing that sheriffs retained as their own the court fines which they collected within their counties.[53]

The judicial powers of the sheriff, so it was held in the eleventh century, were derived from the authority of the alderman or earl. The author of the *Iudex* [54] regards the alderman as head of the judicial system of the shire,[55] and Heming even represents him as holding sessions of the hundred.[56] The earl's third penny of the pleas of the shire and hundred shows that this judicial supremacy is, and has been, no fiction. Yet it has already been shown that by the reign of Ethelred the sheriff is the alderman's representative who takes his place in the shiremote. Indeed, the rapidly decreasing number of aldermen and the corresponding increase in the number of counties governed by a single alderman rendered this inevitable. The lawgivers of Edgar's time probably meant to deal with this situation when they required the alderman as well as the bishop to attend the two annual sessions of the shire.[57] Canute's law repeats the requirement [58] at a time when, instead of the original alderman in charge of each shire, there were some half-dozen earls ruling all the shires in England.[59] There is every reason to believe that in

[49] Kemble, *Cod. Dipl.* nos. 853, 874, 875, 880, 881.
[50] *D.B.* i. 149.
[51] See Liebermann, *Gesetze*, ii. 719 (3 e), 722 (19); also Episcopus, 10, *ibid.* i. 478.
[52] *Quamvis Aluricus vicecomes sedisset in ea villa semper reddebat de ea firmam regis et filii eius post eum* (*D.B.* i. 208).
[53] Kemble, *Saxons in England*, ii. 166.
[54] Iudex, 8, Liebermann, *Gesetze*, i. 475.
[55] See Davis, *E.H.R.* xxviii. 421. Cf. Kemble, *Saxons*, ii. 137.
[56] Heming, *Chartulary*, ed. Hearne, i. 123, 137.
[57] 3 Edgar, 5. 2.
[58] 2 Canute, 18. 1.   [59] Larson, in *Am. H.R.* xv. 725.

the eleventh century, as in the tenth, the sheriff acted in a judicial capacity for his superior.   The earl, however, is to be found personally discharging his proper judicial functions [60] and continues to be recognized as the head of the shiremote.   The Confessor's writs to this body are regularly addressed to the bishop, the earl and the thegns of the shire, and only in a minority of the known cases is the sheriff's name added.[61]   Since the matters which were brought forward there largely concerned financial administration, it can hardly be supposed that the sheriff was absent when his name is not specifically mentioned in the king's writ.[62]   As early as Ethelred's reign, it has been seen, the requirement that the alderman should attend the sessions of the shire was satisfied by the presence of the shireman.   The size of an earldom like Godwine's appears to prove that the earl could not attend regularly, and even among the few extant writs in which a sheriff is named there are some addressed to this official and the bishop, no mention being made of the earl.[63]   It is not questioned that in the absence of the earl the sheriff often presided over the shire assembly.   In a county like Shropshire, where he had superseded the earl in the exercise of other powers,[64] it is probable that he alone presided and that he was actually the " constituting officer " [65] of the county court.

The judicial activity of the sheriff in the hundred, well known through a writ of Henry I., was not a result of Norman rule.   The laws of Ethelred show that there was a lesser reeve in the wapentake [66] and the hundred [67] who directed criminal procedure.   In the reign of Edward the Confessor the *praepositus* is mentioned as

---

[60] Kemble, *Cod. Dipl.* nos. 755, 898 ; Thorpe, *Diplomatarium,* 346, 376 ; *Liber Eliensis,* 139.

[61] Zinkeisen, in *Pol. Science Quarterly,* x. 138.   In the *Ramsey Chronicle,* p. 79, occurs an instance, in Ethelred's time, in which both alderman and king's reeve preside.

[62] In assuming that he was absent Zinkeisen seems to go further than the evidence justifies.

[63] Kemble, *Cod. Dipl.* nos. 869, 870, 858.   With these should no doubt also be included nos. 856, 857, 861, addressed to portreeves of London.

[64] Cf. the sheriff's proclamation of the king's peace (note 88).   The burghal third penny here went not to the earl, but to the sheriff (Ballard, *Domesday Boroughs,* p. 44).

[65] See Stubbs, *Const. Hist.* i. 128 ; Zinkeisen, in *Pol. Science Quarterly,* x. 137, 138.

[66] 3 Ethelred, 3. 1, 2.

[67] The employment of a *gerefa* at the preliminaries of purgation (3 Ethelred, 13) implies that he is at the head of the hundred court, which is the regular tribunal in criminal cases (1 Ethelred, 1. 2 ; 2 Canute, 17, 25. 1).

fixing the term of the case to be tried in the hundred.[68] But it is clear that in this reign the sheriff also held sessions of the hundred-mote. In some hundreds either he or the *motgerefa* presided,[69] a circumstance tending to show that the latter acted under his authority. It was an unusual exemption granted to the cathedral church of Worcester,[70] as well as to some other churches, which had anciently excluded the sheriff from jurisdiction in the hundreds that they possessed. It had become the rule in the hundred of Wormelow near the city of Hereford that the sheriff held three of the twelve sessions in the year.[71] So strongly do these sessions resemble the sheriff's tourn of a later day that they have even been regarded as views of frankpledge.[72] They do well fit into the requirements of the prevailing system of suretyship for keeping the peace,[73] but the evidence to prove the existence of frankpledge in the period under consideration is not strong enough [74] to warrant the conclusion that they represent more than an incipient form of sheriff's tourn which later came to make view of frankpledge tithings. Lastly, it is to be noted that the sheriff's jurisdiction in the hundred extended to the burghers of many towns. The fact that in the reign of King Edward the sheriff was setting the term for the suits of the men of Shrewsbury, which they were bound to observe under penalty of ten shillings,[75] is to be interpreted in the light of the requirement that the *iudices civitatis* of Chester were to attend the

[68] *D.B.* i. 269 b.

[69] Kemble, *Cod. Dipl.* no. 840. Cf. the *gingran* of the *scirman* (Polity, 7, in Thorpe, *Ancient Laws*, p. 426 ; Episcopus, 10, in Liebermann, *Gesetze*, i. 478).

[70] See *D.B.* i. 172.

[71] *D.B.* i. 179. Special attendance at court three times a year is also mentioned as an obligation resting upon the tenants of the bishop of Winchester in his manors about Taunton (Kemble, *Cod. Dipl.* no. 897). These were *placita episcopi* (*D.B.* i. 87 b) held in a hundred which had passed into the bishop's hands.

[72] Liebermann, *Gesetze*, ii. 521, sect. 31 d. Dr. Round (*V.C.H. Hereford*, i. 299, n. 250) makes it clear that the horsemen who accompanied the sheriff on the journey from Hereford to the hundred of Wormelow are to be considered as his guard. He had to pass near the Welsh region of Archenfield in order to hold his court.

[73] See the writer's *Frankpledge System*, p. 113.

[74] See Liebermann, *Gesetze*, ii. 745, sects. 10 d, 11. Cf. *E.H.R.* xxviii. 422. Even if twelfth-century views represent a genuine tradition instead of the reading of a familiar institution back into 2 Canute, 20, it is probable that before 1066 *borh* groups existed which a writer unfamiliar with institutional development might readily identify later as frankpledge tithings. See *The Frankpledge System*, pp. 20-29 ; as to the tithing of Canute's laws, which according to Dr. Liebermann's interpretation becomes a suretyship-tithing to hold and lead every man to plea, *ibid.* pp. 20, 27 ; and as to the comparative elasticity of the Anglo-Saxon *borh* system and that of the frankpledge of Norman days, *ibid.* pp. 29-30.

[75] *D.B.* i. 252.

sessions of the hundred, and for absence without manifest excuse to pay a penalty of the same amount.[76]

The military functions of the sheriff on the eve of the Norman Conquest seem normally to have been confined to matters within his own county. He had a right to claim *inward* or a bodyguard. In the counties of the western border he took the command against Welsh incursions, just as at an earlier time the alderman led the men of the shire to repel the Danes. Military expeditions prior to 1066 are ordinarily treated in the Domesday record as the king's, and the summons is *edictu regis*,[77] yet in Shropshire and Hereford-shire the order comes from the sheriff.[78] There can be no doubt that when in 1056 Leofgar, the unclerkly bishop of Hereford and former chaplain of Earl Harold, took the field with his priests against Griffin, the Welsh king, and was slain along with Elfnoth the sheriff,[79] the latter was in command of a general levy of the shire. It was only after its defeat that Earls Leofric and Harold came to aid the sorely harassed English forces. The well-known story of the negotia-tions of William the Conqueror with Esgar, sheriff of London and Middlesex, seems to prove that upon the latter devolved the duty of defending the city against the Norman advance.[80] The death at Hastings [81] of Godric, the sheriff of Berkshire, tends to show that the shire levies of the near-by counties which came to aid King Harold in his last battle were led by their sheriffs. As Kemble says, they were the natural leaders of the militia and the *posse comitatus*.[82] Yet it seems clear that the troops which, according to the statements in Domesday, the western sheriffs took into Wales were of a different kind. At Shrewsbury and at Hereford the sheriff called out, not indeed as did the king's officials in other towns, a fixed number of men, but apparently as many as he needed. Individual persons were no doubt designated,[83] and the high penalty

---

[76] *D.B.* i. 262 b.   [77] *D.B.* i. 172.

[78] *Cum in Walis pergere vellet vicecomes qui ab eo edictus non pergebat XL. solid. de forisfactura dabat* (*D.B.* i. 252, Shrewsbury). *Si vicecomes iret in Wales cum exercitu ibant hi homines cum eo. Quod siquis ire iussus non iret emendabat regi XL. solid.* (*ibid.* i. 179, Hereford).

[79] *A.-S. Chron.* a. 1056; Florence of Worcester, a. 1056.

[80] See Stenton, *William the Conqueror*, p. 224. His office of staller may, in part, account for his military functions. Cf. Page, *London* (1923), 232-3.

[81] *Abingdon Chron.* i. 491.

[82] *The Saxons in England*, ii. 164. That they acted in the latter capacity may perhaps be assumed from the analogy of the *gerefa* of Athelstan's time and the hundredman of Edgar's reign.

[83] See above, note 78.

of forty shillings, for failure to go, seems fairly good evidence that summons was not general.   Moreover, the custom which made the Welsh of the district of Archenfield near Hereford the advance-guard when the army went into Wales and rear-guard on the return [84] points to arrangements out of keeping with a general summons and haphazard assemblage.

While the sheriff came to possess powers which pertained to the headship of the shire, he by no means lost those derived from his reeveship.   Among these may be named his powers connected with the peace and with police.[85]   The king's peace was given in the reign of the Confessor by the king's hand or writ or else by his *legatus*.   At Chester this *legatus* was the earl, the earl's official or the king's reeve.[86]   At Shrewsbury the king's peace was proclaimed by the sheriff.[87]   The monastic story of Leofstan, the wicked sheriff who violated the sanctuary of St. Edmund's Abbey to seize a criminal, dating from but a short time after the Norman Conquest and pur-porting to come from an earlier period, is good testimony regarding the existence before that event of one of the sheriff's most familiar duties.[88]   In the Institutes of Polity the bishop when exercising jurisdiction over his clergy in capital cases is called Christ's *scirgerefa*.[89]

As fiscal agent of the king, the sheriff had powers which were possessed by the earliest known king's *gerefa*.[90]   The administra-tion of royal rents, dues, services and forfeitures had made the office of reeve one of importance in the state, and they were destined to make the office of sheriff still greater in the next period of its history.   But for King Edward's time finance is still an obscure subject, and the sheriff's place in the scheme is only here and there recorded.   He was closely connected with the *ferm* of the royal demesne lands, the very core of the financial system.   These lands were sometimes in the custody of various farmers or local officials.[91]

---

[84] *D.B.* i. 179.

[85] See 4 Athelstan, 7 ; 6 Athelstan, 8, 8. 2-4.

[86] *D.B.* i. 262 b.

[87] *Ibid.* i. 252.   The same was true in Warwickshire in 1086.

[88] *Memorials of St. Edmund's Abbey* (R.S.), i. 30-32, 112-13 ;   Liebermann, *Ungedruckte anglo-normannische Geschichtsquellen*, pp. 231-2.

[89] Institutes of Polity, 25, in Thorpe, *Ancient Laws*, p. 439.

[90] Liebermann, *Gesetze*, i. 14, Wihtraed, 22.   Here the *gerefa* appears to be a domanial agent of the king.

[91] *Rex tenet Axeminstre . . . Eccha praepositus accommodavit cuidam presbitero unum ferling terrae (T.R.E.: D.B.* i. 100 ; iv. 76) *. . . qui tenebat eam T.R.E. concessit eam (i.e. a haw in Guildford) Toui praeposito villae pro emendatione unius*

Sometimes the sheriff himself was the farmer.[92] It is impossible
to say whether as yet he farmed all such lands in his county and
then leased part of them to bailiffs. The total amount of the king's
annual income from a county prior to 1066 is known in but one case,[93]
and it is not stated in this instance whether it was or was not the
result of a *ferm*. If it is unsafe to assume that counties were farmed
as a whole at the date of the Domesday inquest,[94] much less is it
permissible to make the assumption for the reign of King Edward.
There are, nevertheless, indications which point to such a practice.
The later custom, whereby the king upon granting away a demesne
manor made allowance to the sheriff for a corresponding decrease in
the *ferm* which he paid, has in one case been traced to the Confessor's
time.[95] The alienation of lands belonging to the royal *ferm* seems
regularly to have been effected through the sheriff's agency,[96] and
his enforcement of the *avera* or carrying services, which were an
ancient part of the *feorm* rendered by royal estates,[97] is a third link
in the chain. It is at least clear that sheriffs were the custodians
of many royal manors and that lands confiscated for crime were
being placed in their hands. If they did not as yet manage the
farming system within their respective shires, they were very important factors in that system. The farming together of a whole
hundred or district,[98] and the grouping of royal manors to make up

---

suae forisfacturae (*ibid.* i. 30). . . . *tenuit Godwinus praepositus regis in firma* (*ibid.*
iv. 97). . . . *quas tenuit I. faber T.R.E. qui propter latrocinium interfectus fuit et
praepositus regis addidit illam terram huic manerio* (*ibid.* ii. 2 b).

[92] *Sed quidam vicecomes misit eas ad V hidas per concessionem eiusdem regis*
(King Edward) *quia firma eius eum gravabat* (*D.B.* i. 197). This is further proof
that the king's manors were not *extra comitatum*, as Eyton held. Dr. Round deals
with the question in *V.C.H. Somerset*, i. 395.

[93] Warwickshire (*D.B.* i. 238). The *firma* of three nights from Oxfordshire and
Northamptonshire (*ibid.* 154 b, 219) is clearly that paid in 1086, rather than before
1066, where Liebermann (*Gesetze*, ii. 422) places it.

[94] See *V.C.H. Northampton*, i. p. 277.

[95] Round, *Commune*, p. 73.

[96] *Fuit de firma regis sed tempore Godrici vicecomitis fuit foris missa* (*D.B.* i. 57).
*Homines de hundredo testantur quod praestitum fuit istud manerium per vicecomitem
extra firmam regis Edwardi* (*ibid.* i. 31). *Alwi vicecomes misit haec extra firmam*
(*ibid.* i. 163).

[97] With the *regalium rerum et operum debita* (Kemble, *Cod. Dipl.* nos. 196, 206 ;
Birch, *Chart. Sax.* nos. 335, 351) of the Mercian kingdom compare the *feorm* at
Taunton (Kemble, no. 1084 ; Birch, no. 612).

[98] *De firma in Bertune hundredo regis Edwardi fuit ad dimidiam diem firmae
reddidit in omnibus rebus* (*D.B.* i. 38 b). . . . *I. mansionem quae vocatur Esmaurige
quae tempore regis E. reddebat per annum ex consuetudine XXX. denarios in firma
Axeminstre regis* (*D.B.* iv. 467, fo. 503 b).

jointly the equivalent of the ancient *feorm* of one night,[99] were useful devices in centralizing control in their hands.

The sheriff's administration of the proceeds of the *ferm* is a matter of record for the Confessor's time. The plan, familiar to the student of the Pipe Rolls, by which the sheriff at the king's instance made local disbursements from the proceeds of his *ferm* was already followed. It is recorded that when the *legati* of the king on their journey down the Trent reached Torksey, the boatmen of that town conducted them to York, and the sheriff out of his *ferm* furnished provisions for the journey.[100] Two peculiar Domesday entries concerning Gloucestershire estates, one telling that from a certain manor the sheriff rendered what it brought at farm, the other that a manor rendered what the sheriff wished,[101] seem to refer to the receipt of quantities of produce upon which a money valuation had not yet been placed.

Since the sheriff was a judicial as well as a fiscal official, he was specially concerned with the king's income from judicial sources. These revenues were largely reckoned with the *ferm*, and formed its second ingredient. But some compositions for the greater offences went directly to the king, and had to be accounted for separately. King Edward had on his demesne throughout England three of these more important *forisfacturae* [102] which were *extra firmas*. With them are to be included in various localities other royal *placita* which were not farmed.[103] They preserve an old distinction [104] and constitute the germ of the future pleas of the Crown. The mention of one of these five-pound forfeitures which was paid for breaking into the town of Wallingford by night and which, it is

---

[99] See Dr. Round's account in *V.C.H. Hampshire*, i. 401-2. In Shropshire this plan dated at least from the reign of Ethelred (*D.B.* i. 253 b).

[100] *Ibid.* i. 337. At Wallingford a similar procedure on the part of the local *praepositus* was apparently not unknown (*ibid.* i. 56).

[101] *Ibid.* i. 163.

[102] *Has III. forisfacturas habebat in dominio rex Edwardus in omni Anglia extra firmas* (*ibid.* i. 252). *Rex vero habebat in suo dominio tres forisfacturas, hoc est pacem suam infractam et heinfaram et forestellum. Quicunque horum unum fecisset emendebat C. solidos regi cuiuscunque homo fuisset* (*ibid.* i. 179).

[103] Thus at Oxford *fyrdwite* was emended by paying a hundred shillings to the king (*ibid.* 154 b). In 1086 trespass upon the king's highway at Dover paid a *forisfactura* to the king of the same amount (*ibid.* i. 1). Cf. *rex E. dedit . . . de firma sua solutam ab omni consuetudine praeter forestam custodiendam excepta forisfactura regis sic est latrocinium, et homicidium et heinfara et fracta pax* (*ibid.* i. 61 b).

[104] 1 Ethelred, 1. 14 ; 2 Canute, 12-15. For the amount collected see 1 Canute, 3, 2 ; 2 Canute, 62.

recorded, went not to the sheriff but to the king,[105] implies that in
Berkshire amounts derived from ordinary pleas were included in
the sheriff's *ferm*. The regular income from the hundred court was
farmed along with that of certain estates,[106] and the earl's third
penny of pleas of the hundred,[107] as well as the king's two pence,[108]
had become annexed to definite manors. That all these arrangements required the sheriff's supervision seems probable. The old
rule, that compositions made in court by the holders of bookland
are to be paid only on witness of the king's *gerefa*,[109] makes either
him or the subordinate reeves responsible. It is to be observed also
that the regular process of collecting *geldwite* was through the sheriff's
seizure of land,[110] and that land forfeited to the king for crime was
probably taken into the sheriff's hands.[111] Moreover, the same
official is named as collector of the *forisfacturae* of the king's sokemen outside the royal demesne.[112] Since among these payments
may be included the great ones not included in the *ferm*, as well as
the lesser ones which were so included, the passage seems to show
that the sheriff will account directly to the king's chamberlain[113]
or his representatives for sums which are not farmed. Such an
arrangement as that which attached to one of Earl Harold's manors
the third penny of a whole shire probably implies that judicial
profits included in the *ferm* were aggregated.[114] A Domesday
entry, which records that the borough of Yarmouth along with the
king's judicial income from three hundreds was worth to the Confessor eighteen pounds *ad numerum*,[115] shows that payment was
being made in coin. It also indicates, as in the case of the special
pleas of the Crown, that the sheriff already accounts at the king's
treasury for these portions of his *ferm*.

[105] *D.B.* i. 56 b.
[106] *T.R.E. reddebat XXI. libras de firma istud Wich cum omnibus placitis eiusdem hundred* (*ibid.* i. 268).
[107] *Ibid.* i. 38 b, 86 b, 101, 263 b ; iv. 462, fo. 479 b.
[108] *Huic manerio pertinebant II. denarii de hundredo Conendovre T.R.E.* (*ibid.* i. 253).
[109] 1 Ethelred, 1. 14.
[110] Below, p. 63. Cf. *D.B.* i. 141.
[111] *Anno quo mortuus est isdem rex (Edwardus) fuit ipse forisfactus et dedit illam Merloswen vicecomiti pro reatu regis* (*ibid.* i. 376). The *praepositus regis* (above, p. 14, n. 121) acts in this capacity also, but probably as the sheriff's representative.
[112] *D.B.* i. 189 b.
[113] Larson, *King's Household*, p. 133 ; Poole, *Exchequer in Twelfth Century*, pp. 23-6.
[114] Dorset (*D.B.* i. 75).
[115] See note 118.

D

Other varieties of income seem to have been collected by the same methods and agencies which dealt with the judicial revenues. The customary manorial payments of a few pence a year, made to the sheriff [116] from lands outside the demesne, were probably included in his *ferm*. The sums rendered by boroughs before Norman days were in part farmed and in part accounted for directly.[117] The borough *ferm* might include that from the pleas of various hundreds,[118] and a small borough on the royal demesne might pay its *ferm* as part of a manor.[119] The case of Hereford, where the reeve farmed immediately from the king,[120] seems quite exceptional. The grouping at Warwick of the royal burghal dues with the king's share of the pleas of the county, the lumping of the judicial income from hundreds with the *ferm* of boroughs, and the statement that at Chester a certain *forisfactura* collected by the reeve was made over to the *minister regis* within the city,[121] seem to indicate at so early a time the sheriff's relation to the *firma burgi*.[122] The division of the profits of the coinage at Shrewsbury between king and sheriff, in the proportion respectively of two to one, implies that the latter had here acquired the earl's rights to the third penny of the borough. The sheriff's work in the levy of gelds is proved by the fact that he reduced the assessment of a Cambridgeshire vill from ten to the more usual five hides.[123] According to Heming, when in Canute's day the Danegeld was overdue, the money was paid to the sheriff ; [124] whether the allusion is to the Danegeld itself or merely to the *geld-*

---

[116] See *D.B.* i. 138 b, 139, 140 b.

[117] At Huntingdon, for instance, in the time of King Edward the fisheries were farmed, but the mill and the mint were accounted for separately (Ballard, *Domesday Boroughs*, p. 92) ; cf. Round in *Domesday Studies*, i. 135-7. The seven pounds paid to the king and the earl by the moneyers of Chester when the coinage was changed were *extra firmam* (*D.B.* i. 262 b).

[118] *Walterus de II. partibus burgi Malmesberie reddit VIII. libras regi. Tantundem reddebat ipsum burgum T.R.E. et in hac firma erant placita hundret' de Cicemtone et Sutelesberg quae regi pertinebant (D.B.* i. 64 b). *Gernemwa. Tenebat Rex E. . . . Tunc valebat cum duabus partibus soche de tribus hundretis XVIII. libras ad numerum, pars comitis IX. libras ad numerum (ibid.* ii. 118).

[119] *Langeford . . . Huic manerio reddebat burgum Totheneis XX. solidos ad firmam regis (ibid.* i. 101).

[120] Ballard, *Domesday Boroughs*, p. 92.

[121] *D.B.* i. 262 b.

[122] See Maitland, *Domesday Book*, pp. 204-5.

[123] This was done with the king's consent (*D.B.* i. 197).

[124] Heming, *Chartulary*, i. 278. He seems even then to have seized land to enforce payment just as he did in the reign of William the Conqueror (*D.B.* i. 141).

wite, the sheriff's responsibility in the matter is equally patent.[125] The hundredmen, the regular collectors of the geld,[126] here also appear to be under his direction.

The work of the sheriff in enforcing service due to the Crown is mentioned in more circumstantial detail. If the *inward* which he supervised was not performed, he collected wardpenny in lieu of it.[127] Special *inward* was rendered to the sheriff of Cambridge-shire when King Edward came into the county.[128] When the king was in the city of Hereford the lesser landholders there did the same service at the hall.[129] This obligation is akin to another, that of providing the king with an escort on his progresses. When King Edward went hunting in the neighbourhood of Shrewsbury, the better burghers who had horses formed his guard, and the sheriff sent thirty-six *pedites* to stalk the deer as long as the king was there. When he departed, the sheriff sent with him twenty-four horsemen to conduct him to the first manor-house in Staffordshire.[130] The rendering to the sheriff of special *avera* is also mentioned, significantly, when the king came to the shire.[131] This obligation, which evidently sprang from the ancient king's *feorm*,[132] was in some places an annual one.[133] Manors sometimes jointly performed such service [134] and, like *inward*, it was commuted by a money payment to the sheriff. Some persons acquitted themselves of it by providing a horse once a year or by paying four pence instead.[135] Of a similar nature was the service performed by the burghers of Cambridge, who three times a year furnished their ploughs to the sheriff.[136] The keeping of hayward *in servitio vicecomitis* is also mentioned.[137] At Hereford the sheriff each August summoned certain persons

[125] This is also hinted by the following : *Rex habet I. Burgum quod vocatur Bade quod tenuit Eaditda Regina die qua rex E. fuit vivus et mortuus et reddebat gildum pro XX hid. quando vicecomitatus gildabat (D.B.* iv. 106, fo. 114 b).

[126] Round, *Feudal England*, pp. 53-4.

[127] *Homines huius manerii reddebant Warpennam vicecomiti regis aut custodiam faciebant (D.B.* i. 190). Grants of King Edward (as Kemble, *Cod. Dipl.* no. 862) mention *wardwite*.

[128] The alternative was the payment of 12s. 8d. (*D.B.* i. 190).

[129] *Ibid.* i. 179.   [130] *Ibid.* i. 252.

[131] *Ibid.* i. 190 b.   [132] See above, note 98.

[133] *Inveniebant vicecomiti regis I. averam et V. denarios et unum quadrantem per annum (D.B.* i. 134).

[134] *Reddiderunt VI. denarios vicecomiti vel unam averam et dimidiam (D.B.* i. 133 b) ; *inveniebant III. partes averae et III. denarios vicecomiti (ibid.* i. 141).

[135] *Inquisitio Comitatus Cantabrigiae* (ed. Hamilton), p. 4.

[136] *D.B.* i. 189.

[137] *Inquisitio Comitatus Cantabrigiae*, p. 34.

three days for the cutting and one for the ingathering of hay.[138] Mention of a certain *liber homo* of Gloucestershire who rendered service to the sheriff throughout England [139] goes considerably beyond the obligation of the boatmen of Torksey,[140] which took them only into the next county, and shows how far from home the sheriff's missions might carry a man.

A review of the fragmentary evidence available at once shows how unsatisfactory is the state of our information concerning the position and duties of the sheriff of King Edward's day, but it reveals clearly enough the fact that he had become an important person. Already he may be seen going about his duties attended by an escort of horsemen.[141]   When he presided for the alderman over the shiremote, it became his duty to proclaim the king's commands and the enactments of king and *witan*.[142]   It was he who might be expected to execute a great part of such orders together with the decrees of the shire assembly.   As occasional president of the hundred court, he exercised the customary criminal jurisdiction. He had authority to proclaim the king's peace and to apprehend criminals.   His was the responsibility for local defence, and he led the forces of his shire against Welsh attack or Norman invasion. The enforcement of the levy of Danegeld gave him a significance for vital national interests and influence over the landholders of the shire.   The collection of court fines increased his power over all persons who remained within the king's *soke*.   The exaction of *avera*, *inward* and hayward also occurred in some places not on the royal demesne.[143]   In the capacity of reeve of the royal demesne the enforcement of payments, renders and services brought the sheriff into immediate touch with the everyday interests of men. His custodianship of royal estates involved manorial duties ; [144] but he seems to have been the superior to whom numerous reeves and farmers of demesne manors accounted.   When the king made a visit to the shire the sheriff provided for his safety, convenience and various needs like a household official.   It is not surprising that in Domesday Book his official acts receive attention almost to the

---

[138] *D.B.* i. 179.
[139] *Ibid.* i. 162.
[140] Above, p. 30.
[141] Above, p. 26, note 72.
[142] See *E.H.R.* xxviii. 425.
[143] Maitland, *Domesday Book*, p. 169.   See *D.B.* i. 139, 200, 200 b.
[144] In 1086 the king had at Holborn two *cotselli*, rendering to the sheriff twenty pence a year, whose guardianship had always been entrusted to the sheriff of Middlesex (*D.B.* i. 127).

exclusion of those of the earl, and that in this great record the sheriff of King Edward is shown due respect when there arises occasion to mention him.   In a word, the whole government of the shire was falling into his hands.

The Old English sheriff presents traits which characterize his better-known Norman successor.   The assumption that he was always a considerable landholder within the shire is not supported by the solitary example which has been adduced in its support ; [145] still, in Domesday Book the sheriff of the period before 1066 is usually a landholder.   Excluding from consideration Esgar and Robert fitz Wymarc, the great landholding stallers,[146] there are named in Domesday fifteen of these sheriffs, of whom all but two are at once seen to be in possession of land ; [147] and the remaining two probably belong to the same category.[148]   Some of these personages had land in shires other than those in which they were sheriffs. Some held by direct gift of the king or by special arrangement with him.[149]   Some bought or sold land by the king's permission.   Frequently they are mentioned as having tenants or dependants holding land of them.   Three of them, Alwin or Ethelwine of Warwick, Ezi of Hants, and Toli of Norfolk and Suffolk, gave land to the

---

[145] Tofig Pruda, cited by Kemble (*Saxons*, ii. 166) as a very wealthy sheriff, can hardly be Tofig or Touid, the sheriff of Somerset (Kemble, *Cod. Dipl.* nos. 837, 839). The latter held office as late as 1067 or 1068 (above, note 48), and Godwine (Kemble, *Cod. Dipl.* nos. 834, 835, 836, 838), who was sheriff as late as 1061, must have been his predecessor in office.

[146] See notes 153, 164.

[147] They are Godwin (Berks, *D.B.* i. 57 b, 58 ; Bucks, *ibid.* i. 149), Edwin (lands in Warwick, *D.B.* i. 238 b, 241, and in Oxford, *ibid.* i. 157 b), Alwin (the correct form of the name appears from *Salt Arch. Soc.* ii. 178, 179, to be Ethelwine, and not Elfwine, as Freeman suggested, Warwick, *ibid.* i. 238 b, 241, 242 b ; he also had land in Huntingdon, *ibid.* i. 206 b, and in Gloucester, *ibid.* i. 167), Alwi (probably Elfwig, Gloucester, *ibid.* i. 162 b, 163 ; he seems to be the same Alwi *vicecomes* who held land of William the Conqueror in Oxford, and who was still living in 1086, *ibid.* i. 160 b), Aluric or Elfric (Huntingdon, *ibid.* i. 203, 208 ; see Kemble, *Cod. Dipl.* no. 903), Blacuin (Cambridge, *D.B.* i. 201), Orgar (Cambridge, *ibid.* i. 197), Osward (Kent, below, p. 38), Ezi (Hants, *D.B.* i. 43), Alured or Alfred (Dorset, *ibid.* i. 83 ; see Kemble, *Cod. Dipl.* no. 871), Heche or Heccha (lands in Devon, *D.B.* iv. 301, 306, 389 ; compare *ibid.* i. 109, 109 b, 111 b), Merloswein (Lincoln, *ibid.* i. 376), and Toli (Suffolk, *ibid.* ii. 299 b, 409 b ; Norfolk, *ibid.* ii. 211 b).

[148] Norman, the sheriff (probably the person mentioned in Kemble, *Cod. Dipl.* nos. 863, 904, as sheriff of Northampton), is named as having the commendation of lands in Suffolk, where a man of his also held lands (*D.B.* ii. 312 b, 334 b).   In the reign of King William he was a tenant of Robert Malet in the same county (*ibid.* ii. 327).   Edric, sheriff of Wilts, is mentioned only because of his trespass upon the king (*ibid.* i. 72 b), which seems to mean occupation of his lands.

[149] King Edward leased land to Alwin (*D.B.* i. 167).   Ezi held a half-hide *in paragio* of the king (*ibid.* i. 43).

church for the good of their souls.[150]   On the other hand, it is evident that it was not solely the Norman sheriff who despoiled the church.   One reads that Ævic in Canute's time occupied lands of the church of Worcester,[151] and Heming speaks of unjust reeves and royal collectors as great robbers of this church.[152]   Godric of Berkshire had acquired for the term of three lives the land of Fifhide, belonging to the church of Abingdon, by means which the monks much resented.[153]

Nor were these officials guiltless of the sharp practice and the usurpation of the rights of the Crown for which some later sheriffs were famous.   More than one of them stands accused of making free with the king's lands.[154]   Complaints against Godric of Berkshire are more numerous in Domesday than those against any other pre-conquest sheriff.   In one place he found pasture for his horses at the expense of the royal demesne ; in another he invaded the king's rights by ploughing with his own ploughs a hundred and twenty acres of Crown lands ; [155] he even granted a half-hide of land belonging to the king's demesne farm to be held as long as he should be sheriff by a certain girl for teaching his daughter orphrey-work.[156] But there are complaints against others.   Osward of Kent gave to Elfstan, a reeve of London,[157] parts of a manor of the farm of King Edward, so that they were lost to the manor, and he removed *extra manerium* six acres of land and wood which in 1086 were still alienated.[158]   Orgar of Cambridgeshire held a portion of royal domain which he placed *in vadimonio* [159] without the king's permission.   The placing of land *extra firmam* by the sheriff was doubt-

---

[150] The grant of Alwin was made by concession of the king and upon testimony of the whole county (*D.B.* i. 238 b).   That by Ezi was made after the death of King Edward and was questioned by the Normans (*ibid.* i. 43).   Toli's gift was to the church of St. Edmund (*ibid.* ii. 211 b).   Ulf, the portreeve of London, and his wife gave land and a wharf to Westminster Abbey (*Westminster Domesday*, MS., Westminster Abbey, f. 506).

[151] Heming, *Chartulary*, i. 277.

[152] *Ibid.* ii. 391.   The Danish invasion, unjust reeves and collectors, and Norman violence are given in chronological order as the three great robbers.

[153] *Chron. Abingdon*, i. 491.   Domesday even relates that King Edward gave to Robert fitz Wymarc land of the church of Hereford to hold as a canon and that the latter made it over to his son-in-law, who held it at the king's death (*D.B.* i. 252 b).   Esgar gained a firm hold upon one of the manors of the monks of Ely, and even appeal to the king failed to have any effect (*Liber Eliensis*, p. 217).

[154] Freeman, *Norm. Conq.* iv. 781.

[155] *D.B.* i. 57 b.

[156] *Ibid.* i. 149.

[157] *Praestitit . . . Alestan* (*D.B.* i. 2 b).

[158] *D.B.* i. 2 b.

[159] *Ibid.* i. 197.

less not always with the king's permission. On the other hand, the Anglo-Saxon sheriff has been accused of encroaching on private rights and of taking land into a royal manor so that his income might be increased while the *ferm* that he paid remained the same.[160]

The complexity and variety of the powers of the Old English sheriff were the peculiar sources of his usefulness to an undeveloped system of government. His has well been called a "generic office".[161] Furthermore, the array of places in which his duties were performed is striking. Some duties might even take a sheriff into another shire or his servants throughout England. It may be added that his action represents more than one authority within the state. In compelling collection of Danegeld he even follows the behest of the *witan*. In a very real sense he was at once the officer of the earl, the king, and the nation.

His relations to the earl and the king respectively constitute an interesting chapter of constitutional history. For some purposes, as shown above, he was the earl's subordinate or deputy. It has been questioned whether, in an age when the earls constituted the strongest political power in the state, the sheriff did not come to represent their choice and interest. Freeman suggests, for instance, that the presence in Herefordshire of the Norman earl Ralph may explain the apparent appointment of the Norman Osbern as sheriff.[162] The title *vicecomes*, according to all indications, was used as the equivalent of *scirgerefa* only after the Norman Conquest, when English documents were being turned into Latin by clerks familiar with the Norman *vicomte*.[163] On the other hand, the sheriff continued to be the king's *gerefa*, over whom the king's control may safely be assumed to have been as great as the laws show it was over his reeves in general. The occasional holding of the office by the household dignitary known as the staller[164] points to the control

[160] *V.C.H. Buckingham,* i. 220.
[161] Adams, *Origin of the English Constitution,* p. 5.
[162] *Norm. Conq.* ii. 345, n. 3.
[163] *Am. H.R.* xiv. 469 ; Stubbs, *Const. Hist.* i. 127, n. 4.   At an earlier time *iudex comitatus* occurs (*ante,* p. 23 ; cf. p. 18).
[164] Esgar or Ansgar, the staller, is accepted not only by Kemble (*Saxons,* ii. 165, n. 2) but also by Round (*Geoffrey de Mandeville,* p. 353) as sheriff of Middlesex. Robert fitz Wymarc (Freeman, *Norm. Conq.* ii. 345, n. 3 ; iv. 736-8 ; Round, *Feudal England,* p. 331) was certainly sheriff of Essex in the reign of William the Conqueror (*D.B.* ii. 98). It is practically sure that he held the same position before the Conquest (*V.C.H. Essex,* i. 345). Kemble (*l.c.*) and Freeman (*Norm. Conq.* iv. 757) believe that the naming of Eadnoth the staller in a writ in the position usually

of appointments by the king himself. The usage by which the sheriff often holds land of the king again shows a close personal relation. At least one of King Edward's sheriffs in addition to the stallers who held the office seems to belong to the court circle.[165] The sheriff, moreover, was the king's personal agent in providing for his necessity and convenience during his progresses, in proclaiming his peace, and in collecting his revenues. The Domesday jurors report the loss to the king arising from the laxness of former sheriffs in administering lands at farm; and one hears how King Edward gave directions to a sheriff in regard to the assessment of a vill.[166] On the other side, a charter of Canute deals with a case in which the reeves of Devonshire oppressed the lands of a church, and the earl complained to the king, possibly also to the *witan*.[167]

The dualism in the government of the Anglo-Saxon shire was not exactly what it has been understood to be. During the recorded history of the sheriff's activity he does not stand purely for central or royal power as against the local influence of the earl. Though directly representing the king in various matters, he was the judicial, it may even be the military, agent of the earl. Administration, as Stubbs held,[168] was in the hands of " a national leader and a royal steward ", but this view of the situation leaves out of the account the fact that the latter exercised some powers of the former and thus tended to assume control of the entire shire government. In the age of great earls it would be futile to regard the abdication of their functions as other than voluntary. There could hardly have existed an active opposition between the interests of the sheriff and those of the earl. No doubt much depended on the influence of the latter in the *witan* to determine national policy. So long as the earl collected the third penny of the shire and the borough he could well leave to the sheriff the actual performance of his duties. Weak kingship seems before the Conquest to be gaining for the

occupied by the sheriff shows him to have been sheriff of Hampshire. There are quite enough authenticated cases of this usage to raise a presumption in favour of the correctness of the view. On the same ground Freeman (*Norm. Conq.* ii. 345, n. 3) assumes that Osbern was sheriff of Hereford.

[165] Merloswegen, sheriff of Lincoln, witnessed charters along with the king's stallers, his two brothers-in-law, Harold and Tostig, the two archbishops, and two bishops (Kemble, *Cod. Dipl.* nos. 806, 808).

[166] Above, note 123.

[167] Kemble, *Cod. Dipl.* no. 729. Although the authenticity of this document is not beyond question, the procedure is sufficiently established by no. 1289.

[168] *Const. Hist.* i. 127.

sheriff what strong kingship will strengthen after the Conquest. The shrievalty is one of a very few centralizing institutions which are to be found in Anglo-Saxon England.   Through it the monarchy of the future will not only direct much of its administration, but will control the activities of public assemblies in shire and hundred.   It contains the germs of a close co-ordination between local and central institutions, which is destined to give rise to the most distinctive features of the medieval English constitution.

III

THE BARONIAL SHRIEVALTY, 1066–1100

FOR a generation after the government of England was fully assumed by Norman officials, about five years subsequent to the Battle of Senlac, the sheriff's power was at its highest. This was the golden age of the baronial shrievalty, the period during which the office was generally held and its tradition established anew by the Conqueror's comrades in arms. The strength of William of Normandy was in no small measure derived from this latter fact. The sheriff in turn profited from the vast access of power which the turn of events and the insight of experience had brought to the king. With the exception of the *curia regis*, the greatest institution at the king's disposal was now the shrievalty. It is the aim of the present chapter to trace the activity and development of the office in this period for which no systematic detailed study of the subject now exists.[1]

There was a strong likeness between the English sheriff and the Norman *vicomte*, and the conquerors naturally identified the one with the other.[2] As the English of the chancery gave place to Latin, *vicecomes* became the official designation; the title *viceconsul* is sometimes found.[3] In the Norman-French of the period the sheriff

---

[1] Stubbs treats the Norman shrievalty in an incidental fashion, covering only its barest outlines (*Const. Hist.* i. 127-8, 295, 299, 425-30). Dr. Round in his various works throws much light particularly upon its financial and genealogical aspects (*Feudal England*, pp. 328-31, 422-30; *Commune*, pp. 72-5; *Geoffrey de Mandeville*, especially appendix P; and numerous chapters in the *Victoria History of the Counties of England*). Mr. Stenton (*William the Conqueror*, pp. 420-4) has treated briefly but with insight and originality the changes in the office brought by the coming of the Normans. The best brief account of the constitutional position of the Norman shrievalty is by Dr. George B. Adams, *The Origin of the English Constitution*, pp. 72-5.

[2] On the Norman *vicomte* in the time of William the Conqueror see C. H. Haskins, *Norman Institutions*, 1918, ch. i.

[3] *D.B.* iv. fo. 312 b.

is the *vescunte*,[4] a name which in the legal language of later times becomes viscount. The employment of Normans in the office gave effect to their administrative ideas. Changes in the shire system soon made the sheriff, like the *vicomte*, the head of government in his bailiwick. At first sight he seems a *vicomté* rather than a *scirgerefa*.[5] Yet the Conqueror did not bodily transplant the Norman office.[6] The legal basis of his shrievalty was that of Edward the Confessor. The history, character and tradition of the English county were very different from those of the Norman *vicomté*. The Norman official had greater advantages and importance in the capacity of sheriff than in that of *vicomte*. The greatest change, moreover, was in the new power behind the sheriff.

It was in accordance with the position claimed by King William as the heir of King Edward that he retained in office a number of English sheriffs, temporarily demanded by administrative necessity. Edward's sheriffs who had served during the few months of Harold's rule seem to have been considered in rightful possession of their shires unless they had resisted the invasion. Godric, the sheriff of Berkshire, who fell fighting with Harold, is mentioned in Domesday Book as having lost his sheriffdom,[7] presumably, as Freeman suggested,[8] because the office was regarded as *ipso facto* forfeit when its occupant moved against William. Osward, the sheriff of Kent, also lost his office,[9] and the proximity of his shire to the place of conflict as well as the known hostility of the Kentishmen to William [10] suggests the same explanation. Esgar, sheriff of Middlesex, who as staller seems to have commanded against the Normans after the battle of Hastings, was not only superseded by a Norman in his office [11] and his lands,[12] but is said to have suffered lifelong imprisonment.[13] In regions more remote from the conflict Englishmen remained in office. Their names, therefore, throw light on Harold's

[4] *Leis Willelme*, 2, 1 ; 2, 2 a, in Liebermann's *Gesetze*, i. 492, 494.

[5] This is well brought out by Mr. Stenton, *William the Conqueror*, p. 422.

[6] The personnel of the two offices was of course different. Roger of Montgomery, viscount of the Hiemois (Ordericus Vitalis, *Hist. Eccles.* ii. 21), became an earl in England.   [7] *D.B.* i. 57 b.

[8] *Norm. Conq.* iv. 729. Godric's lands were seized and granted to a Norman with the exception of the single hide given to his widow for the humble service of feeding the king's dogs (*D.B.* i. 57 b ; cf. Freeman, iv. 37).

[9] *D.B.* i. 2 b.

[10] Ordericus Vitalis relates that after the battle of Hastings they came to terms with William and gave hostages (*Hist. Eccles.* ii. 153).

[11] See note 50.

[12] See *D.B.* i. 129, 139 b.   [13] *Liber Eliensis*, p. 217.

last campaign.   Edric was still sheriff of Wiltshire in 1067 [14] and Touid or Tofig of Somerset apparently as late as 1068.[15]   Alwin or Ethelwine of Warwickshire [16] and Robert fitz Wymarc [17] both remained in office ; and the latter, if not the former as well, was succeeded by his son.   Marloswein or Maerleswegen, whom Harold had left in charge of the north,[18] retained his position in Lincolnshire until he joined the Danes in their attack on York.[19]   The names of several others who continued in office are probably [20] to be added.   There is evidence that the families of Toli,[21] the Confessor's sheriff of Norfolk and Suffolk, and Elfric, his sheriff of Huntingdon,[22] enjoyed King William's favour.   Few of Edward's sheriffs are known, but their importance to William and his attitude towards them are evident.

[14] Round, *Feudal England*, p. 422 ; Davis, *Regesta*, i. no. 9.

[15] Davis, *ibid.* nos. 7, 23.

[16] Alwin appears as sheriff in a document which Eyton ascribes to the year 1072 (*Salt Arch. Soc.* ii. 179).   He was permitted to acquire land by special licence of the Conqueror (*D.B.* i. 242 b).   His son Thurkil seems to have been sheriff of Staffordshire (*Salt Arch. Soc.* ii. 179 ; Davis, *Regesta*, i. no. 25).   His style, Turchil of Warwick (*D.B.* i. 238), suggests that he may have succeeded to the shrievalty of his father (Freeman, *Norm. Conq.* v. 792).   He became an important tenant-in-chief (*D.B.* i. 240 b ; Ballard, *Domesday Inquest*, p. 100).

[17] Robert fitz Wymarc had been staller to King Edward, and is said to have sent to William the news of Stamford Bridge (Freeman, *Norm. Conq.* iii. 413, n. 3).   He was succeeded by his son, Swein of Essex, before 1075 (Davis, *Regesta*, i. nos. 84-6).   Eyton dated his death or superannuation 1071-2 (*Shropshire Arch. and Nat. Hist. Society Publications*, ii. 16).

[18] Gaimar, *Estoire des Engles* (Rolls Series), l. 5235.

[19] *A.-S. Chron.* a. 1067, 1069 ; see Davis, *Regesta*, i. no. 8.

[20] Cyneward (Kinewardus) was sheriff in Worcestershire, but mention of him in 1072 (Heming, *Chartulary*, ed. Hearne, i. 82 ; Thorpe, *Diplom.* p. 441) hardly proves his occupation of the office at that time, as Mr. Davis (*Regesta*, i. no. 106) assumes. See Freeman, *Norm. Conq.* v. 763.   The statement of William of Malmesbury (*Gesta Pontificum*, p. 253) that Urse was sheriff when he built the castle at Worcester, which was before 1069, makes it probable that the English sheriff was superseded by Urse d'Abetot at an earlier date.   The names of Swawold, sheriff of Oxfordshire in 1067 (Parker, *Early History of Oxford*, Oxford Historical Society, p. 301 ; Davis, *Regesta*, i. no. 18), and of Edmund, sheriff of Hertfordshire (*ibid.* no. 16), suggest that they may be sheriffs of King Edward who were not displaced.   One Edwin, who had been the Confessor's sheriff in an unknown county, was probably retained for a time (*D.B.* i. 238 b, 241): *H. tenet de rege et III. hidas emit ab Edwino vicecomite* (*ibid.* i. 157 b).

[21] Toli seems to have died about 1066.   His successor, Norman, may have been the same person as King Edward's sheriff of Northampton (Kemble, *Cod. Dipl.*, nos. 863, 904).   As to Norman's shrievalty in East Anglia see *D.B.* ii. 312 b ; Davis, *Regesta*, i. no. 41 ; Round, *Feudal England*, pp. 228-30.   Toli's widow was still a tenant in Suffolk in 1086 (*D.B.* ii. 299 b).

[22] Elfric's wife and sons were permitted to retain the manor he had held (*D.B.* i. 203).   This Aluric may have been the same as Aluric Godricson, named in 1086 as formerly sheriff of Cambridgeshire (*ibid.* i. 189).

But changes in the shrievalty were rapid. By 1071 Englishmen continued in the office are rare.[23] By 1068 there were Norman sheriffs in fortress cities like London and York, and apparently in Exeter and Worcester.[24] Furthermore, gradual changes in the constitution of the shire added greatly both to the power and the dignity of the office. Whether or not the bishop for a time continued as a presiding officer of the county court,[25] the establishment of separate ecclesiastical courts[26] soon turned his interest in another direction. The earldom also quickly lost its old significance.[27] Domesday Book still carefully records the earl's rights and perquisites, but to all appearances no earl remains except in Kent and a few counties of the extreme west and north.[28] In Kent the sheriff was certainly the creature of the king rather than of Earl Odo.[29] In the palatinates of Chester[30] and Durham[31] the sheriff was long to be the official of the earl and of the bishop respectively. The Montgomery earls in Shropshire,[32] and probably for a short time the

[23] Moreover, Swein of Essex and Thurkil of Warwick (above, notes 16, 17), despite their names, are to all practical intents Norman barons.

[24] See below, p. 59 and notes.

[25] The present writer does not believe with Mr. Davis (*Regesta*, i. 7) that mention of the bishop's name in writs to the county court demonstrates his actual presidency of that body. There is too much evidence of the sheriff's activity. See pp. 55-56.

[26] See Liebermann, *Gesetze*, i. 485.

[27] In the counties of Derby, Nottingham and Lincoln the earl is mentioned in 1086 as if still existent (*D.B.* i. 280 b, 336 b). In Yorkshire the earl may recall persons who have abjured the realm, and proclaim the king's peace (*ibid.* i. 298 b). In Worcester the earl is still said to have the third penny (*ibid.* i. 173 b). But there is no earl.

[28] This striking result was due to the merger of the earldom of Wessex with the Crown, the extinction of the earls of the house of Godwin, the disappearance of Edwin and Morcar by 1071, and finally the revolt of 1075, leading to loss of rank for Roger fitz Osbern and Ralph Guader, the heads of two newly created earldoms, and to the execution of Waltheof, the last surviving English earl.

[29] Concerning Haimo, the sheriff, see note 47. He was in office before the arrest of Odo in 1082, and held the position for years. His family and that of his brother, Robert fitz Haimo (note 71), remained loyal to William Rufus during the great feudal revolt of 1088, in which Odo was involved.

[30] The earl of Chester held of the king the whole shire except what belonged to the bishopric (*D.B.* i. 262 b).

[31] The bishop of Durham had his own sheriff at least as early as Ranulf Flambard's time (Lapsley, *Durham*, pp. 80-1. Compare Symeon of Durham, ii. 209).

[32] Freeman, *Norm. Conq.* iii. 501; Davis, *England under Normans*, p. 517. Earl Roger held Shrewsbury and all the demesne which the king had held in the county. To him obviously, and not to the king, goes the *ferm* of three hundred pounds one hundred and fifteen shillings for the city, demesne manors, and pleas of the county and hundreds (*D.B.* i. 254). Compare the farming of the earl's judicial revenues in Cheshire (*ibid.* i. 262 b). The sheriff at Shrewsbury was the earl's official (Davis, *l.c.*). The shrievalty was successively held by the two husbands

Fitz Osbern earls in Herefordshire,[33] and Count Robert of Mortain in Cornwall,[34] appointed and controlled the sheriff. · In the reign of William Rufus the sheriff of Northumberland was the relative and steward of Earl Robert Mowbray.[35] But elsewhere the subordination of the sheriff to the earl was ended. The burghal third penny generally passed from the earl's into the king's hands,[36] and, as if to emphasize the change, it was occasionally regranted to a sheriff.[37] Except in rare cases like those just mentioned, and soon limited to the palatinates, earls after 1075 did not as such hold administrative office.[38] It was the sheriff and not the earl [39] who had charge of public justice and the maintenance of the peace,[40] and the earl's military headship of the shire was at an end. The conquest of Carlisle from the Scots in 1092 was followed by the appointment of a sheriff.[41] Soon after 1066 a county was

---

of Roger's niece, Warin the Bald and Rainald (Ordericus Vitalis, *Hist. Eccles.* iii. 29 and n. 6 ; *D.B.* i. 254-5).

[33] Heming (*Chartulary*, i. 250) regards Radulf de Bernai (*D.B.* i. 181), the sheriff, as the henchman of William fitz Osbern ; but this could only have been previously to 1075.

[34] Robert held of the king, his brother, almost the whole shire. Thurstin, the sheriff, held land of him (*D.B.* iv. 204 b, 234, 507 b), and as *Tossetin vicecomes* witnessed one of his charters (*Monasticon Anglicanum*, vi. pt. 2, p. 989). Mr. Davis thinks (*Regesta*, i. p. xxxi) that Cornwall could not have been a palatinate as late as 1096, when Warin, the sheriff, is addressed by the king in a writ of the form (*ibid.* no. 378) usually addressed to county courts.

[35] Davis, *England under Normans*, p. 105 ; *A.-S. Chron.* a. 1095. Roger the Poitevin, son of Roger of Montgomery, had a *vicecomes* when his brother Hugh was earl (*Monasticon*, iii. 519 ; Farrer, *Lancashire Pipe Rolls and Early Charters*, 272), apparently in the region between the Ribble and the Mersey (Freeman, *William Rufus*, ii. 57). It is to be observed, however, that the heads of feudal baronies sometimes had *vicecomites* of their own. See Round, *Calendar*, no. 1205 ; also " Some Early Sussex Charters ", in *Sussex Archæological Collections*, vol. xlii. ·

[36] This was true of the burghal third penny at Bath (*D.B.* i. 87), and in the boroughs of Wiltshire (*ibid.* i. 64 b), and must have held for Worcester (note 27) and Stafford (*D.B.* i. 246). Bishop Odo has revenues at Dover which appear to be derived in part from the third penny which Earl Godwin has held (*ibid.* i. 1), but he is not rightfully entitled to Godwin's portion of certain dues at Southwark (*ibid.* i. 32). The record concerning Northampton and Derby shows that the third penny might not be appropriated without grant (*ibid.* i. 280 b).

[37] Baldwin was the recipient of the third penny at Exeter, Hugh of Grantmesnil at Leicester (see Ballard, *Domesday Boroughs*, p. 37, n. 6) and Robert of Stafford at Stafford (*D.B.* i. 246).

[38] The old practice of conferring the third penny upon them and of naming them in writs to the county court has become mere form.

[39] For the theory of the Anglo-Saxon period see *Eng. Histor. Rev.* xxxi. 27.

[40] Below, pp. 55, 60.

[41] Davis, *Regesta*, i. no. 478 ; *Monasticon Anglicanum*, i. 241.

being called a *vicecomitatus* or sheriffdom.[42]   Unobscured by any greater official, the sheriff now stands out as the sole head of the shire.

The importance and power of the Norman shrievalty were further enhanced by a tenure of office usually long and by a personnel of remarkable character.   The removability of the sheriff was still an effective principle, the usefulness of which by no means ended with its application to the cases of English sheriffs who fought for Harold.   William dismissed from the office Normans of no little importance.[43]   Yet the *crementum*, the sum of money occasionally paid for the privilege of farming the shire,[44] seems to represent a bid for the appointment.   The influence of feudal usage was also strong.   It has been held justly that William I. could not have dismissed sheriffs wholesale as did Henry II. without risking a feudal rebellion.[45]   The Norman viscounty was, in some instances, hereditary.[46]   The sheriff was appointed for no specified term, and the tendency of the age was to treat offices like fiefs.

Personal claims to the king's friendship or gratitude did much to lengthen the tenure of office.   The leading sheriffs of the Conqueror often held office for life, and some of them survived until the reign of Henry I.[47]   A few who stood especially high in the king's favour

[42] Herman's *Miracula Sancti Eadmundi*, written about 1070, has *Aerfasto duarum Eastengle vicecomitatuum episcopo* (Liebermann, *Ungedruckte anglonormannische Geschichtsquellen*, p. 248).   In the Domesday inquest for Bedfordshire appears the expression, *Omnes qui iuraverunt de vicecomitatu* (D.B. i. 211 b) ; and in the record of the judgement in the case of Bishop Wulfstan against Abbot Walter, 1085–6, we read *iudicante et testificante omni vicecomitatu* (Heming, *Chartulary*, i. 77).

[43] Among these was Froger, sheriff of Berkshire (*Chron. Abingdon*, i. 486, 494).   About 1072 Ilbert lost the shrievalty of Hertfordshire (D.B. i. 133).   For the date compare Round, *Feudal England*, pp. 459–61, with Liebermann, *Gesetze*, i. 485.   Swein of Essex lost his place, to be followed by Ralph Bainard (D.B. ii. 2 b).   This was before 1080 (Davis, *Regesta*, i. no. 122).   The latter by 1086 (D.B. ii. 1 b) had been superseded by Peter of Valognes, who was sheriff of Essex (*V.C.H. Essex*, i. 346).   Peter, Swein and Ralph were all Domesday tenants-in-chief.

[44] See below, p. 64.

[45] Stenton, *William the Conqueror*, p. 423.

[46] See Haskins in *Am. H.R.* xiv. 470 [*Norman Institutions*, p. 47].

[47] Haimo, who has been identified as son of Haimo Dentatus, slain at Val-es-Dunes (cf. *E.H.R.* xxxiii. 150, n. 47), and who was a distant relative of William the Conqueror and *dapifer* both to him and to William Rufus, is mentioned as sheriff of Kent about 1071 (Bigelow, *Placita Anglo-Norm.* p. 8) and also in 1086.   Farrer shows (*Itinerary of Henry I.*, nos. 584A-586) that he was not superseded in the period 1078–83 as Davis, no. 188, indicates (no. 98 shows that he was sheriff in 1077) ; he seems to have remained in office until his death, which Mr. Davis shows was in 1099 or 1100 (*ibid.* nos. 416, 451).   He was succeeded both in his household office (*Monasticon Anglicanum*, v. 100, 149 ; *E.H.R.* xxvi. 489) and his shrievalty

held great household offices at court.[48]   Another group are known

(*Monasticon*, i. 164 ; iii. 383 ; Round, *Calendar*, no. 1378) by another Haimo, who was undoubtedly his son.

Roger Bigod, probably son of a knight closely attached to the fortunes of the Conqueror (cf. *E.H.R.* xxxiii. 150, n. 47), became the greatest noble in East Anglia and *dapifer* to William II.   He was sheriff of Norfolk by 1069, sheriff of Suffolk for two different terms prior to 1086, as well as under Henry I., and Domesday sheriff of both counties.   For his share in the rebellion of 1088 he apparently lost his estates temporarily, and surrendered his office for a time to Herbert, the king's chamberlain, but he served as sheriff later than 1091 and probably until his death, which occurred in 1107.   Ralph de Bellfago was sheriff of Norfolk at some time in the period 1091–1102 (*Ramsey Chron.* 228).

Urse d'Abetot, or, more properly, d'Abbetot (Dupont, *Recherches historiques et topographiques sur les compagnons de Guillaume le Conquérant*, 1907, pt. 2, p. 7), a trusted agent of the Norman kings for a period of forty-five years or more following the Conquest, was the brother of Robert the despenser of the Conqueror (Heming, *Chartulary*, i. 268) and William II. (Davis, *Regesta*, i. no. 326).   He became the greatest lay landholder in Worcestershire, of which county he was sheriff apparently (note 20) from 1068.   He is still mentioned as sheriff about 1110 (Liebermann, *Gesetze*, i. 524), and at his death before Sept. 1113 (Farrer, no. 290) he was succeeded by his son (note 63).

Edward of Salisbury, a great landholder in the southern and south-western counties and another (cf. *E.H.R.* xxxiii. 150, n. 47) *curialis*, was sheriff of Wiltshire in 1081, and possibly as early as 1070.   He seems to have been sheriff so late as 1105 (*E.H.R.* xxvi. 489-90).   His daughter Matilda married the second Humphrey de Bohun, who shared his vast possessions with his son, Walter of Salisbury (*Monasticon*, vi. 134, 338, 501).

Baldwin de Meules (Meulles : Dupont, pt. 2, 49), or Baldwin de Clare, son of Count Gilbert of Brionne (Ordericus Vitalis, ii. 181), one of the guardians of the Conqueror's minority, was delegated to build a castle at Exeter after the revolt of 1068 (*ibid.*).   A great landholder, he enjoyed the rare distinction of having a castle of his own (*D.B.* i. 105 b), which was situate at Okehampton.   He was sheriff of Devon by about 1070 (Davis, no. 58), and without doubt held the office until his death a little before 1096 (Round, *Feudal England*, p. 330, n. 1).

Durand of Gloucester, another Domesday sheriff, served for fifteen years or more (note 62) preceding his death.

Hugo de Port, sheriff of Hampshire possibly as early as 1070 (Davis, no. 267), and a great landholder, seems to have held office until in 1096 he became a monk (*ibid.* no. 379).   He was sheriff of Nottingham also in the period 1081-7 (*Monasticon*, i. 301).

[48] As to Haimo and Roger Bigod, see note 47.

Robert d'Oilly, sheriff of Warwickshire in and before 1086 (*V.C.H. Warwick*, i. 279 ; Davis, *Regesta*, i. nos. 104, 130, 200), his shrievalty beginning about 1070 (*ibid.* no. 49), was constable under William I. and William II. (*ibid.* p. xxxi).

Robert Malet, son and probably successor in office (note 82) of a well-known follower and sheriff of the Conqueror (see p. 162 and Dupont, *Recherches historiques et topographiques*, pt. 2, pp. 13-14), sheriff of Suffolk from 1070 (Davis, no. 47) to at least 1080 (*ibid.* no. 122), and an important tenant-in-chief in several shires, was the king's great chamberlain (Round, *Geoffrey de Mandeville*, p. 180).

Aiulf, the chamberlain, Domesday sheriff of Dorset (note 82), and in the reigns of William II. and Henry I. sheriff of Somerset (Davis, nos. 315, 417 ; *Montacute Chart.*, Somerset Record Soc., p. 120), was a tenant-in-chief both in Dorset (*D.B.* i. 82 b) and Wiltshire (*ibid.* 75), and probably at court a deputy to Robert Malet.

Edward of Salisbury is believed to have been a chamberlain of Henry I. (*E.H.R.* xxvi. 489-90).

E

to have been in his special employment at the *curia* or elsewhere.[49] To practically all of these he made large grants of land *in capite*, usually in several shires. Similar grants prove his friendship for a still larger group.[50] With the exception of a very few of whom

[49] These are Urse d'Abetot (Heming, *Chartulary*, ii. 413; Round, *Feudal England*, p. 309; Davis, *Regesta*, i. nos. 10, 416, 422; see also below, p. 59 and note 130), Edward of Salisbury (notes 48, 49; Davis, nos. 247, 283), Hugo de Port (*ibid.* nos. 207, 220), Baldwin of Exeter (above, note 47), Hugo de Grantmesnil (note 58), and Peter de Valognes (Davis, no. 368). The last named was the Domesday sheriff of Essex and Hertfordshire, and tenant-in-chief both in these shires and in Lincolnshire, Norfolk and Suffolk. His wife, Albreda, was the sister of Eudo the *dapifer* (*Monasticon*, iii. 345; iv. 608). He was sheriff of Hertfordshire about 1072 (note 43), and still sheriff of Essex in the reign of William II. (Davis, nos. 436, 442). Hugh de Beauchamp was sheriff of Buckinghamshire in the reign of William II. (Davis, no. 370), at whose court he was employed (*ibid.* nos. 419, 446, 447). Hugh of Buckland witnessed writs of William II. (*ibid.* nos. 444, 466), and in 1099 was delegated to execute a judgement of the king's court (*ibid.* no. 416).

[50] Geoffrey de Mandeville, sheriff of London and Middlesex from the Conquest (Round, *Geoffrey de Mandeville*, p. 37, n. 2, p. 439; Davis, *Regesta*, i. nos. 15, 93), though not at the date of Domesday (*D.B.* i. 127; Davis, *ibid.* no. 306), and at some period of his career sheriff of Essex and Hertfordshire (Round, *ibid.* pp. 141-2), is well known as a landholder in eleven different shires. Cf. Page, *London* (1923), 233-34.

Hugh fitz Grip, sheriff of Dorset, was dead by 1086, but his wife was a tenant-in-chief, holding some forty manors (*D.B.* i. 83 b).

Ralph Bainard, a Domesday tenant-in-chief in Essex, Norfolk and Suffolk (*D.B.* ii. 68, 247, 413), a pre-Domesday sheriff of Essex (Davis, no. 93), possibly of London as well (*ibid.* no. 211), and his brother, Geoffrey Bainard, a noted adherent of William II. (Freeman, *William Rufus*, ii. 63), who, in the reign of the latter, seems to have been sheriff of Yorkshire (Davis, nos. 344, 421, 431; *E.H.R.* xxx. 283-4), bear the name of a well-known baronial family; as does Ralph Taillebois, sheriff of Bedfordshire and Hertfordshire (*V.C.H. Buckingham*, i. 220), who died before 1086 (*D.B.* i. 211 b), and Ivo Taillebois, *dapifer* to William II. (Davis, nos. 315, 319, 326), *tenens* in Norfolk, and presumably sheriff of Lincolnshire before 1086 (*E.H.R.* xxx. 278).

Hugh fitz Baldric, sheriff of Yorkshire from 1069 to about 1080 (*E.H.R.* xxx. 282), and also sheriff of Nottinghamshire, was a Domesday *tenens* not only in these shires but also in Hampshire (*D.B.* i. 48, 356) and Lincolnshire.

Ansculf de Picquigny, sheriff of Buckinghamshire (*D.B.* i. 148 b) and Surrey (*ibid.* i. 36), also deceased before 1086, was father of the prominent Domesday baron, William de Picquigny.

William de Mohun, sheriff of Somerset in 1084 and 1086, and probably for a considerable period (Maxwell-Lyte, *History of Dunster*, pp. xiii and 3), was a great landholder and founder of a well-known house.

Durand of Gloucester (*D.B.* i. 168 b, 186 b), though himself not a great tenant, represents an important family interest.

Robert of Stafford (Davis, no. 210 and app. xxvi; see *D.B.* i. 225, 238, 248 b) held much land of the Crown.

Picot, the notorious sheriff of Cambridgeshire, one of the barons who attended the *curia regis* in the time of William II. (*Deputy Keeper's 29th Rep.*, app. p. 37), who was in office as early as 1071 (Davis, no. 47), and as late as some date in the period 1090-98, was a tenant-in-chief in his own shire (*D.B.* i. 200).

Eustace of Huntingdon, of almost equally evil memory, sheriff by 1080 (Davis,

little is recorded,[51] and a very few in the counties still under an earl,[52] the known sheriffs [53] at or near the date of Domesday, some twenty in number, are all tenants-in-chief [54] of the Crown, and as a rule great tenants-in-chief. Four of them left heirs, who within two generations became earls.[55] The baronial status of the shrievalty is thus well established. As important barons or household officials a number of them frequently appear at meetings at the *curia regis*,[56] even as *vicomtes* usually attended the duke's *curia* in Normandy.[57] Rank, importance, or official position, moreover, entitled the sheriff of more than one English shire to a place in this Norman body.[58]

The greater power and prestige of the Norman as compared with

---

no. 122) and superseded by 1091 (*ibid.* nos. 321, 322, 329), was a Domesday *tenens* in Cambridgeshire and Northamptonshire as well as in Huntingdonshire.

William of Cahaignes, sheriff of Northamptonshire under both William I. and William II. (*ibid.* nos. 283 b, 288), was also a Domesday tenant-in-chief (*D.B.* i. 201 b).

[51] Ranulf of Surrey (*D.B.* i. 32), Roger of Middlesex (*D.B.* i. 127), and Gilbert (*D.B.* i. 20 b), who may be sheriff of Sussex or *vicomte* of the honour of Pevensey. There was an " O ", sheriff of Surrey in the period 1070–98 (Westminster Domesday, Westminster Abbey, f. 161).

[52] Rainald, formerly sheriff of Shropshire (*D.B.* i. 181), Gilbert or Ilbert of Hereford (notes 149, 212), and Thurstin of Cornwall (note 34).

[53] The counties whose sheriffs I am unable to name are Berkshire, Oxford, Leicester, Rutland, Derby, Cheshire and Northumberland. It seems impossible to tell how long Froger, the first Norman sheriff of Berkshire, remained in office.

[54] See notes 47, 50. Haimo, one of the smallest landholders among these, had in Kent three whole manors and parts of others (*D.B.* i. 14), lands in Essex, besides (*ibid.* ii. 54 b). Durand, another small tenant, had lands in the south-west (*D.B.* iv. fo. 8 b), as well as in Gloucestershire (*ibid.* i. 168 b) and Herefordshire (*ibid.* i. 179).

[55] Hugh, second son of Roger Bigod ; Patrick, grandson of Edward of Salisbury ; Miles of Gloucester, grandnephew of Durand ; and Geoffrey de Mandeville, grandson of the sheriff of the same name.

[56] This appears in connexion with the trial of Bishop William in 1088 : see *Columbia Law Review*, xii. 279.

[57] Haskins, *Norman Institutions*, p. 47.

[58] Robert d'Oilly, the constable, and Robert Malet, the chamberlain (above, note 48), both appear at William's *curia* in Normandy (Davis, *Regesta*, i. nos. 199, 207), as do also Hugo de Port and Baldwin of Exeter (*ibid.* nos. 125, 220). Hugo de Grantmesnil appears in attendance even before the conquest of England (*ibid.* no. 2). In 1050 along with his brother Robert he founded the monastery of St. Evroul. Present at Hastings, he was employed by the Conqueror about 1068 to hold Hampshire. Subsequently he received an important post at Leicester (Ordericus Vitalis, *Hist. Eccles.* ii. 17, 121, 186, 222). He was a great landholder in the midlands in 1086, and appears as witness to one of the writs of William II. (Davis, no. 392). The language of Ordericus (*vicecomes Leyrecestrae*, iii. 13 ; *praesidatum Leyrecestrae regebat*, iii. 270) and his possession of the third penny at Leicester (note 37) indicate that he was sheriff (Freeman, *Norm. Conq.* iv. 232). He died in the habit of a monk, Feb. 22, 1093 (Ordericus Vitalis, iii. 453). His son Ivo, who succeeded to his English possessions, was one of the four lords of Leicester and *municeps et vicecomes et firmarius regis* (*ibid.* iv. 169).

the Anglo-Saxon sheriff are evident. No longer was he a man of moderate means, overshadowed by the nobility and prelates of the shire ; on the contrary, he was often himself the greatest man in all his region and not infrequently a benefactor of the church.[59] Since no official superior stood between him and the king, he enjoyed great freedom of action. As a baron and a personal adherent of the king, he combined the prestige of a local magnate and the status of a trusted official. He was, as it were, a sheriff of King Edward, who through royal favour had grown into a great landholder and a prominent king's thegn. The effective control exercised over the office by the early Norman kings [60] is thus largely explained, though its basis could not be expected to survive the generation which followed the Conqueror at Hastings.

The hereditary nature of some of the Norman shrievalties is well understood,[61] but the known instances are not numerous. The families of Roger dé Pistri and of Urse d'Abetot each supplied four sheriffs, the former in Gloucestershire,[62] the latter in Worcester-

[59] Peter of Valognes and his wife founded the priory of Binham (*Monasticon,* iii. 345 ; iv. 608), Roger Bigod that of Thetford (*ibid.* v. 148-9), Ivo Taillebois the monastery of Spalding (*ibid.* iii. 215, 217), Picot the priory of Barnwell (Miss Norgate, *England under the Angevin Kings*, ii. 463). Hugo de Grantmesnil endowed the monastery of St. Evroul (Ordericus Vitalis, *Hist. Eccles.* ii. 14 ff.), and later gave it some of his English property (Davis, *Regesta*, i. no. 140). Robert d'Oilly endowed the church at Abingdon (*Chron. Abingdon*, ii. 12-15). Warin gave land to the monastery of Shrewsbury (*Monasticon*, iii. 518), Haimo to the church of St. Andrew at Rochester (Davis, *Regesta*, i. no. 451), and Hugh fitz Baldric tithes to the abbey of Preaux (*ibid.* no. 130). Baldwin of Exeter and both his sons who succeeded him were benefactors of Bec (Round, *Feudal England*, table facing p. 473). Geoffrey de Mandeville founded the priory of Hurley (Round, *Geoffrey de Mandeville*, p. 38), and also gave land to St Peter of Westminster for his wife's soul (Davis, *Regesta*, i. no. 209), Durand to St. Peter of Gloucester *pro anima fratris sui Rogerii* (*D.B.* i. 18), Thorold to St. Guthlac of Croyland *pro anima sua* (*ibid.* i. 346 b), Rainald to the church of St. Peter *pro anima Warini antecessoris sui* (*D.B.* i. 254).

[60] See Adams, *English Constitution*, p. 72.

[61] Stubbs, *Const. Hist.* i. 295.

[62] Roger de Pistri was sheriff of Gloucester as early as about 1071 (Davis, *Regesta*, i. no. 49). His brother Durand, the Domesday sheriff, seems to have succeeded him before 1083 (*ibid.* 186). After the death of Durand about 1096 (Round, *Feudal England*, p. 313), his nephew, Walter fitz Roger (*D.B.* i. 169), better known as Walter of Gloucester, became sheriff, although Durand's son Roger, who seems to have succeeded to his lands, lived until 1107. Walter is mentioned as holding the office in 1097 (Davis, *ibid.* no. 389), and again in 1105-6 (*Monasticon*, i. 544). He evidently served for many years, for his son Miles, who was sheriff in 1129, was confirmed in possession of the family lands but shortly before (Farrer, *Itinerary*, no. 578) and still owed a sum which he had recently engaged to pay for the land and *ministerium* of his father (*Pipe Roll*, 31 Henry I. p. 77). Miles was constable of England until he was superseded in Stephen's time by Walter de Beauchamp. He was created by Matilda earl of Hereford (Round, *Geoffrey de Mandeville*, pp. 263, 285).

shire.[63]   The power of these families, already strong through their
local baronial standing, was further increased by the fact that in
each case the custody of a castle was held together with the
shrievalty.[64]   Baldwin of Exeter, another great tenant-in-chief
and custodian of Exeter castle,[65] was succeeded as sheriff of Devon
in turn by two of his sons.[66]   The Grantmesnil and Malet shrievalties seem to have passed from father to son,[67] but both sons were
ruined in consequence of their adherence to Duke Robert of Normandy in the early years following the accession of Henry I.[68]
Haimo was succeeded both as *dapifer* and as sheriff of Kent by his
son Haimo,[69] and his son Robert [70] is no doubt the Robert fitz
Haimon who was sheriff of Kent in the earlier years of Henry I.[71]
Ralph Taillebois and Ivo Taillebois seem both to have been sheriffs
of Bedfordshire before the Domesday inquest.[72]   Swein of Essex
and probably Turchil of Warwickshire were hereditary sheriffs of
a slightly earlier date.[73]   The surname of Walter of Salisbury indicates that he succeeded Edward, his father.[74]   Henry de Port,
sheriff of Hampshire in 1105, was the son, though not the immediate
successor, of Hugo de Port.[75]   The second Geoffrey de Mandeville
in the time of King Stephen greatly increased the strength of his
newly acquired earldom by regaining the three shrievalties held by
his grandfather in the days of the Conqueror.[76]   By this time such

[63] Urse d'Abetot held the Worcestershire shrievalty from about 1068 (above,
note 20).   The office passed at his death, about 1113, to his son Roger, and after
the latter's disgrace to Walter de Beauchamp, the husband of Urse's daughter
(Round, in *D.N.B.*, art. "Urse d'Abetot", and in *V.C.H. Worcester*, i. 263).   Walter's
son, William de Beauchamp, still held the position in the reign of Henry II.

[64] Below, p. 59.

[65] Baldwin was the patron of the church of St. Mary within the castle (*Devonshire Association for Advancement of Science*, xxx. 27).

[66] Round, *Feudal England*, p. 330, n. 37 ;   *V.C.H. Devon*, i. 555.

[67] See notes 48, 58, 82.   [68] Ordericus Vitalis, *Hist. Eccles.* iv. 167.

[69] Above, note 47.   [70] See Davis, *Regesta*, i. no. 451.

[71] At some time in the period 1103–9 (*Monasticon*, iii. 383 ; Round, *Calendar*,
no. 1377).   He was still prominent in 1130 (*Pipe Roll*, 31 Henry I., pp. 95, 97).
This cannot be Robert fitz Haimon, the conqueror of Glamorgan, and brother of
the elder Haimo (William of Jumièges, Migne, *Patrolog. Lat.* cxlix. 898), who was
injured and lost his reason in 1105 (*E.H.R.* xxi. 507-8).   He left no son.

[72] *D.B.* i. 209, 209 b.   Ivo exacted the sheriff's *crementum* for demesne manors.
See note 50.

[73] Above, notes 16, 17.

[74] Walter, moreover, was the father of Patrick, earl of Salisbury (*Monasticon*,
vi. 338, 501), sheriff of Wiltshire in the seventh year of Henry II.   But see below,
p. 78, n. 27.

[75] Davis, *Regesta*, i. nos. 377, 379 ;   *E.H.R.* xxvi. 489-90.

[76] Round, *Geoffrey de Mandeville*, pp. 141-2,

power has long been regarded as a menace to the state. In the great majority of counties there is no life tenure nor hereditary succession, and sheriffs follow each other in more rapid succession.[77] The sheriff was in so many known instances surnamed from the chief town of his shire that this usage has been assumed to be the rule.[78] The title of Swein of Essex affords almost the only case of a different usage for this period.[79] Sometimes a sheriff was placed over two counties, but this double tenure in nearly every case seems to have been of brief duration.[80] The Conqueror and his sons limited the hereditary sheriff to one shire.[81] Occasionally a sheriff held two shires in succession.[82] Hugh of Buckland, one of the new *curiales* of William Rufus,[83] who in the reign of Henry I. was *carus regi* and sheriff of eight shires,[84] held three of these by or before 1100.[85] New men will in the future be utilized to check the influence of the powerful sheriff with baronial interests. The participation

[77] For the sheriffs of Lincolnshire and Yorkshire see *E.H.R.* xxx. 277 ff. ; for the sheriffs of Essex and Hertfordshire prior to 1086, above, notes 43 and 50. In Warwickshire also the succession was comparatively rapid. In London, Geoffrey de Mandeville (note 50), Ralph Bainard (Davis, no. 211), and Roger (*D.B.* i. 127) all served before 1086.

[78] See Round, *Feudal England,* p. 168, where a list of instances is given. To this may be added Durand of Gloucester (*D.B.* i. 168 b) as well as Peter of Oxford, who belongs to the reign of William II. (*Chron. Abingdon,* ii. 41). Urse d'Abetot appears as *Urso de Wircestre* (*D.B.* i. 169 b).

[79] Yet Turchil de Warewicscyre appears in Thorpe, *Diplomatarium,* p. 441.

[80] Mr. Round has shown that the Domesday reference to Urse d'Abetot in Gloucestershire (i. 163 b) does not prove that he ever had this shire along with that of Worcestershire (*V.C.H. Worcester,* i. 263). Roger Bigod, the famous sheriff of Norfolk, was sheriff also of Suffolk at various times (note 47). Ralph Taillebois, who died before 1086, served both Bedfordshire (*D.B.* i. 218 b) and Hertfordshire (*V.C.H. Buckingham,* i. 220), but in Hertfordshire Edmund was sheriff at the opening of the reign (Davis, no. 16), and Ilbert probably before 1072 (above, note 43). Concerning the length of time during which Ansculf held the shrievalties of Buckinghamshire (*D.B.* i. 148 b) and Surrey (*ibid.* i. 36), and Geoffrey de Mandeville those of Essex and Hertfordshire (see note 50), there is no definite information. Hugh fitz Baldric, sheriff of Yorkshire (note 50), was also sheriff of Nottinghamshire in 1074 (*E.H.R.* xxx. 282).

[81] Miles of Gloucester, however, was sheriff of Staffordshire and Gloucestershire, 1128–30 (*Pipe Roll,* 31 Henry I., pp. 72, 76).

[82] Aiulf, sheriff of Dorset in and before 1086 (*D.B.* i. 83), was in office in the period 1082–4 (Davis, *Regesta,* i. no. 204), and was sheriff of Somerset before 1091 (*ibid.* nos. 315, 316), and also (above, note 48) in the reign of Henry I. William Malet, sheriff of Yorkshire from 1067 to 1069 (*E.H.R.* xxx. 281), seems to have been sheriff of Suffolk before April 1070 (Round, *Feudal England,* pp. 429-30).

[83] Above, note 49. Ordericus Vitalis (*Hist. Eccles.* iv. 164) mentions him only as one of the men *de ignobile stirpe* raised from the dust by Henry I.

[84] *Chron. Abingdon,* ii. 117.

[85] He held Bedfordshire (Davis, *Regesta,* i. nos. 395, 471), Hertfordshire (*E.H.R.* xxvii. 103), and Berkshire (below, note 112) in the reign of William II.

in the rebellion of 1088 by two such officials doubtless recalled the dangerous revolt of Norman *vicomtes* in 1047.[86]

The perquisites of the office, both legitimate and other, were probably greatest in the generation following the conquest of England. The view that the Danegeld was farmed and constituted the sheriff's greatest source of profit [87] is untenable,[88] but there are indications in Domesday that the farming of the king's lands and the local pleas yielded a handsome margin.[89] How the oppressive sheriff might turn his power to financial advantage will appear later. The fact that so great a tenant as Urse d'Abetot might apparently gain exemption from the relief of 1095 [90] hints what influence at court might do. Sheriffs are mentioned as having certain lands for the term of their office.[91] The reeveland [92] as well as certain pence pertaining to the shrievalty, which Edward of Salisbury received,[93] might add to the sheriff's profits, though the latter and probably the former were held subject to certain official obligations.

The Domesday sheriff had personal agents or *ministri*. Among these may possibly be under-sheriffs, for the spirited denunciation written by the monk of Ely indicates that Picot of Cambridge had such a subordinate.[94] It is clear that among these *ministri* were reeves, and there is a presumption that by 1086 the sheriff was the head of the royal and public reeves of the shire. The *ministri regis* are sometimes seen to perform the same duties as reeves,[95] and the

[86] William of Malmesbury, *Gesta Regum*, ii. 286.
[87] Stubbs, *Const. Hist.* i. 412.
[88] Round, *Feudal England*, pp. 499-500.
[89] Below, p. 64.   [90] Round, *Feudal England*, p. 313.
[91] A manor in Dorset held by Aluric, presumably the sheriff in the time of King Edward, is held by Aiulf of the king as long as he shall be sheriff (*D.B.* i. 83); *Quam terram dederat Ilbertus cuidam suo militi dum esset vicecomes* (ibid. i. 133).
[92] *D.B.* i. 181 ; iv. 29 ; Maitland, *Domesday Book*, p. 169.
[93] *D.B.* i. 69.
[94] *Gervasius . . . irae artifex, inventor sceleris, confudit fas nefasque ; cui dominus eius dictus Picotus tamquam caeteris fideliori pro sua pravitate totius comitatus negotia commiserat.* The account ends with the story that St. Etheldreda and her sisters appeared and punished Gervase with death for his offences against this church (*Liber Eliensis*, p. 267). At the inquest of several shires taken at Keneteford the sheriffs of Norfolk and Suffolk were represented by a deputy (Davis, *Regesta*, i. no. 122).
[95] *De his ii hidis nec geldum nec aliquod debitum reddiderunt ministri regis* (*D.B.* i. 157 b, Oxfordshire). Certain customs which the king formerly had at Gloucester neither he nor *Rotbertus minister eius* now has (ibid. i. 162. *Hanc forisfacturam accipiebat minister regis et comitis in civitate* (ibid. i. 262 b, Chester). According to *Leges Henrici*, 9, 10 a (Liebermann, *Gesetze*, i. 556), the *ministri regis* are officials who farm the local pleas.

ministri vicecomitis have the same functions.[96] The sheriff of the period is known to have had reeves with fiscal duties.[97] Since the authority of the sheriff regularly extended to manors of the royal demesne,[98] it follows that the king's reeve of Domesday was his subordinate. This is attested by fairly convincing evidence.[99] The dependence of the hundredmen upon the sheriff is shown by the fact that in Devonshire they as well as king's reeves were collectors of the king's *ferm*, including the portion derived from the pleas of the hundred.[100] In Norfolk one of the hundred-reeves had for more than a decade held land *per vicecomites regis*.[101] Finally, Mr. Ballard's conclusion,[102] that except at Hereford and Dover the borough *praepositus* of Domesday was the sheriff's subordinate, appears to be well founded.

Under the early Norman kings the sheriff's judicial position was most important and his independence in judicial matters

[96] The *ministri* of Roger Bigod increased a render to fifteen and later to twenty pounds (*D.B.* ii. 287 b, Suffolk). The Conqueror granted a hundred to the abbot of Evesham, *quod nullus vicecomes vel eorum ministri inde se quicquam intromittant vel placitent vel aliquid exigant* (Davis, *Regesta*, app. xiii). At the Domesday inquest for Hampshire the *ministri regis*, contrary to the testimony of the men of the shire and the hundred, declare that a certain piece of land belongs to the king's *ferm* (*D.B.* i. 50).

[97] The Domesday sheriff of Wiltshire was responsible for the *ferm* collected by reeves, and when there was a deficiency had to make it good (*D.B.* i. 69). Roger Bigod as sheriff of Suffolk warranted to a reeve a free man who had been joined to the *ferm* of *Brunfort* (*ibid.* ii. 282). William II. enjoined a sheriff to make reparation for wrong done by his reeve Edwy and his other *ministri* (*Chron. Abingdon*, ii. 41). Haimo's agents who seized some of Anselm's property during his absence from England are mentioned by the latter as *vestri homines* (epist. lvii. Migne, *Patrolog. Lat.* clix. 233).

[98] Maitland, *Domesday Book*, p. 167; see also *E.H.R.* xxi. 31, note 97.

[99] A *praepositus regis* claimed land for pasturing the king's cattle, but was met by the witness of the shire that he might have it only through the sheriff (*D.B.* i. 49, Hants). A sheriff made certain estates reeveland for the *praepositi regis* (*ibid.* i. 218 b). Moreover, these officials are mentioned as taking part in the collection of the *ferm* (*ibid.* iv. fo. 513 b). Roger Bigod is shown to have been closely associated with the act of the *praepositus regis* in his shire who seized unto the king's hand the land of an outlawed person (*D.B.* ii. 176 b; cf. *ibid.* ii. 3). According to *D.B.* iv. fo. 513, the *ferm* of a manor was rendered *praeposito regis de Winesford*, who seems to be the ordinary official of the manor (*D.B.* i. 179 b).

[100] *Comes [de Moritonio] habet i. mansionem quae vocatur Ferdendella . . . De hac mansione calumniantur hundremani et praepositi regis xxx. denarios et consuetudinem placitorum ad opus firme Ermtone mansione regis* (*D.B.* iv. fo. 218). The reeve who held the hundredmote was apparently a dependent of the sheriff in the time of King Edward (*E.H.R.* xxxi. 28).

[101] *D.B.* ii. 120. The land had been given to the reeve originally by Earl Ralph, who was overthrown in 1075.

[102] *The Domesday Boroughs*, pp. 45-7. Certainly this was true at Canterbury, for the sheriff, Haimo, held this city of the king (*D.B.* i. 2).

greatest. The usage which in the reign of Henry I. regarded the sheriff as solely responsible for holding the sessions of the hundred and the shire was evidently not new.[103] According to Domesday Book the sheriff holds local courts even in Herefordshire,[104] which for a time has probably been a palatinate, and in Shrewsbury,[105] where the earl's authority over sheriff and shiremote is still great.[106] The essence of one of the very greatest franchises is exemption of a hundred from the jurisdiction of the sheriff and his reeves.[107]   In separating ecclesiastical from secular jurisdiction the Conqueror forbade any sheriff or reeve or *ministri regis* to interfere in matters which belonged to the bishop.   If any one contemns the bishop's summons three times the *fortitudo et iustitia regis vel vicecomitis* are to be invoked.[108]   In all but most exceptional causes the Norman sheriff for a time must have been the justice.[109]   To commission some one else required a special exercise of the royal prerogative. The pleas of the Crown, the income from which was not farmed, and went to the king *in toto*,[110] as well as the ordinary causes triable in

---

[103] The writ of 1109–11 (Liebermann, *Gesetze*, i. 524) establishes no new principle in this regard, but merely directs the sheriff how these sessions are to be held.

[104] Of the Welsh of Archenfield we read, *si vicecomes evocat eos ad siremot meliores ex eis vi. aut vii. vadunt cum eo. Qui vocatus non vadit dat ii. solid. aut unum bovem regi et qui de hundret remanet tantundem persolvit* (*D.B.* i. 179).

[105] *Siquis burgensis* (of Shrewsbury) *frangebat terminum quem vicecomes imponebat ei emendabat x. solid.* (*D.B.* i. 252).

[106] Above, note 32. See also Davis, *England under Normans*, p. 517.

[107] *E.H.R.* xxxi. 28. See also above, note 96. The church of St. Mary of Worcester had a hundred with similar liberty (*D.B.* i. 172 b), and the exclusion of the sheriff from the hundred of Hornmere, held by the monastery of Abingdon (*Chron. Abingdon*, ii. 164), was of long standing.

[108] Liebermann, *Gesetze*, i. 485 ; Stubbs, *Sel. Charters*, p. 85.

[109] The king's court is in the main " only for the great man and the great causes " (Pollock and Maitland, i. 108).

[110] The usual five-pound *forisfacturae* (*E.H.R.* xxxi. 32-3), which were *extra firmas*, the king had everywhere on his demesne in Worcestershire from all men (*D.B.* i. 172), and in Kent from all *allodiarii* and their men. The list in the last-named county (*D.B.* i. 2) included the felling of trees upon the king's highway. For grithbreach in Kent in certain cases eight pounds was paid, and in Nottingham (*ibid.* i. 280) the same amount for impeding the passage of boats down the Trent or for ploughing or making a ditch in the king's highways toward York.   Manslaying on one of the four great highways (*Leis Willelme*, 26, Liebermann, *Gesetze*, i. 510) counted as breach of the king's peace. In Yorkshire (*D.B.* i. 298 b) and Lincolnshire (*ibid.* i. 336 b) the king was entitled in twelve hundreds, the earl in six, to eight pounds for breach of peace given by the king's hand or seal. At Oxford the housebreaker who assailed a man (*ibid.* i. 154 b), and in Berkshire the man who broke into a city by night (*ibid.* i. 56 b), paid five pounds to the king. Burghers in some towns (*ibid.* i. 154 b, 238) who failed to render the due military service paid the same amount, although sums collected for various other offences in boroughs were often less. In Cheshire the lord who neglected to render service

the shire and hundred, seem to be dealt with by him and his subordinates. It has been shown, however, that as early as the reign of William Rufus there were special royal justices locally resident.[111] Hugh of Buckland, sheriff of Berkshire in this reign, seems to combine the two offices,[112] but they are already separable.

The sheriff's position as head of the judicial system of the shire is the central fact in Norman local government. It involved numerous duties and responsibilities. The law of the king's court being as yet unformed and fitful in operation, the most important law-declaring body was still the county court.[113] A strong sheriff could exert a decided influence upon customary law.[114] His control tended towards uniformity of practice. About 1115 the observances of judgement, the rules of summons, and the attendance in the counties convened twice a year are said to be the same as those in the hundreds convened twelve times a year.[115] In the one instance in which Domesday affords data for comparison the sum collected for absence from the hundred is the same as that for absence from the shire.[116] All this means activity for the sheriff and the reeves under him.[117] The two great sessions of each hundred held annually

---

toward repairing the bridge and the wall of the city (*D.B.* i. 262 b) incurred a *forisfactura* of forty shillings, which is specifically stated to have been *extra firmas*. On a Berkshire manor *latrocinium* is mentioned among the great *forisfacturae* (*ibid*. i. 61 b). The *murdrum* fine (*Leis Willelme*, 22, Liebermann, *Gesetze*, i. 510) was already being collected (Davis, *Regesta*, i. no. 202) in the Conqueror's reign. Half the goods of the thief adjudged to death in some places went to the king (*D.B.* i. 1); for certain offences a criminal's chattels were all confiscated. According to the *Leis Willelme* (2, 2 a-2, 4, Liebermann, *Gesetze*, i. 494-5) the *forisfactum regis* of forty shillings in the Mercian law and that of fifty shillings in Wessex belong to the sheriff, while in the Danelaw the man with sake and soke who is impleaded in the county court forfeits thirty-three ora, of which the sheriff retains ten for the king.

[111] Davis, *England under Normans*, p. 520. As to the local justiciar of the twelfth century see Round, *Geoffrey de Mandeville*, pp. 106-9. A writ of William II., directed to his *iudicibus*, sheriffs, and officials (Davis, *Regesta*, i. no. 393), seems to show the change.

[112] *Et Berchescire vicecomes et publicarum iusticiarius compellationum a rege constitutus* (*Chron. Abingdon*, ii. 43).

[113] Vinogradoff, *English Society in the Eleventh Century*, p. 91.

[114] Davis, (*England under Normans*, 522) suggests that the sheriff's influence contributed to the great diversity of local judicial usage. Cf. Stubbs, *Const. Hist.* i. 393, n. 4.

[115] *Leges Henrici Primi*, 7, 4-7, 8, Liebermann, *Gesetze*, i. 553-4.

[116] Above, note 104. Compare *Rex habet in Dunwic consuetudinem hanc quod duo vel tres ibunt ad hundret si recte moniti fuerint et si hoc non faciunt forisfacti sunt de ii. oris* (*D.B.* ii. 312).

[117] Thus a writ of Henry I. addressed to Roger Bigot and *omnibus ministris de Suthfolcia* directs them to permit a vill of St. Benedict of Ramsey to be quit of

to make view of frankpledge [118] presumably met in this period under the sheriff's presidency,[119] no less than in the reign of Henry II.[120] Sentence of outlawry was pronounced by the sheriff in the county court,[121] and Mr. H. W. C. Davis [122] has found indications that in the time of the Conqueror the forest law was sometimes enforced in the same way. It is usually assumed that this machinery was turned to financial oppression in the king's interest during the reign of Rufus.[123] So far as one may judge, it was through the sheriff's jurisdiction that the king's financial claims were enforced.[124] Nothing but the sheriff's power could have enabled Ranulf Flambard to drive and supervise " his motes over all England ". To the sheriff in the shiremote [125] were communicated the king's grants, proclamations, and administrative orders. About him turned the administrative as well as the judicial system of the shire.

The sheriff might be directed by royal writ to reserve certain cases to the king's court,[126] and he was sometimes commissioned to assume its judicial powers, as were *vicomtes* in Normandy.[127] The mention of a resident justice in the shire [128] shows, on the other hand,

shires and hundreds and of all other pleas except *murdrum* and *latrocinium* (*Ramsey Chartul.* i. 249). There is evidence that the sheriff summoned men to the shiremote (note 104).

[118] *Leges Henrici*, 8, 1-8, 2, Liebermann, *Gesetze*, i. 554 ; cf. *Leis Willelme*, 25, *ibid.* i, 511.

[119] Dr. Liebermann even believes that this was true in the reign of Edward the Confessor (cf. *E.H.R.* xxxi. 29, note 28), when the sheriff is known to have held sessions of the hundred. See the present writer's *Frankpledge System*, pp. 113-14.

[120] Assize of Clarendon, § 9, Stubbs, *Sel. Charters*, p. 144.

[121] *Siquis pro aliquo reatu exulatus fuerit a rege et comite et ab hominibus vicecomitatus* (*D.B.* i. 336). Since there was no longer an earl the presidency of the sheriff follows.   [122] *Regesta*, i. p. xxxi.

[123] Stubbs, *Const. Hist.* i. 327 ; Freeman, *William Rufus*, i. 344.

[124] *E.H.R.* xxxi. 33 ; see below, pp. 62-3.

[125] See W. H. Stevenson, *E.H.R.* xxi. 506-7. Of a grant addressed in the familiar form, *Willelmus rex Anglorum, Gilleberto de Britteville et omnibus fidelibus suis, Francigenis et Angligenis, de Berkascire*, the Abingdon chronicler (*Chron. Abingdon*, ii. 26) says : *rex Willelmus iunior . . . concessit istas ad comitatum Berkascire inde litteras dirigere*. Dr. Liebermann finds evidence (*Trans. R.H.S.*, N.S. viii. 22) that the coronation charter of Henry I. was to be read in every shire court in the kingdom ; cf. Davis, *England under Normans*, p. 119, n. 4.

[126] See the writ of William II. to the sheriffs in whose shires the abbot of Evesham held lands (Davis, *Regesta*, i. no. 429 ; *Monasticon*, ii. 22).

[127] See Davis, *Regesta*, i. nos. 117, 132 ; Haskins, *Norman Institutions*, p. 46.

[128] See the case of Hugh of Buckland dating from the reign of William II. (above, p. 56). A charter of William I. which mentions the sheriffs and justiciars of Devon has been explained by Mr. Davis (*Regesta*, i. no. 59) as probably a variant of later date. The charter of Henry I. to London (*Gesetze*, i. 525) not only shows that the sheriff and *iustitiarius* are two different persons, but shows that the function of the latter was *ad custodiendum placita coronae meae et eadem placitanda*.

that some other agent of the king might be entrusted with judicial functions which the sheriff had formerly discharged. During the Conqueror's reign a sheriff is known in but one instance to have sat alone as a commissioned royal justice; [129] but the earliest known eyre, some time in the period 1076–9, was held before two sheriffs [130] along with other barons. Precepts of William II. order sheriffs to dispose of certain assigned cases.[131] Through such royal mandates the sheriff first came into contact with that royal inquest for ascertaining facts which constituted the original form of the jury. The king's writ enjoining such procedure might come direct to the sheriff [132] or to a person serving as the king's justice at whose instance the sheriff sometimes acted.[133]

The military functions of the sheriff in the period under consideration were derived both from English and from Norman usage. The principle of the general levy provided a fighting force exceedingly useful in an emergency, though inferior to that yielded by the system of knight service now imported from Normandy. The sheriff of King Edward led both the shire levies and the special forces sent by the boroughs.[134] Vestiges of such arrangements still appear in Domesday Book.[135] Florence of Worcester mentions the military service rendered by Urse d'Abetot against the rebellious earls in 1074 in terms which suggest that he commanded a general levy.[136] Robert Malet, sheriff of Suffolk, was one of the leaders of

[129] *Yale Law Journal*, xxiii. 506.

[130] Round, *Feudal England*, p. 329. Urse d'Abetot may have sat as justice in his own shiremote under the presidency of Geoffrey of Coutances (Davis, *Regesta*, i. no. 230; compare no. 184).

[131] To do right to the abbot of Westminster concerning the churches of Scotland (Davis, no. 420) or to summon three and a half hundreds to deal with a case concerning the rights of the abbot of Ramsey (nos. 448, 449). Humphrey the Chamberlain, in the latter case, seems to be acting as sheriff.

[132] *Hist. Monast. St. Augustini* (R.S.), pp. 353–4, 356; Davis, *Regesta*, i. no. 448.

[133] See the case in which Picot and Odo of Bayeux were concerned; below, p. 72.

[134] *E.H.R.* xxxi. 30.

[135] The Welsh of the district of Archenfield, who in King Edward's time served under the sheriff of Hereford, number 196 in 1086. They are required to make expeditions into Wales only when the sheriff goes (*D.B.* i. 179). To this service *in exercitu regis* they are so firmly bound that if one of them dies the king has his horse and arms (*D.B.* i. 181). At Taunton all were under obligation to go *in expeditione* with the bishop's men (*D.B.* iv. fo. 174). The quota demanded of boroughs was usually fixed at a comparatively small figure. See Maitland, *Domesday Book*, p. 155, n. 8.

[136] *Wulfstan cum magna militari manu et Angelwinus Eoveshamensis abbas cum suis ascitis sibi in adiutorium Ursone vicecomite Wigorniae et Waltero de Laceo cum copiis suis et cetera multitudine plebis* (Florence of Worcester, a. 1074).

the king's forces which put down the revolt of 1075 in East Anglia.[137]
The *inward*, which in the Confessor's time was rendered in the west
and midlands under the sheriff's direction,[138] still abounds in the
Domesday period.[139]   In Kent the tenants of certain lands guarded
the king for three days when he came to Canterbury or Sandwich.[140]
The Norman *vicomte*, on the other hand, was keeper of the king's
castles,[141] and the earlier sheriffs of the Conqueror often appear in
this capacity.[142]   William Malet held the castle at York, and in
1069 unsuccessfully defended it against the Danes.[143]   The story
of the excommunication of Urse d'Abetot shows that he was the
builder of the castle at Worcester ; [144] he was also its custodian,[145]
a post to which his daughter's husband, Walter de Beauchamp, and
his grandson, William de Beauchamp, succeeded in turn.   The
custodianship of the castle at Exeter likewise became hereditary in
the family of Baldwin, the sheriff who erected it.[146]   The constable-
ship of Gloucester was attached to the shrievalty at least as early
as the time of Walter of Gloucester.[147]   There is evidence of such
an arrangement elsewhere,[148] although sheriffs were not necessarily

[137] Davis, *Regesta*, i. no. 82.
[138] *E.H.R.* xxxi. 29, 35.
[139] See, for example, *D.B.* i. 132 b, 190.
[140] *Ibid.* i. 1.   This obligation was commuted in one Kentish district by rendering for each *inward* two sticks of eels, and in another by a payment of twelve pence
for each *inward*.
[141] See Haskins, *Norman Institutions*, p. 46.
[142] This suggests that William Peverel (Ordericus Vitalis, *Hist. Eccles.* iv. 184),
in whose hands the castle of Nottingham was placed when it was built in 1068,
may have been sheriff.
[143] *Habuit Willelmus Malet quamdiu tenuit castellum de Euruic . . . Dicunt
fuisse caisitum Willelmum Malet et habuisse terram et servitium donec fractum est
castellum* (*D.B.* i. 373).   Florence of Worcester (Engl. Hist. Soc.), ii. 4, adds
details.   William was on a military expedition apparently in the fenland at
the time of his death (Dupont, *Recherches historiques et topographiques*, pt. 2,
p. 14).
[144] William of Malmesbury, *Gesta Pontificum*, ii. 253.
[145] Round, *Geoffrey de Mandeville*, pp. 313-14 ; *D.N.B.* art. "Urse d' Abetot ".
[146] Ordericus Vitalis, *Hist. Eccles.* ii. 181 ; Round, *Geoffrey de Mandeville*,
p. 439 ; above, p. 51.
[147] His son Miles in the reign of Henry I. held its custody *sicut patrimonium
suum* (Round, *ibid.* p. 13, n. 1 ; *Monasticon*, vi. 134).   Walter also had charge of
the castle of Hereford.
[148] It has not been proved that Geoffrey de Mandeville held the tower of London,
but both his son and grandson did so (Round, *Geoffrey de Mandeville*, pp. 37-8, 166).
Similarly the shrievalty of Wiltshire in the twelfth century included an hereditary
custodianship.   In Dorset Hugh fitz Grip cleared ground for work on the castles
(*D.B.* i. 75), and the sheriff at Lincoln performed a similar service (*ibid.* i. 336).
The same was true at York and apparently at Gloucester and Cambridge.   See
below, note 249.

custodes castelli.[149]   When Roger Bigod rebelled in 1088 he seized
Norwich Castle,[150] so as sheriff he was hardly its guardian.   Both
he and Hugh de Grantmesnil, however, must have been materially
strengthened in this revolt by the resources of their office.   After
the failure of the movement in the north Durham Castle was delivered
to the sheriffs of Lincolnshire and Yorkshire.[151]   During this rebellion the sheriffs also took possession of the men, lands and property
of Bishop William of Durham,[152] one of the rebels.

   The retirement of the earl left the sheriff the authority for
keeping the peace and administering matters of police within his
bailiwick.   At Shrewsbury, in a region where the sheriff had been
exceptionally prominent, it was he and not the earl who proclaimed
the king's peace in the time of King Edward.[153]   After the earl has
disappeared throughout the greater part of England, the Domesday
inquest for Warwickshire shows that this function belongs to the
sheriff,[154] and an entry for Yorkshire proves that the realm may
be abjured before him, and that he has the power of recalling and
giving peace to a person who has thus made abjuration.[155]   The
sheriff's well-known power of arresting malefactors [156] was extended
when he was made responsible for enforcing the forest laws.[157]   This
phase of his activity can hardly have been new,[158] but the severity of
Norman forest regulations [159] certainly gave it new significance.   A

[149] *Custodes castelli* are mentioned in Sussex (*D.B.* i. 21).   Robert the despenser,
brother of Urse d'Abetot, held the castle and honour of Tamworth (Round, *Geoffrey
de Mandeville*, p. 314).   Gilbert the sheriff of Herefordshire had the castle of Clifford
to farm, but it was actually held by Ralph de Todeni (*D.B.* i. 173).   Robert d'Oilly,
castellan of Oxford in the reigns of William I. and William II., was sheriff of
Warwickshire (*Monasticon*, i. 522 ; *Chron. Abingdon*, ii. 12).

[150] *A.-S. Chron.* a. 1088 ; W. Malmesbury, *Gesta Regum*, ii. 361.

[151] *E.H.R.* xxx. 282-3.   They were possibly former sheriffs.

[152] *Monasticon*, i. 245.   [153] *D.B.* i. 252.   [154] *D.B.* i. 172.

[155] *Si vero comes vel vicecomes aliquem de regione foras miserint ipsi eum revocare
et pacem ei dare possunt si voluerint* (*ibid.* i. 298 b).

[156] *E.H.R.* xxxi. 30-31.

[157] Mr. Davis (*Regesta*, i. p. xxxi) has established such a responsibility.   Not
only does the sheriff of Kent serve on a commission to judge forest offences
(*ibid.* no. 260), but a precept of the king to his sheriff and liegemen of Middlesex
forbids any one to hunt in the manor of Harrow which belongs to Archbishop
Lanfranc (*ibid.* no. 265).   In the Confessor's time the guarding of the forest might
be a manorial duty for which commutation was made by money payment (*D.B.*
i. 61 b).   So in the reign of the Conqueror (*D.B.* i. 180 b, Herefordshire), *Willelmus
comes misit extra suos manerios duos forestarios propter silvas custodiendas.*   Mr.
Davis associates foresters with the enforcement of forest law only by the time of
William Rufus.

[158] See II. Canute, 80, 1, Liebermánn, *Gesetze*, i. 366-71.

[159] See *A.-S. Chron.* a. 1087 ; Freeman, *Norm. Conq.* v. 124-5.

letter of Bishop Herbert de Losinga implores the lord sheriff and God's faithful Christians in Norfolk and Suffolk to seek and give up those who have broken into his park at Homersfield and killed a deer.[160] There are in the archives of Westminster Abbey letter forms,[160a] dating apparently from the time of Abbot Gilbert and addressed to the Sheriffs of Middlesex, Surrey and Essex, by which the abbot claims right of sanctuary with life and members for an offender who has sought the altar of St. Peter and the body of King Edward. The sheriff's duties were further increased through the enactment of the Conqueror providing that he was to deal with those who contemned the authority of the episcopal court.[161] A writ of Henry I., addressed in 1101 to the shiremote of Lincolnshire, and presumably sent to other shires, orders the sheriff and certain notables to administer to the king's demesne tenants the oath to defend the realm against Robert of Normandy.[162]

The sheriff was the recipient of royal mandates of many varieties. The king's writs, whether addressed directly to the sheriff or to the county court to be published by the sheriff,[163] imposed special administrative no less than judicial duties. They attest the prerogative powers of the Norman kingship and reveal the shrievalty as an arm of a central executive. Notices to shiremotes of royal grants of lands or privileges [164] incidentally warrant the surrender by the sheriff and reeves of part of the king's rights. Sheriffs made livery of lands,[165] and placed grantees in possession of customs or privileges by writ or order of the king.[166] To the usual clause of the king's writ-charter forbidding any one to disturb the grantee [167] may sometimes be added another restraining the sheriff or another

<hr>

[160] Goulburn and Symonds, *Herbert de Losinga*, pp. 170-72 ; *Epistolae Herberti Losingae*, etc., ed. Anstruther, Caxton Soc. 1846, p. 71.

[160a] Westminster Domesday, f. 79, 79 d.

[161] Above, p. 55.   [162] *E.H.R.* xxi. 506-9.

[163] Of a mandate of the Conqueror in the usual form confirming its lands to the church of Abingdon it is said, *Quarum recitatio litterarum in Berkescire comitatu prolata plurimum et ipsi abbati et ecclesiae commodi attulit* (*Chron. Abingdon*, ii. 1).

[164] See Davis, *Regesta*, i. nos. 160, 162, 176, 209, 210, 212, 245. Nos. 244, 277, 289 give possession with *sac* and *soc.*

[165] The sheriff of Yorkshire gave possession of land to Bishop Walcher *per brevem regis* (*D.B.* i. 298). See also *ibid.* i. 167, and Davis, *Regesta*, i. no. 442. In some places an act of livery must have been usual when the writ was read. In the Domesday inquest, as, for instance, i. 36, 50, 62, 164, both the men of the shire and the hundred seem to doubt that a grant of land has been made, because they have never seen the king's writ nor act of livery.

[166] Davis, *Regesta*, i. no. 87.

[167] *Ibid.* nos. 14, 17, 85, 243, 244, 294.

officer from doing so,[168] or else ordering the sheriff to see that no injustice is done in the matter.[169] A common method of enforcing the decision of the king's court, especially when held locally by a royal justice, was by writ to the sheriffs.[170] A form of peremptory command bids the sheriff see that a given person shall have certain property or rights, and let the king hear no further complaint on the matter.[171] The sheriffs may be ordered to seize the property of rebels or other persons under the royal ban.[172] Henry I. commands the sheriffs of Kent and Essex to prohibit fishing in the Thames before the fishery at Rochester on pain of the king's *forisfactum*.[173] William I. causes Lanfranc and Geoffrey of Coutances to summon the sheriffs and tell them in the king's name to restore lands, the alienation of which had been permitted by bishops and abbots.[174] William II. orders the sheriffs of the shires wherein the abbot of Ramsey has lands to alienate none of his demesne without the king's licence.[175] The Conqueror's writ to William de Curcello, presumably sheriff of Somerset, enjoins that payment of Peter's pence shall be made at next Michaelmas by all thanes and their men, and that William, together with the bishop, is to make *inquisitio* concerning all who do not pay and to take them in pledge.[176]

The sheriff has charge of the king's property and of his fiscal rights. Land at the king's farm may be *in manu vicecomitis*,[177] and the sheriff often holds land which is *in manu regis*.[178] Lands which the king holds in demesne are mentioned as having been officially received by the sheriff.[179] The sheriff has the custody of land which has fallen to the king through forfeiture.[180] He seizes

[168] As in Round, *Calendar*, no. 1375.

[169] *Monasticon*, ii. 18; Davis, *Regesta*, i. no. 104.

[170] *Ibid.* i. nos. 129, 230, 288 b.

[171] *Ibid.* no. 329.

[172] Above, p. 60.

[173] *Monasticon*, i. 164.

[174] Davis, *Regesta*, i. no. 50.

[175] *Ibid.* no. 329.

[176] *Cal. of MSS. of the Dean and Chapter of Wells*, Hist. MSS. Commission, i. 17; Davis, *Regesta*, i. no. 187. Pledge was not to be taken upon the bishop's land until the matter came before him.

[177] *Modo est in manu vicecomitis ad firmam regis* (*D.B.* ii. 5).

[178] A part of *Blontesdone* held by Edward the sheriff is *in manu regis* (*ibid.* i. 74); *modo custodit hoc manerium Petrus vicecomes in manu regis* (*ibid.* ii. 1). Of the half hundred and borough of Ipswich it is said, *hoc custodit Roger Bigot in manu regis* (*ibid.* ii. 290).

[179] *Rex tenet in dominio Rinvede . . . Quando vicecomes recepit, nisi x. hidae. Aliae fuerunt in Wilt* (*D.B.* i. 39). Cf. *Quando Haimo vicecomes recepit* (*ibid.* i. 2b).

[180] *Hoc invasit Berengarius homo Sancti Edmundi et est in misericordia regis. Hic infirmus erat. Non potuit venire ad placitum. Modo sunt in custodia vicecomitis* (*ibid.* ii. 449). *Quas tenuit i. faber T.R.E. qui propter latrocinium interfectus fuit et praepositus regis addidit illam terram huic manerio* (*D.B.* ii. 2 b).

land for failure to render service due [181] or to pay geld [182] or gavel,[183] and he brings action against a person who has invaded lands *de soca regis*.[184]   One reads at times of the king's saltpans as in his charge [185] and of boroughs as held by him.[186]   It is his business to see that the king's estates of which he is guardian are kept properly stocked with plough oxen,[187] and he is the custodian of the peasants who till the land.[188]   Through an application of the doctrine of seisin the profits from pleas are said to be *in manu vicecomitis*.   Bishop Odo sued the sheriff of Surrey in order to obtain the third penny of the port dues at Southwark.[189]   Control of the king's lands also means control of their issues.   It is this which in the past has made the sheriff an attendant upon the royal progresses.[190]

The innate financial genius of the Norman, together with the unusual opportunities which the period afforded for increasing the royal income, renders the sheriff's fiscal functions of striking importance both to the king and the realm.   The early development of direct taxation in England as compared with the Continent has been pronounced one of the most remarkable facts of English history.[191] Here the sheriff appears both as the agent of a dominant central power and also as its main support.

A *firma comitatus* existed at least in one case before 1066.   It

---

[181] See below, p. 70.

[182] *Hanc terram sumpsit Petrus vicecomes . . . in manu eiusdem regis pro forisfactura de gildo regis* (*D.B.* i. 141).

[183] *. . . ille gablum de hac terra dare noluit et Radulfus Taillgebosc gablum dedit et pro forisfacto ipsam terram sumpsit* (*D.B.* i. 216 b).

[184] Picot was the sheriff and Aubrey de Vere the trespasser (*ibid.* i. 199 b).

[185] *Ibid.* ii. 7 b ; cf. Ellis, *Introduct. to Domesday Book*, p. xli.

[186] Thus Haimo held Canterbury of the king (*D.B.* i. 2).   The see of St. Augustine and Abbot Scotland were in 1077 reseised of the borough of Fordwich which Haimo held (*Hist. Mon. S. Augustini*, p. 352).   See also above, note 178.

[187] *D.B.* ii. 1, 2 ; see also *V.C.H. Essex*, i. 365.

[188] The services of the sokemen whom Picot lent Earl Roger to aid him in holding his pleas (*D.B.* i. 193 b) were regarded as lost to the king.   Richard fitz Gilbert in Suffolk held as appurtenant to one of his manors certain *liberi homines* formerly acquired by agreement with the sheriff (*ibid.* ii. 393).   In Buckinghamshire the sokeman who has land which he can give and sell nevertheless *servit semper vicecomiti regis* (*D.B.* i. 143, 143 b).   The sheriff's custodianship of some cottiers at Holborn was of longer standing (*D.B.* i. 127).   When in 1088 William of St. Calais was proclaimed a rebel the villeins on his Yorkshire manors were seized or held to ransom by the sheriff (*Monasticon*, i. 245).   On a Gloucestershire manor of the royal demesne the sheriff is said to have increased the number of villeins and bordars (*D.B.* i. 164).

[189] *D.B.* i. 32.   Ranulf the sheriff, apparently overawed, let the matter go by default.

[190] Above, p. 33.

[191] Vinogradoff, *English Society in the Eleventh Century*, p. 140.

F

is known that by 1086 there are instances of the payment by the sheriff of one sum for the royal revenues of the county which are farmed.[192]   The number of such cases casually mentioned suggests that this may long have been the rule in counties where any of the king's lands are held at *ferm*.   Not only is there a *ferm* of Wiltshire,[193] but the sheriff is said to be responsible for the *ferm* collected by reeves, and must make good the amount which is due from them.[194]   The annual *ferm* from Warwickshire [195] and from Worcestershire [196] consists both of the *firma* of demesne manors and of the *placita comitatus*, as in the days of the Pipe Rolls.   Indeed the *Leges Henrici* will speak of the *soke* of sheriffs and royal bailiffs comprised in their *ferms*.[197]   Northamptonshire and Oxfordshire [198] each pays a lump sum in commutation of a *ferm* of three nights. Geoffrey de Mandeville held London and Middlesex for an annual *ferm* of £300, Essex for £300, and Hertfordshire for £60.[199]   William de Mohun, sheriff of Somerset, likewise accounted for a fixed sum ; [200] in Shropshire, which has become a palatinate, the earl in 1086 received one *ferm* for the demesne lands and the pleas of the county and hundreds.[201]   The *augmentum* or *crementum* mentioned in Domesday [202] appears to be a premium paid by the sheriff in excess of the regular *ferm* for the privilege of farming the shire, the equivalent of the *gersoma* of the Pipe Roll of Henry I.[203]

There are various other evidences of the sheriff's activity as

[192] Round, *Commune*, pp. 72-3.
[193] *Hanc terram tenet Edwardus (de Sarisberie) in firma de Wiltescira iniuste ut dicit comitatus (D.B.* i. 164).   [194] Above, note 97.
[195] £145 *ad pondus*, to which are added certain customary payments, partly in the nature of commutation, *xxiii. libras pro consuetudine canum, xx. solidos pro summario et x. libros pro accipitre et c. solidos reginae pro gersuma (D.B.* i. 238).
[196] . . . *reddit vicecomes xxiii. libras et v. sol. ad pensum de civitate et de dominicis maneriis regis reddit cxxiii. libras et iiii. solidos ad pensum.   De comitatu vero reddit xvii. libras ad pensum, et adhuc x. libras denariorum et de xx. in ora pro summario. Hae xvii. librae ad pensum et xvi. librae sunt de placitis comitatus et hundredis et si inde non accepit de suo proprio reddit (D.B.* i. 172).
[197] *Leges Henrici*, 9, 10 a, Liebermann, *Gesetze*, i. 556.
[198] *D.B.* i. 154 b, 219.   For Oxfordshire the amount is £150.
[199] Round, *Geoffrey de Mandeville*, pp. 141-2, 166-7.
[200] Round, *Commune*, p. 73.   [201] Above, note 32.
[202] In Oxfordshire £25 *de augmento* is mentioned (*D.B.* i. 154 b).   Edward of Salisbury paid £60 *ad pondus* as *crementum* (*ibid.* i. 64 b).   The *gersuma* of Domesday is smaller, and seems to be in theory a gift.   Oxfordshire (*D.B.* i. 154 b) paid a hundred shillings as the queen's *gersuma*.   In Essex a *gersuma* of the same amount was paid by a manor or borough to the sheriff (*ibid.* ii. 2 b, 3, 107).   See below, note 205.   Six manors in Herefordshire rendered twenty-five shillings *gersuma* at Hereford (*ibid.* i. 180 b).
[203] *Pipe Roll*, 31 Henry I., pp. 2, 52, 73.

head of the *ferm* of the shire. Of this the pleas of the hundred formed an important source,[204] the income from which might regularly be included in the *ferm* of lands.[205] There are instances in which the sheriff annexes the revenue from a hundred court to that of a royal manor [206] or borough.[207] Moreover, Maitland's inference that the sheriff lets boroughs to *ferm* [208] has been justified by more recent research. The case of Worcester and the familiar example of Northampton [209] by no means stand alone. The facts collected by Mr. Ballard make it clear that the sheriff was ordinarily accountable for borough renders.[210] In the Domesday inquest the sheriff appears as a witness to facts concerning the *ferm*,[211] and sometimes he himself farms royal estates,[212] though in most cases they are farmed by some one else. The sheriff is frequently mentioned as letting such lands to farm,[213] and the person who holds

[204] Both the two pence of the king and the third penny of the earl derived from Appletree hundred, Nottinghamshire, are *in manu et censu vicecomitis* (*D.B.* i. 280). Because seven of the hundreds of Worcestershire had been exempted from his control the sheriff lost heavily in *ferm* (*ibid.* i. 172). Swein of Essex had been granted from the pleas of one hundred in Essex a hundred shillings, from those of another twenty-five (Ballard, *Domesday Inquest*, p. 70).

[205] *Vicecomes inter suas consuetudines et placita de dimidio hundred recepit inde xxxiiii. libras et iv. libras de gersuma* (*D.B.* ii. 2, Essex). *De hac mansione calumpniantur hundredmanni et praepositus regis xxx. denarios et consuetudinem placitorum ad opus firme Ermtone mansione regis* (*ibid.* iv. fo. 218).

[206] *T.R.E. reddebat vicecomes de hoc manerio quod exibat ad firmam. Modo reddit xv. libras cum ii. hundred quos ibi apposuit vicecomes* (*ibid.* i. 163, Gloucester).

[207] *Ibid.* i. 162. The income from three hundreds had been combined with that of the borough of Winchcombe.

[208] *Domesday Book*, p. 209.

[209] *Ibid.* pp. 204-5. Mr. Ballard has remarked that this is the only case in Domesday in which burgesses appear to farm a borough (*Domesday Boroughs*, p. 92). It has been pointed out, however (*V.C.H. Northampton*, i. 277), that it was a century before they acquired the privilege of farming directly of the Crown. As to the *ferm* of the city of Worcester, see note 196.

[210] *Domesday Boroughs*, pp. 44-5. The sheriff is mentioned as increasing a borough render. There is allusion to the time when he received a borough upon entering office (*D.B.* i. 2, Canterbury ; i. 280, Northampton). He is said to account for the burghal third penny. The collection of the *census domorum* at Worcester (*D.B.* i. 172), of the poll tax at Colchester (*ibid.* ii. 106 b), of the port dues at Southwark (*ibid.* i. 32), and of toll in many places (*D.B.* i. 209 ; Davis, *Regesta*, i. no. 201) seems to be the work of his agents.

[211] *D.B.* i. 248 ; ii. 446 b.

[212] Thus Gilbert the sheriff of Herefordshire held at farm the *castelleria* and borough of Clifford (*D.B.* i. 183). Harkstead manor in Essex was farmed by Peter of Valognes (*D.B.* ii. 286 b). Urse d'Abetot personally accounted for the *ferm* of certain manors in Worcestershire (*D.B.* i. 172, 172 b).

[213] *Hoc manerium cepit W. comes in dominio et non fuit ad firmam. Sed modo vicecomes posuit eum ad lx. solidos numero* (*D.B.* i. 164). *Durandus vicecomes dedit haec eadem Willelmo de Ow pro lv. libris ad firmam* (*ibid.* 162). See also below, notes 217, 220.

them under him may be regarded as holding at the king's *ferm*.[214] William II. let the hundred of Normancros to the monks of Thorney for a hundred shillings, payable annually to the sheriff of Huntingdonshire.[215] Extensive districts were sometimes administered collectively. There was a *ferm* of the king's rights for the Isle of Wight.[216] The *ferm* for a whole group of estates might be collected through a head manor,[217] a plan necessarily followed when great groups of manors in the south jointly paid the amount of a day's *ferm* in commutation of the ancient food-rent rendered to the king.[218] A money economy prevails except in the case of certain old renders which seem to have been added to *ferms*,[219] and sometimes a cash value is set on these. Two Domesday passages record the payment of borough *ferms* to the sheriff about Michaelmas or Easter,[220] although only the latter of these dates corresponds with one of the known terms for the half-yearly payment of Danegeld.[221]

Other fiscal duties of the sheriff are occasionally mentioned in Domesday Book. The revenues from the special pleas of the Crown, such as *murdrum* and the five-pound *forisfacturae*, though

[214] *Reddit per annum xvi. libras ad pensum et quando Baldwinus vicecomes recepit hanc qui tenet eam ad firmam de rege reddebat tantumdem* (*D.B.* iv. fo. 83 b).

[215] Davis, *Regesta*, i. 453.

[216] *D.B.* i. 38 b.

[217] Briwetone and Frome together rendered the *ferm* of one night *cum suis apenditiis* (*D.B.* iv. fo. 91). Robert holds *Bedretone in firma Wanctinz* (*ibid.* i. 57, Berks.) Four hides of land lying in a Gloucestershire manor are *ad firmam regis* in Hereford (*D.B.* i. 163 b). *Ad hoc manerium apposuit vicecomes tempore W. comitis Walpelford* (*D.B.* i. 179 b).

[218] See Round, *Feudal England*, p. 109 ff.

[219] Such as sheep, hawks, sumpter horses, food for the king's dogs, wood for building purposes (*D.B.* i. 38 b, Dene), salt, corn and honey. Thus, Domesday has: *dimidiam diem de frumento et melle et aliis rebus ad firmam regis pertinentibus. . . . De consuetudine canum lxv. solidi* (i. 209 b); *ii. denarios et theloneum salis quod veniebat ad aulam* (*ibid.* i. 164); *Ilbertus vicecomes habet ad firmam suam de Arcenefeld consuetudines omnes mellis et ovium* (*ibid.* i. 179 b). See also notes 195, 196. Domesday Book (iv. fo. 91) mentions *firmam unius noctis cum appenditiis*.

[220] Roger Bigot gave Ipswich to farm for £40 at Michaelmas (*D.B.* ii. 290). At Colchester the burghers of the king each year, fifteen days after Easter, rendered two marks of silver which belonged to the *firma regis* (*ibid.* ii. 107). The reeves on the lands of Worcester made certain money payments at Martinmas and in the third week of Easter (Heming, *Chartulary*, i. 98-9). The burghers of Derby rendered corn to the king at Martinmas (*D.B.* i. 280).

[221] Mr. Round (*Domesday Studies*, ed. Dove, i. 91) points out the coincidence between the earlier of these periods and the usual time of the meeting of the great council at Winchester, the seat of the treasury. He holds that the final annual accounting of the collectors of the Danegeld was at Easter. The payment of Peter's pence was at Michaelmas (above, p. 62).

not included in the *ferm*, were collected by the sheriff.[222] The collection locally of the pence for the maintenance and wages of the king's levies [223] probably fell under his supervision. Picot had from the lawmen of Cambridge, as heriot, eight pounds, a palfrey and the arms of one fighting man ; and Aluric Godricson, when he was sheriff, had twenty shillings as the heriot of each lawman.[224] From the reign of King Edward the sheriff or the king's reeve in Suffolk had the commendation or half the commendation of men on certain lands.[225] It is recorded that in the counties of York, Nottingham and Derby the thane with more than six manors gave a relief of eight pounds to the king, while the thane with six manors or less paid three marks of silver to the sheriff.[226] There is reason to hold that the sheriff had charge of the collection of the Danegeld,[227] and he is mentioned as responsible for port dues collected.[228] Anselm complains that during his absence from England the agents of Haimo took toll of the archbishop's property at Fordwich.[229] At Holborn the king had two cottiers who rendered twenty pence a year to the sheriff.[230] Numerous persons in Hertfordshire, not on the royal demesne, rendered to the sheriff pence in lieu of *avera* or in addition to *avera*.[231] At Cambridge the sheriff had exacted of the burghers nine days' service with their ploughs instead of the three days formerly required. Moreover, the *inward* which he claimed, like the *avera*, might be commuted by a money payment.[232] From three manors which Queen Edith held in Surrey the sheriff had £7 on account of *adiutorium* which was due from the men when she

---

[222] Above, note 110 ; *E.H.R.* xxxi. 32-3. *Averam et viii. denarios in servitio regis semper invenerunt et forisfacturam suam vicecomiti emendabant* (*D.B.* i. 189 b).

[223] See *D.B.* i. 56 b ; ii. 107. It is to be noted that William Rufus made this a systematic means of extortion (Stubbs, *Const. Hist.* i. 327).

[224] *D.B.* i. 189.   [225] *D.B.* ii. fos. 312 b, 334, 334 b.

[226] *D.B.* i. 280 b, 298 b.

[227] *E.H.R.* xxxi. 34-5. The collectors of the Danegeld were reeves of the class usually under the sheriff's control. His responsibility is assumed by Stubbs (*Const. Hist.* i. 412) and by Mr. Round (*Feudal England*, p. 170), although one of the instances cited by the latter (*Chron. Abingdon*, ii. 160) shows that in the reign of Henry I. there was a collector of the geld for Berkshire who was not the sheriff. The evidence of the Pipe Roll of Henry I. seems to establish the usage also for an earlier period. The Abingdon chronicler (*ibid.* ii. 70) gives well-nigh conclusive evidence for the period when Waldric was chancellor, namely (Round, *Feudal England*, pp. 480-81) just before November 1106. The geld was to be collected in Oxfordshire *per officiales huic negotio deputatos*. From this payment the abbey was acquitted by a mandate of the king directed to the sheriff.

[228] Above, p. 63.   [229] Epist. lvi. Migne, *Patrolog.* clix. 233.

[230] *D.B.* i. 127.   [231] *E.H.R.* xxxi. 35-6.

[232] Above, note 140.

had need.[233]   The royal service called also for outlays of the produce or money in the sheriff's hands.   The sheriff of Yorkshire in 1075 received Edgar the Atheling at Durham and let him find food and fodder at the castle on his route as he travelled to meet King William on the Continent.[234]

The Norman sheriff is famous for his extortion and oppression. The vague words of Domesday sometimes suggest that *ferms* may as yet be increased without the king's consent, and there is abundant evidence [235] that during the Conqueror's reign the sheriff and his agents exacted such additions.   The old *firma unius noctis* paid by a group of manors in the southern counties, and worth about £70 in the time of King Edward,[236] had risen by 1086 to £105.[237] Norman prelates [238] and barons [239] were very ready to farm the king's lands, and the English Chronicle [240] complains that the king let his lands " as dearest he might," and that they went to the highest bidder.   With *ferms* sometimes in excess of the value of lands,[241] the chronicler may well declare that the king " cared not how iniquitously the reeves extorted money from a miserable people." [242]   That the sheriff at the head of the system reaped his

[233] *D.B.* i. 30 b.
[234] *A.-S. Chron.* a. 1075. At an earlier time the sheriff had provided the sustenance of the king's *legati* in going by water from Torksey to York (above, p. 30). The king's reeves at Wallingford met the expense of the burghers in the king's service with horses and by water *non de censu regis sed de suo* (*D.B.* i. 56).
[235] *Quando Rog. Bigot prius habuit vicecomitatum statuerunt ministri sui quod reddent xv. libras per annum quod non faciebant T.R.E. Et quando Robertus Malet habuit vicecomitatum sui ministri creverunt eos ad xx. libras. Et quando Rog. Bigot rehabuit dederunt xx. libras, et modo tenet eos* (*D.B.* ii. 287 b). Roger Bigot had increased the *ferm* of Ipswich to £40, but finding it would not yield that amount he pardoned £3 (*ibid.* ii. 290 b). Mr. Round maintains (*Geoffrey de Mandeville*, pp. 101, 361) that in the twelfth century the amount collected from a given manor was always the same.
[236] Round, *V.C.H. Hampshire*, i. 401.
[237] Round, *Feudal England*, p. 113. Under Edward the Confessor a one night's *ferm* collected from a group of Hampshire manors was £76 : 16 : 8. Under the Normans this was increased to £104 : 12 : 2, and in Wilts and Dorset to about £105 (*V.C.H. Hampshire*, i. 401).
[238] The bishop of Winchester farmed Colchester (*D.B.* ii. 107 b) and the archbishop of Canterbury held the borough of Sandwich, which yielded a *ferm* of £40 (*D.B.* i. 3).
[239] For instance, Hugo de Port (*D.B.* i. 219), Hugh fitz Baldric (*ibid.* i. 219 b) and William of Eu (*ibid.* i. 162).
[240] a. 1087.
[241] Ballard, *Domesday Inquest*, pp. 221-2 ; *V.C.H. Hampshire*, i. 414. The collection of the old *ferm* from a manor which had lost lands and the increase of *ferms* is well shown in the case of royal demesne lands in Gloucestershire (*D.B.* i. 163).
[242] *A.-S. Chron.* a. 1087.

harvest is shown by the *crementum* which he paid.[243]   He might exact from those to whom he let the king's lands a *gersuma* or bonus over and above the amount of the *ferm* due to him.[244]   In Bedfordshire this was called *crementum*.[245]   To such sheriffs, holding warrant from the king to suppress unruly peasants, one may with some confidence ascribe the beginning of the amercement imposed upon the vill or group which fails to produce absconding members guilty of crime, in other words the basic administrative element in frankpledge suretyship ; and the origin of the *murdrum*, the imposition laid upon the hundred which fails to produce the slayers of men whose English ancestry cannot be proved, is attributable to the same agency.

The sheriff stands accused of bad stewardship and greed in trespassing upon the king's rights,[246] in wasting the property in his charge, and in depriving individuals of their property.   Two manors in Dorsetshire had lost a hundred shillings in value through the depredations of Hugh fitz Grip.[247]   Sheriffs are credited with the loss of men and animals on the manors of the royal demesne,[248] and with the destruction of houses, usually to make room for a castle, a practice which led to a decline of population in some towns.[249]   Norman sheriffs showed little regard for private rights

[243] Above, note 202.
[244] In Essex the *gersuma* exacted from a borough or manor in several instances amounted to £4 (*D.B.* ii. 2, 2 b, 107 b), but £10 was collected from one manor (*ibid.* ii. 3).   Mr. Ballard (*Domesday Boroughs*, p. 45) interprets the hawk and £4 of *gersuma* paid by the burghers of Yarmouth to the sheriff as a gift to propitiate him.
[245] *D.B.* i. 209, 209 b.   The *crementum* rendered by a manor here usually consisted of a certain sum of money plus an ounce of gold for the sheriff annually.   To one of the demesne manors in this shire the king granted Ralph Taillebois the right to add other demesne lands to offset the burden of the amount thus imposed.
[246] Thus Ralph Taillebois gave to one of his own knights land which he had seized for non-payment of gavel (*D.B.* i. 216 b).   *Superplus invasit Picot super regem* (*D.B.* i. 190).
[247] *D.B.* iv. 34.
[248] Loss of plough oxen on Essex manors is charged to sheriffs, especially to Swein and Bainard (*D.B.* ii. 1, 2).   Urse d'Abetot so wasted men that they could no longer render the salt due (*D.B.* i. 163 b).
[249] The Domesday inquest for Lincoln states that certain houses beyond the metes of the castle have been destroyed, but not by the oppression of sheriffs and their *ministri*, as if the reverse were the rule (*D.B.* i. 336 b).   Such destruction occurred at Dorchester, Wareham and Shaftesbury from the accession of Hugh fitz Grip to the shrievalty (*D.B.* i. 75) ; and a *destructio castellorum* occurred at York in 1070, for which another sheriff, Hugh (*ibid.* i. 298 b), was responsible.   At Cambridge (*ibid.* i. 189) and Gloucester houses were taken down for the same purpose (*ibid.* i. 162).

of property.[250]   Domesday Book records complaint that some of
them have unjustly occupied the lands of individuals.[251]   In one
instance the shire testified that land taken by the sheriff for non-
payment of Danegeld had always been quit of the obligation.[252]
Violent imposition of *avera* and *inward* is mentioned several times in
Bedfordshire, and land was taken even from a former sheriff because
he refused *avera vicecomiti*.[253]   Demands upon burghers were sometimes so great that they fled.[254]   The exactions of Picot at Cambridge
are among the worst recorded.[255]   Through fear of him the men of
Cambridge are related to have wrongfully decided a lawsuit in his
favour.[256]

Best known of all are the grievances of the churches and monasteries.   The spoliation of ecclesiastical possessions by the followers
of the Conqueror was due to the policy of the king, as well as to the
rapacity of the baronage.[257]   But the plundering of the sheriff was
sometimes almost systematic.   The wholesale seizure of the lands
of the church of Worcester by Urse d'Abetot is notorious,[258] and the
best of evidence shows that they were permanently retained.[259]
Evesham and Pershore, the other great monasteries of this county,

[250] Freeman says (*Norm. Conq.* iv. 728) of one of these officials who robbed
various persons of their possessions, " he seems to have acted after the usual manner
of sheriffs ".

[251] Froger of Berkshire held certain lands which he had placed at the king's
ferm absque placito et lege (*D.B.* i. 58).   Ansculf unjustly disseised William de
Celsi (*ibid.* i. 148 b).   Ralph Taillebois wrongfully occupied the lands of others
(*ibid.* i. 212, 217 b).   Eustace of Huntingdon appropriated the burghers as well as
the lands of Englishmen (*ibid.* i. 203, 206, 208).

[252] *Ibid.* i. 141.

[253] *Ibid.* i. 132 b.

[254] Ballard, *Domesday Boroughs*, p. 87.

[255] See above, p. 53, n 94.   Picot also imposed service with carts and appropriated
some of the common pasture, building upon this land his three famous mills ;
whereby several houses were destroyed, as well as a mill belonging to the abbot of
Ely and another belonging to Count Alan (*D.B.* i. 189).

[256] Below, p. 72.

[257] The Conqueror undertook to subject the monasteries to feudal service by
compelling them to provide a certain number of knights in war or to surrender part
of their lands.   Out of 72 manors which Burton Abbey originally possessed, over
40 were lost (*Salt Arch. Soc.* v. pt. 1, p. 1).   King William quartered 40 knights on
the Isle of Ely, towards the support of whom the abbot gave in fee certain lands to
leading Normans, among whom were Picot the sheriff and Roger Bigot (*Liber
Eliensis*, p. 297).   It is said that William Rufus demanded 80 knights (*Monasticon*,
i. 461).   Mr. Round (*Feudal England*, pp. 296-301) shows the process by which
a number of abbeys established knights' fees.   Haimo, sheriff of Kent, was one of
the *milites* of the archbishop of Canterbury, to whom he had given lands (*D.B.* i. 4).

[258] Heming, *Chartulary*, i. 253, 257, 261, 267-9 ; Freeman, *Norm. Conq.* v. 761,
764-5.

[259] Round, *Feudal England*, pp. 169-75.

also suffered heavy losses at Urse's hands.[260]   Others acted in a similar spirit.[261]   The invective directed by the monk of Ely against the greed and impiety of Picot of Cambridge in appropriating lands of St. Etheldreda deserves to be a classic.[262]   It was well for the prelate to have influence with the sheriff.[263]   The story that the sheriff, departing from York with an imposing retinue, met the laden wains of Archbishop Aldred as they entered the city and ordered the seizure of their contents,[264] at least expresses a twelfth-century churchman's conception of this official.

William the Conqueror, though powerful and not devoid of a sense of justice, made little progress with the perennial medieval problem of honest local government.   There was no appeal from the sheriff except to the king or his duly accredited representative ; this made it practically impossible for any but men of the greatest influence to oppose the head of the shire.   In Aldred's case, just cited, the archbishop is said to have obtained restitution through a direct appeal to King William.[265]   The clause in royal charters commanding the sheriff to see that no injustice is done the grantee is much more than form.[266]   When the king's justice convened a

[260] Freeman, *Norm. Conq.* v. 765.   Evesham lost 28 out of 32 newly acquired properties.   These were seized by Bishop Odo at a *gemot* of five shires which he held, and a large part of them soon given over to Urse and his associates (*Chronicon Abbatiae de Evesham*, pp. 96-7 ; *D.B.* i. 172).   Mr. Davis (*Regesta*, i. no. 185) shows that Urse retained a hide belonging to the abbot of Evesham after four shires had adjudged the whole manor to the abbot.

[261] Froger, like his Anglo-Saxon predecessor, won evil renown by holding property of the monastery of Abingdon (*Chron. Abingdon*, i. 486).   Peter of Valognes made aggression upon the property both of St. Paul's (*Domesday Studies*, ii. 540) and of the abbey of St. Edmund's (Davis, *Regesta*, i. nos. 242, 258).   Eustace of Huntingdon deprived the abbot of Ramsey (*D.B.* i. 203) of burgesses, and violently seized lands of the abbey, which for a long time he handed over to one of his knights (*Chron. Ramsey*, p. 175).   Ralph de Bernai, with the aid of Earl William fitz Osbert (*D.B.* i. 181 ; Freeman, *Norm. Conq.* v. 61), also took lands from the church of Worcester (Heming, *Chartulary*, i. 250).

[262] *Liber Eliensis*, p. 266.

[263] During his exile Anselm wrote to Bishop Gundulf of Rochester to urge upon Haimo and his wife the restoration of a market belonging to the archbishop which had been seized by a neighbour (epist. lxi. Migne, *Patrolog.* clix. 235).

[264] See Raine, *Historians of the Church of York* (R.S.), ii. 350-53.   If the story is true the sheriff was William Malet.

[265] The same procedure is implied in the instance wherein William Rufus orders the sheriff of Oxford to right the injuries done by his subordinates to the monks of Abingdon (*Chron. Abingdon*, ii. 41).   Anselm wrote to Haimo that on his return to England his goods ought to have been freed according to the king's precept, and asked the sheriff to restore what his subordinates had seized at Sandwich and Canterbury, *ne me facere clamorem ad alium cogatis* (epist. lvi. Migne, *Patrolog.* clix. 233).

[266] One form of notifying the sheriff of a royal grant prescribed that if injury

local court within the shire [267] the sheriff took a lower place. The bishop of Bayeux, presiding in the shiremote of Cambridgeshire, not only refused to accept the recognition of a jury alleged to be intimidated by Picot, but ordered the sheriff to send them and another twelve to appear before him in London.[268] In taking the Domesday inquest the *barones regis* placed upon oath the sheriff as well as others. Domesday records the contested claims or questionable conduct of the sheriff himself, though usually of a sheriff no longer in office. Machinery has been fashioned which may call him to a reckoning.[269] But the Domesday inquest was never repeated, and the mission of royal justices to the county was as yet unusual. Where the king was not directly concerned the sheriff was left to do much as he pleased. Strength and loyalty were his great qualifications. An over-display of the former might be condoned so long as the latter was assured. The spirit of feudality remained, despite striking manifestations of royal power.

By the early years of the twelfth century the long process of reducing the sheriff's power was under way. It is not improbable that the ministry of Ranulf Flambard took the first steps in this direction. William Rufus had his experience with rebellious sheriffs, and the calling out of an army of 20,000 foot soldiers in 1194 served as further reminder of the military possibilities of the office.[270] The employment of local justiciars was a device which might take from the hands of such sheriffs the control of the pleas of the Crown. The baronial opposition to Henry I. brought further changes. By this reign the sheriff seems to be castellan only when he inherits the position. The hereditary shrievalty still exists in

---

be done the grantee, the latter is to make complaint to the king, who will do full right. See *Monasticon*, ii. 18 ; Davis, *Regesta*, i. no. 104. Another form of writ enjoined the sheriff to see that in matters affecting the royal grant no injustice was done. See above, p. 62.

[267] He might convene several hundreds (see note 131), a shire court, or several shires. Odo of Bayeaux is said to have presided in a *gemot*, at which were present three or more sheriffs (Davis, *Regesta*, i. app. xxiv).

[268] *Textus Roffensis*, Eng. Histor. Soc., 150-51 ; Bigelow, *Placita Anglo-Normannica*, pp. 35, 36 ; Stenton, *William the Conqueror*, pp. 434-5.

[269] In the *Leis Willelme*, 2, 1, Liebermann, *Gesetze*, i. 492-3, possibly written in the first third of the twelfth century, but perhaps as old as 1090, the sheriff may be convicted before the justice for misdeeds to the men of his bailiwick.

[270] Florence of Worcester, using a formula of the reign of Henry I., tells that when in 1085 the king of Denmark threatened an invasion of England King William brought over troops from Normandy, and sending throughout England *episcopis, abbatibus, comitibus, baronibus, vicecomitibus ac regis praepositis, victum praebere mandavit*. Cf. note 223.

some shires, but the feudal danger may soon be met by placing a group of shires in the hands of a new officer whom the king has raised from the dust.

A strong local official under the king's direction, whose activity epitomized shire government and whose business was administration, was a novelty in a feudal age. The king had other agents to whom he entrusted special judicial and military functions, and in some measure fiscal functions as well, but the fact that some sheriffs were given duties of this sort at the *curia* indicates that the king's servants there were not usually of superior administrative ability. The sheriff's personal prestige, and a feudal status which might even give him a seat in the king's great council, imparted to his office a dignity and a substantial quality which eight centuries have not effaced. Some modification of the functions of the Anglo-Saxon shrievalty came through Norman usage, fiscal efficiency, and the introduction of new feudal dues and services, but the strong combination of powers in the sheriff's hands was nearly all wielded by his English predecessor. The disappearance of the earl hardly added functions which the sheriff had not already performed. The fiscal system which supported the Norman monarchy was largely English, although the sheriff's ideas of financial administration were Norman, as was the practice which made him keeper of the king's castles. Functions incident to ecclesiastical jurisdiction were actually lost. The new life infused into the office which made it powerful came through the energy of the Norman kings and their enhanced views of the royal prerogative. In a manner astonishing to the student of old English polity they assume their own right to do justice, and to that end depute sheriffs or other agents. In the course of general administration the king's direction of their activity is equally prominent. The writ which follows the form of the Confessor's announcements to the shire court assumes initiative. Through it the king issues positive commands to sheriffs, and even lays down rules for their guidance which have all the force of the older English laws.

The need of loyal local officials on the part of a feudal ruler permitted the shrievalty to assume the semblance of a viceroyalty, but its holder was subject to this strong means of control supplemented by the local law and custom of the shire, and usually by his vassalage to the king. The dread agent of Norman monarchy, fitting counterpart of the grim Conqueror, under whose administration the peasant was oppressed by excessive rents, the monastery deprived of its lands, and every one subjected to the danger of wanton oppression, seems a heartless adventurer. But he was no instrument of feudal anarchy. Despite his feudal interests, personal attachment to the king and the rewards which it brought committed him to the cause of strong monarchy. His profits in holding the shire were a buttress to the king's authority. His authority over both hundred and shire prepared for the rule of the common law at a later time, and apparently led to the system by which vills came to be represented in the shiremote and hundredmote.[271] His view of frankpledge kept him in personal touch with the hundredmote. The public nature of this body could not be jeopardized through the encroachment of feudal lords so long as the income from its pleas formed an integral part of the sheriff's *ferm*. The strong local position of the sheriff, sometimes supplemented by command of the castle, made him powerful to enforce judicial decrees or royal orders affecting even the strongest lords of his county.[272] His check upon the political power of feudalism and his preservation of the old communal assemblies to render important service to later generations, to say nothing of his maintenance of law and order and his great services to administration in general, demand for the Norman sheriff lasting gratitude.

[271] See *Leges Henrici*, 7, §§ 4-8, Liebermann, *Gesetze*, i. 553-4.

[272] The defection of Earl Roger in 1075 was due in part to the fact that the king's sheriffs had held pleas on his lands (Adams, *Political History of England*, p. 61).

IV

THE PROBLEM OF CENTRAL CONTROL UNDER
HENRY I. AND STEPHEN, 1100–1154

THE lay officials employed by William the Conqueror, whether in central government or in the management of the shires, were regularly of baronial status. These were not the only officials, for there was an important group of curial bishops, and a force of trained clerks was utilized in the work of the chancery, if not also in that of the treasury. But the dominant element in early Norman administration in England was baronial. The first apparent impulse in an opposite direction may be due to the feudal disorders of the reign of William Rufus. At any rate it is clear that this king appointed some special agents to carry into effect his novel measures and policies. By 1106, the date of the battle of Tinchebrai, the second generation of the feudal nobility in which the Conqueror placed dependence had in no small measure proved wanting. The barons who loyally supported Henry I. in the early and troublous years of his reign seem to have enjoyed his undying favour. They remained a powerful influence in government. In numerous instances, however, their sons did not attain the same position. The new men, who aided the king at the crisis of the reign, and who sometimes acquired the confiscated lands of the rebels, henceforth became more and more prominent. Within twenty years a remarkable circle of these persons held the great offices of state and at the same time served as sheriffs, a combination of functions which had not been infrequent in the days of baronial control. The best illustration, therefore, of the change from the earlier to the later type of administrative staff is afforded by the personnel of the shrievalty in the reign of Henry Beauclerc.

Before Tinchebrai, as indeed in the Conqueror's time, one finds the names of obscure sheriffs ; [1] but well after 1100 the heads of a dozen shires were still either sheriffs of the Domesday period or their sons. William of Cahagnes was in office in the earliest years of the reign,[2] Roger Bigod apparently until his death in 1107,[3] Edward of Salisbury possibly until about the same date,[4] Urse d'Abetot some years longer,[5] and Aiulf the king's chamberlain until fairly late in the reign.[6] Devon, formerly in the hands of William,[7] son of Baldwin of Exeter, about 1107 was passing to his brother Richard fitz Baldwin. Haimo the *dapifer* and Robert,[8] sons of Haimo the *dapifer* of the Conqueror, for fifteen years or more served as sheriffs of Kent. In the place of Robert of Stafford was his son Nicholas,[9] and in that of Hugh de Port in Hampshire his son Henry.[10] Ivo de Grantmesnil probably held his father's position in Leicestershire [11] until in 1102 he suffered forfeiture for his rebellion. Finally,

[1] These include " P.", sheriff of York before 1104 (*Monasticon*, vi. pt. iii., 1178) ; Roger, sheriff of Huntingdon (*Ramsey Chartul.*, R.S., i. 238), who may well be the same as Roger, sheriff of Surrey (Farrer, *Itinerary*, no. 86); Helgot, sheriff of Nottingham before 1108 (Blythe Chart., Harl. MS. 3759, fo. 120) ; Alfred of Essex (*Cartul. Monast. S. Iohannis de Colecestre*, p. 27 ; *Chron. Abingdon*, ii. 57-60) ; William of Oxford (*ibid.* ii. 84, 93), who was a tenant-in-chief (Farrer, no. 302) ; Foucher, sheriff of Shropshire (*ibid.* nos. 38, 51) ; Roger Picot of Northumberland (*Monasticon*, vi. 144), who stands in place of Robert Picot, sheriff in 1095 (Davis, *Regesta*, no. 51). Geoffrey, sheriff of Buckingham (*Monasticon*, i. 165), may belong to this period, but possibly only to the preceding reign. Richard son of Gotse, sheriff of Nottingham (*ibid.* vi. part iii. 1179), and apparently of Derby also (Farrer, *Itinerary*, no. 38), is a better-known figure.

[2] *Ibid.* nos. 44, 123. The succeeding sheriff of Northamptonshire, Robert de Paville, is mentioned between 1104 and 1106 (*ibid.* no. 147).

[3] A writ of the period 1102–6 (*Ramsey Chartul.*, R.S., i. 249) shows that he was sheriff of Suffolk ; the form of address in various writs (Farrer, *Itinerary*, nos. 78, 79) indicates that in 1105 he also held Norfolk.

[4] See *E.H.R.* xxvi. 490. Active at the *curia* as early as 1170 (*Selby Coucher Book*, i. 11-12 ; *Monasticon*, iii. 499), and still earlier if we accept *Hist. Monast. Selebiensis*, p. 9. Cf. n. 27 below.

[5] Roger, his son and successor, was sheriff of Worcestershire at some time 1110–September 1113 (Farrer, *Itinerary*, no. 290A).

[6] Domesday sheriff of Dorset, holding Somerset also in the preceding reign (*E.H.R.* xxxiii. 151, n. 48) ; sheriff of both counties in this reign before the death of Urse d'Abetot (*Monasticon*, i. 44, no. 67). Cf. p. 81, n. 53, below.

[7] Certainly sheriff in 1107 (*Montacute Chartul.*, Somerset Record Soc., p. 121), hardly in 1100 (Farrer, *Itinerary*, no. 32).

[8] Haimo appears as sheriff quite early in the reign (*Monasticon*, i. 164, no. 14 ; Farrer, no. 21), also within the period July 1107–July 1108 and in the period 1114–16 (Farrer, *Itinerary*, nos. 202, 359) ; Robert at some time within the period 1103-9 (Round, *Calendar*, no. 1377).

[9] Sheriff in the preceding reign (Davis, no. 456), mentioned as in office in a document possibly as late as 1117 (*Monasticon*, vi. part ii. 1043, no. 7).

[10] Davis, nos. 377, 379 ; Round, *Calendar*, no. 154 ; Farrer, no. 37.

[11] Bateson, *Leicester Records*, p. xiii ; Ordericus Vitalis, ed. Prevost, iv. 169 : *municeps erat et vicecomes et firmarius regis*.

another hereditary shrievalty had towards the end of the late reign fallen to Walter of Gloucester, who was destined to hold it about twenty years longer and to become the king's constable.[12]

Even before 1106 the king had counteracted the influence of baronial officials of doubtful or more than doubtful loyalty by the employment of new men. At least two sheriffs who owed all to the royal favour were a heritage from the reign of William Rufus, and both continued to rise. Osbert, formerly known as the priest, retained the shrievalty of Lincolnshire, and before 1107 was also entrusted with that of Yorkshire.[13] His marked material prosperity [14] and his tenure in both these positions until his death show that he enjoyed the king's especial favour.   Hugh of Buckland, an important curial, justice, and sheriff of Berkshire, Bedfordshire and Hertfordshire before the close of the late reign, 15 was much esteemed by the king, and before Tinchebrai held in addition to these counties [16] at least three others, including the shrievalty of London and Middlesex.[17]   It is possible that the remaining two of the eight attributed to him a little later are to be counted also at this time, but there is no certainty as to their identity.[18]   Richard de Belmeis, who despite his earlier employment in the service of Robert of Belesme [19] remained loyal to the king in 1102, was made administrator of the Shropshire palatinate after its forfeiture and placed in a position which is described both as that of steward [20] and of

[12] Mentioned as sheriff in this period (Farrer, *Itinerary*, nos. 277, 290A ; *Chron. Abingdon*, ii. 105) and as constable in 1115 (Farrer, *Itinerary*, no. 361).

[13] *Selby Coucher Book*, i. 27-8.

[14] For his grant to Selby Abbey see *Selby Coucher Book*, i. 6-7 ; Farrer, *Early Yorkshire Charters*, i. 355. He was sometimes a witness to the king's writs.

[15] *Chron. Abingdon*, ii. 43 ; Davis, no. 395 ; *E.H.R.* xxvii. 103 ; cf. Page, *London* (1923), 235-6.

[16] Mr. William Page (*London*, 207, n. 6), suggests Oxfordshire and Sussex. But there are several known sheriffs of Oxfordshire, and it would be difficult to date Hugh's possible shrievalty.

[17] Essex at Christmas, 1101 (Farrer, *Itinerary*, no. 32 ; cf. *Monasticon*, i. 164, no. 15, and vi. 105) ; Buckinghamshire before 1107 (*Chron. Abingdon*, ii. 106-7 ; cf. 98, 99), certainly by 1104 ; London and Middlesex before the summer of 1107 (*ibid.* ii. 56), possibly as early as 1103 (Round, no. 1377), clearly before William of Mortain suffered forfeiture after Tinchebrai (St. Albans Chartulary, Cotton MS., Otho D. iii. fo. 73).

[18] Hertfordshire in 1106 or 1107 (*Liber Eliensis*, p. 298) ; Berkshire (*Monasticon*, i. 523).

[19] *Annales Monastici*, R.S., ii. 43.

[20] *Brut y Tywysogion*, anno 1106 ; Ordericus Vitalis, ed. Prevost, iv. 275. According to Eyton, *Antiquities of Shropshire*, ii. 193, he was the successor of Rayner, the Domesday sheriff. It is probable, then, that Foucher, the sheriff of 1102, was his subordinate, and indeed both sheriffs are named in *Monasticon*,

sheriff. His elevation in 1108 to the see of London and his appearance among the king's great officials [21] are further proofs of his standing at court.

The period between 1106 and 1110 was marked by the rise of several more sheriffs of the same class and by the displacement of some of the baronial sheriffs in shires which they had long held. Gilbert the knight, whose shrievalty in Surrey begins not later than 1107, in 1110 had also the counties of Cambridge and Huntingdon.[22] Hugh of Leicester, the seneschal of Matilda, daughter of Earl Simon of Senliz, in her widowhood, is represented as sheriff and also as benefactor of the church at Daventry in Northamptonshire.[23] Hugh almost certainly held the shrievalty of Northamptonshire before the death of Earl Simon in 1109,[24] and apparently held that of Warwickshire by about 1108.[25] His surname, moreover, indicates that he was best known as sheriff of a third county, which he is known to have held at some time after 1109 and before 1116.[26] Indeed he may have been made sheriff of Leicester, through the influence of Robert of Meulan, not long after the disgrace of the Grantmesnils. Other baronial shrievalties had vanished, one in Wiltshire by 1107,[27] and another in Hampshire by 1110. At the latter date William of Pont de l'Arche, already prominent in the king's curial service, held both counties.[28] Since Hugh of Buck-

---

vi. pt. ii. 1043. Farrer, *Itinerary*, no. 437, possibly shows Bishop Richard in control as late as 1121. Owen and Blakeway, *Hist. of Shrewsbury*, i. 73, n. 2, make Payn fitz John his successor. Fulco is named as sheriff in the period 1120–22 (Farrer, *Lancashire Pipe Rolls*, p. 272).

[21] As one of the judges who sat at the treasury at Winchester (*Chron. Abingdon*, ii. 116). He had been reeve of Chichester in 1107 (Farrer, *Itinerary*, no. 106).

[22] Farrer, *Itinerary*, nos. 252, 267 ; Round, *Commune*, pp. 121-3. Sheriff of Surrey before the death of Roger Bigod (Robinson, *Gilbert Crispin*, pp. 141-2 ; Westminster Chartul., Cotton MS., Faustina A. iii. fo. 67).

[23] *Monasticon*, v. 178-9 ; cf. Farrer, *Itinerary*, no. 311.

[24] *Monasticon*, vi. part iii. 1273, no. 33 ; cf. Farrer, *Itinerary*, no. 219.

[25] He seems to be the " H. sheriff of Warwick " of *Monasticon*, vi. part ii. 1043, no. 4. His son Ivo in 1130 held land of the earl of Warwick (*Pipe Roll*, 1130, p. 108) in Warwickshire.

[26] After the death of Earl Simon and while Geoffrey Ridel was living, *i.e.* before 1120 (Chartul. of St. Andrews, Northampton, Cotton MS., Vespasian E. xvii. fo. 17 d). But he was called Hugh of Leicester while Roger, son of Urse d'Abetot, held the shrievalty of Worcestershire (*Register of Priory B. Mariae Wignorensis* (Camden Soc.), p. 30 a), in other words, at some time between 1110 and 1116 (p. 76, n. 5 ; p. 79, n. 33).

[27] Walter Hosate by this date (Farrer, *Itinerary*, no. 173) holds the old shrievalty of Edward of Salisbury.

[28] *E.H.R.* xxxv. 392, no. 25. He was employed upon the king's special business in Hampshire by 1106, possibly by 1103 (*ibid.* p. 391, nos. 21, 23).

land is mentioned in the tenth year of the reign as sheriff of his eight shires,[29] it seems clear, therefore, that no less than seventeen shires were at that date under the control of six of the king's trusted agents, who were all new men.

This high centralization, obviously the result of unusual stress, was probably soon relaxed somewhat, for in the lifetime of Hugh of Buckland another sheriff is mentioned in Hertfordshire.[30] Certainly this was the case after his death and that of Osbert, both of which seem to have occurred in 1115.[31] Fewer counties were now likely to be controlled by one person, and a succession of sheriffs of lesser rank becomes traceable in various localities. In Oxfordshire alone do sheriffs of some rank seem to supersede obscure persons. Here Thomas of St. John and Richard de Monte, who on occasion appear at the great *curia*, alternately and for at least one year jointly,[32] held the office for approximately the whole of the decade following 1107. But the number of hereditary shrievalties held by great barons continued to decline despite the well-known succession in Worcestershire, first of the son, then of the son-in-law of Urse d'Abetot.[33] Well before 1120 Kent had passed from the family of Haimo and was in the hands of William of Eynesford, distinctly a man of the new order.[34] William Bigod, who succeeded his father as *dapifer*,[35] was made sheriff of Suffolk and probably of Norfolk [36] also in recognition of the services of the elder Bigod. Yet before William went down with the White Ship in 1120 both of these counties had been held by two other sheriffs of

[29] *Chron. Abingdon*, ii. 117.

[30] Rannulf, *Matthew Paris*, vi. 36 ; Farrer, *Itinerary*, no. 361.

[31] Farrer, *Early Yorkshire Charters*, ii. 305-6 ; Farrer, *Itinerary*, no. 361 ; Robinson, *Gilbert Crispin*, p. 138.

[32] Thomas was sheriff first. As to his possessions in England and Normandy, see *Pipe Roll*, p. 3, and Round, *Calendar*, no. 724. The joint shrievalty was in the eleventh year of the reign (*Chron. Abingdon*, ii. 119). Richard was sheriff for several years after this (*ibid.* pp. 89, 120), then apparently Thomas again (Farrer, *Itinerary*, no. 316).

[33] See *D.N.B.*, *s.n.* ' Urse d'Abetot '. Walter de Beauchamp, the successor of Urse's banished son Roger, was placed in possession of the lands of the latter possibly as early as 1114 (Brit. Mus. Add. MS. 28024, fo. 149 ; Farrer, *Itinerary*, no. 335), clearly by 1116 (*ibid.* no. 319).

[34] Sheriff of London at some time during the reign (*Chron. Ramsey*, p. 249) ; mentioned as sheriff of Kent in a writ possibly as early as 1114 (Elmham, *Hist. Monast. Sti. Augustini*, R.S., pp. 365-6), and according to Farrer's chronology (*Itinerary*, nos. 385, 386, 554) sheriff both in 1118 and 1127.

[35] *Monasticon*, v. 148.

[36] *Ramsey Chartul.* i. 245, 249.

G

good family but of lesser rank.[37] Nicholas of Stafford, although called by the title of sheriff in 1130,[38] could not have held the ancestral office for seven years preceding that date.[39] In all these, as in the two earlier cases, the son of a Domesday sheriff in his lifetime [40] made way for a member of the new ruling circle. Moreover Aiulf the chamberlain was after a time superseded in Dorset and Somerset. By 1123, at the latest, only three sheriffs of the older type remained. Walter of Gloucester in Gloucestershire, Walter de Beauchamp in Worcestershire and Richard fitz Baldwin in Devon each held the county ruled by two of his family before him. Richard, like his successor of 1128–30, appears also to have held Cornwall.[41]

For the period of almost two decades following 1110 the usual difficulties attending the study of the sheriff are very much increased. But despite the paucity of documents and the increasing use at the chancery of a form of writ which mentions sheriffs only by title and not by name, some rather striking data are attainable. If the long tenure and wide authority of Gilbert the knight are without parallel, it is more than a coincidence that William of Pont de l'Arche, who rose to high position through administrative service, governed in the fiscal year 1127–8 the same two shires [42] which were in his charge seventeen years before. Moreover the Radulf *vicecomes*, witness of a confirmation made to Abingdon by the count of Meulan in the eighth year of the reign, seems to be Ralph Basset, a *vicinus* and

[37] Ralph de Beaufeu preceded Robert fitz Walter (*Chron. Ramsey*, p. 267), whose shrievalty may date as early as 1114 (Chartul. of St. Benet's Holme, Cotton MS., Galba E. ii. fo. 31).

[38] *Pipe Roll*, p. 82.

[39] Below, p. 83.

[40] Haimo the *dapifer* was only recently deceased in 1130 (*Pipe Roll*, pp. 64, 66) and his brother Robert was still living (pp. 95, 97). Walter of Salisbury, son of Edward, is mentioned at that date, and Henry de Port was a royal justice holding pleas at Dover (*Pipe Roll*, p. 65).

[41] *H. rex Angliae Willelmo episcopo Exoniae et Ricardo filio Bald' vicecomiti et omnibus fidelibus suis de Divescira et Cornwall*: Inspeximus of 22 Edward I. in Cotton MS. xvii. 7, part ii. (before 1118). Mr. Round has shown that just prior to 1130 Richard acted as an itinerant justice (*E.H.R.* xiv. 420-22). Cornwall, however, had as sheriff at some time prior to 1130 Frewin or Frawin, who seems to have been removed for his part in the killing of the six sons of Tochi beyond St. Michael's Mount (*Book of Fees* (1920), part i. p. 441 and references).

[42] In 1130 (*Pipe Roll*, p. 36) he still appears as sheriff of Hampshire, accounting for the old farm and for the *auxilium civitatis* of the third year before (p. 40; cf. Farrer, *Itinerary*, no. 551). The princely sums still due to the king in Wiltshire identify him as William the predecessor (*Pipe Roll*, p. 16) of Warin in that shrievalty (Farrer, *Itinerary*, nos. 350, 418), and link the earlier and later periods of his tenure there.

especial friend of this monastery [43] and a prominent official at the *curia*. It is quite possible that he is Ralph, the sheriff of Warwickshire, predecessor of Geoffrey de Clinton, who at some time about 1125 witnessed a grant of the latter to Kenilworth, although in one of the charters of the same series the name of the sheriff of this county is Roger.[44] Hugh of Leicester in 1129, as in earlier years, was in possession of the shires of Northampton and Leicester.[45] He seems to be the same person as Hugh de Warelville who held these counties until Easter 1130.[46] He had for a time acted as sheriff of Lincolnshire.[47] Moreover, in the Pipe Roll of 1130 Hugh de Warelville accounted for Sussex.[48] There is no doubt that Robert fitz Walter, a tenant-in-chief, was sheriff of Norfolk and Suffolk for nearly, if not quite, fifteen years before 1129.[49] Odard of Bamborough, who held of the king a barony in his own county,[50] occupied the same position in Northumberland for as long a period,[51] probably retaining office until his death just before 1133. Warin, who may have been an exchequer official,[52] succeeded Aiulf in Dorset before 1118 and in Somerset by 1123.[53] During the fiscal year 1128–9 he held Wiltshire as a third county.[54] William of Eynesford was employed for a long term which was divided among various counties.[55] Aubrey de Vere, heir of a Domesday landholder and chamberlain, was sheriff of London and Middlesex probably

[43] *Chron. Abingdon*, ii. 103 ; cf. pp. 105, 170, 188. He was in King Henry's service before 1103 (*ibid.*).

[44] *Monasticon*, vi. 221 ; *Collections for a Hist. of Staffordshire* (Salt Arch. Soc.), ii. 195.

[45] *Pipe Roll*, p. 81.

[46] *Ibid.* p. 85.

[47] *E.H.R.* xxiii. 725, no. 2 ; Farrer, *Itinerary*, no. 464 ; *Collections for a Hist. of Staffordshire*, ii. 203. In 1130 Hugh paid a sum which he still owed (*Pipe Roll*, p. 81) *pro separatione comitatus Lincolniae*.

[48] The sheriff of this county at a time prior to Dec. 1125 was William fitz Ang' : Reg. de Bello (Exch. Misc. Books, Augmentation Office, no. 56, fo. 65 a).

[49] Sheriff in the period 1114–16 (*Chron. Ramsey*, p. 267) ; in office until Michaelmas 1129 (*Pipe Roll*, p. 90) ; mentioned at many intermediate points.

[50] Round, *Ancient Charters*, Pipe Roll Soc. x. 33.

[51] Sheriff in 1115 or 1116 (Farrer, *Itinerary*, no. 346) and also appearing in the Pipe Roll of 1130.

[52] See *ibid.* pp. 16, 23, for the record of his being pardoned his danegeld.

[53] Brit. Mus. Add. Charter 24979, part iv. ; Farrer, *Itinerary*, no. 487.

[54] *Pipe Roll*, p. 12. He also accounted for the new farm of Wiltshire in 1129–30, but it is uncertain if he now held Somerset.

[55] London (*Chron. Ramsey*, p. 249), Kent (above, p. 79, note 34), and, 1128–30, Essex (*Pipe Roll*, p. 52). But there was a William de Eynesford *senex* (*ibid.* p. 65).

before 1113 [56] and as late as 1120 ; also of Essex, presumably in the years just preceding 1128.[57]   He and Robert fitz Walter were probably the only sheriffs of this group who could make pretensions to being of good family.   Maenfinin Brito, sheriff of the counties of Bedford and Buckingham from 1125 to 1129,[58] was a rising person who had probably acquired considerable possessions.   Even excluding the various sheriffs of London [59] and of Lincolnshire, the list points to the dominance of the new class of local officials who often serve for long periods.   Some rule wider territories than those ever entrusted to the feudatory sheriffs, and some are professional sheriffs.

The firm entrenchment of the new order in its position is further shown by the fact that officials of this type are occasionally succeeded by near relatives.   Gilbert at his death in 1125 made way in all three of his shires, as Mr. Round has shown,[60] for Fulk, his nephew. Anschetill de Bulmer, Osbert's successor, originally reeve of the North Riding,[61] and later the possessor through royal grant of some of the lands of former rebels, retained Yorkshire until his death in 1129,[62] when it passed to his son, Bertram, the sheriff of 1130. William of Buckland for a time had two of the counties which his father had ruled,[63] and the services of his family still find recognition in 1130 in the fact that he farms Windsor.[64]   Richard de Heriz,[65] after an interval of some years, was followed as sheriff of Nottinghamshire and Derbyshire by his son, Ivo de Heriz,[66] in office 1127-9.

---

[56] *Monasticon*, vi. 155 ; Farrer, *Itinerary*, no. 470.   At one time he acted jointly with Roger, *nepos* of Hubert (*Cal. of Letter Books*, ed. Sharpe, Book C, 221), identified by Mr. Round as the father of Gervase of Cornhill.

[57] Brit. Mus. Add. MS. 14847, fo. 39 ; Add. Charter 28313 ; Farrer, *Itinerary*, no. 575.

[58] Mr. Round shows (*E.H.R.* vi. 438) that in 1166 H. fitz Meinfelin, presumably his son, held fifteen knights' fees.   His predecessor in both counties was Richard of Winchester (*Pipe Roll*, p. 100).

[59] For the list see Round, *Commune*, pp. 121-3 ; Farrer, *Itinerary*, nos. 267-8 ; also Page, *London*, 204-7.   Here the change was from the older kind of sheriff to the burgess type.   For the sheriffs of Lincolnshire, see *E.H.R.* xxx. 280-81.

[60] *Commune of London*, pp. 121-2.

[61] Farrer, *Yorkshire Charters*, ii. introd. p. vi.

[62] *E.H.R.* xxx. 285.

[63] Berkshire in 1119 (*Chron. Abingdon*, ii. 160), and apparently both just before and after that year ; also Hertfordshire, on account of which shire in 1130 he still owed £29 *pro defectu* covering a period of a half-year (*Pipe Roll*, p. 127).

[64] *Ibid.* p. 126.

[65] Richard, son of Gotse, whom he succeeded before Mar. 1114 (Farrer, *Itinerary*, no. 290), seems to have held both counties.

[66] Superseded by 1130 (*Pipe Roll*, pp. 6, 7).   See Farrer, *Itinerary*, no. 610.

Furthermore, Serlo de Burg, who first appears in the royal service as custodian of the property of the archbishop of York,[67] in later years as justice,[68] the type of person much employed in administering King Henry's demesne, held these two shires [69] prior to 1127, and seems to have purchased them, in or just before 1129, for his son, Osbert Silvanus.[70]   The king, having assured himself that administrative power shall not be in the hands of the nobles who have defied authority, is content to let it rest with new families which have proved their capacity.   The administrative motive is present in these days even when the personal or political seems to prevail.

This holds good even where hereditary or baronial shrievalties remain.   Miles of Gloucester, sheriff as well as constable of the realm, after the decease of his father, Walter, from 1128 to 1130 [71] held Staffordshire [72] along with the old family county.   Moreover, he and Payn fitz John, another border lord prominent at the *curia*, and sheriff of Shropshire and Herefordshire,[73] were royal justices both in this region and in Pembrokeshire.[74]   The story told in the *Gesta Stephani* [75] of their oppressive rule from the Severn to the sea shows that they were not lacking in energy.

The occasional letting of a county for a *gersoma*, the equivalent of the later fine *pro comitatu habendo*, is a further mark of the new administrative system.   Men who held even the highest positions at court were in some instances permitted to purchase them.   The *gersoma* represented the consideration for which a sheriff received his office with its opportunities for emolument.   The arrangement was sometimes made for a period of five years, the original recorded instance of this being the shrievalty of Robert de Stanley in Staffordshire,[76] which apparently covers the interval between 1123 and 1128.

[67] *Pipe Roll*, p. 31.   Presumably in the period 1114–19.
[68] *Ibid.* p. 35.   Osbert, his nephew, sat with him.
[69] *Ibid.* p. 31.   Cf. *Monasticon*, vi. part iii., 1180.
[70] He owes in 1130 *xx Marcas argenti pro ministerio Osberti filii sui* (*Pipe Roll*, p. 31).   Osbert also held a knight's fee of the king (*ibid.* p. 9).   He owed only a half-year's new ferm at Michaelmas 1130 (*ibid.* p. 7).
[71] Farrer (*Itinerary*, no. 578) dates between 1127 and 1129 the writ confirming him in his father's lands.   The Pipe Roll of 1130 indicates that he had been sheriff two years.
[72] *Ibid.*   Apparently for two years preceding Michaelmas 1130.
[73] Farrer, *Itinerary*, nos. 547, 690 ; *Gesta Stephani*, R.S. p. 16 ;   Owen and Blakeway, *History of Shrewsbury*, i. 73, n. 2.
[74] *Pipe Roll*, pp. 74, 78, 136 ;   Miles also in Hants (p. 38).
[75] P. 16.
[76] *Pipe Roll*, p. 73.

The payment made was heavy and varied in individual cases.[77] The arrangement was discontinued when a sheriff was replaced, and he then paid according to the proportion of the specified time he had been in office.[78] Fulk, the nephew of Gilbert, left office owing eighty pounds of *gersoma*,[79] presumably the amount due for the four years he held his various counties. Whenever applied, this whole procedure was obviously in conflict both with indeterminate office-holding and with hereditary expectancy.

The evidence of curial control over the shrievalty in and just before 1130 is both varied and convincing. Further proofs which may be cited are the increased activity of itinerant justices and the obvious subordination of the sheriffs' fiscal activities to a strong exchequer. Sheriffs were beginning to come and go in comparatively rapid succession,[80] as in the time of Henry II. The last of the hereditary sheriffs in the south-west was superseded.[81] Some were mulcted at the exchequer for negligence. One whose administrative incapacity is fairly well established by the Pipe Roll of 1130[82] had lately been dismissed from office. Furthermore, it is clear that special curial agents who had long held the position were being shifted from county to county,[83] very much as was done at some later periods of administrative reorganization.

A final and still more potent consideration appears in the personnel of the shrievalty for the fiscal year 1129-30. The king's household and other state officials now serve as sheriffs in larger

[77] For Oxfordshire in 1130, 400 marks, term unspecified (*Pipe Roll*, p. 2); for London, 120 marks for the year (*ibid.* p. 144).

[78] William de Eynesford for holding the counties of Essex and Herts a year gave 20 marks, a fifth of the sum which had been specified for a quinquennial period (*ibid.* pp. 52-3); Hugh de Warelville 20 marks for holding Leicestershire and Northamptonshire a half-year of the five for which he was to pay 200 marks (*ibid.* p. 85).

[79] *Ibid.* p. 44.

[80] Apart from Lincolnshire and London and Middlesex, the shrievalty of Nottinghamshire and Derbyshire now exemplifies this tendency. In Berkshire Anselm *vicomte* of Rouen, in office 1127-9, succeeded Baldwin fitz Clare (*Pipe Roll*, pp. 122, 124). In Kent Ansfrid, probably the former *dapifer* of the archbishop of Canterbury (Rochester Chart. Dom. Ax. p. 102), could have held office but one or two years preceding Ruallo, sheriff 1120-23.

[81] Richard fitz Baldwin served apparently in 1126 or 1127 (Farrer, *Itinerary*, no. 532). For 1128-30 Geoffrey de Furnell held both Devon and Cornwall.

[82] Restold, sheriff of Oxfordshire (*Pipe Roll*, p. 2).

[83] William of Pont de l'Arche from Wiltshire (above, p. 78) to Berkshire, which he holds with Hampshire, 1129-30; William de Eynesford from Kent to Essex (above, p. 79, note 34); Aubrey de Vere still earlier from London to Essex (above, pp. 81-2); Hugh of Warelville in 1130 holds Sussex after giving up various other counties.

number than in the two preceding reigns. The few feudal figures
of importance who are still sheriffs are so employed. Such are the
two hereditary sheriffs, Miles the constable and Walter de Beauchamp, who sits in the seat of Urse d'Abetot and is probably the
king's dispenser besides.[84] The second Robert d'Oilly, another
constable, seems to be the Robert who for the past year has been
sheriff of Oxfordshire.[85] Of the three leading chamberlains who are
sheriffs, only one, Aubrey de Vere, belongs to the hereditary class.
The Norman, Geoffrey de Clinton, who has Warwickshire, has
risen by long service at the *curia*,[86] and has been treasurer [87] as well
as chamberlain, and also itinerant justice in many counties. William
of Pont de l'Arche, lately married to the daughter of William
Mauduit, is now paying a large sum for his *ministerium curiae*,[88]
and a little later will apparently be treasurer.[89] William d'Aubigny
the Breton, a staunch supporter of the king in 1106,[90] now the
husband of a daughter of Roger Bigod [91] and sheriff of Rutlandshire,
has been the king's justice.[92] Bertram de Bulmer as well as Warin
appears to have a post at the exchequer, and Osbert Silvanus has
apparently sat at the king's pleas.[93] Excluding from consideration
the sheriffs of London, who this year number four, and the unknown
sheriff of Somerset, a county apparently lost from the Pipe Roll of
1130, there are only seven sheriffs of English counties [94] who seem

---

[84] This office after his decease in 1133 was held along with his lands by his son
William (Farrer, *Itinerary*, no. 705 ; cf. no. 497).

[85] The sheriff of 1130 owes 26 gold marks *de debito patris sui pro pecunia
Widonis de Oilly* (*Pipe Roll*, pp. 1-2). Robert d'Oilly is represented (*D.N.B. s.n.*
" Robert d'Oilgi ") as *civitatis Oxnefordiae sub rege preceptor*, and he and his
wife Edith, in legend at least, as resident in the castle here (*Monasticon*, vi. 251),
when they founded the abbey of Osney. He was the founder of the church of
St. George within the castle of Oxford (Farrer, *Itinerary*, no. 603). I believe that
he first links this shrievalty with the custodianship of the castle which had still
earlier been in his family.

[86] For his appearance at the *curia* about 1110 see *Ramsey Chartul.* i. 241-2, ii.
83. Mr. Round (*The Ancestor*, xi. 156-7) traces his origin to the neighbourhood
of St. Lô, where he had a castle.

[87] *Monasticon*, vi. 220, 221.

[88] *Pipe Roll*, p. 37.

[89] So accepted by the editors of the Oxford edition of the *Dialogus* (pp. 20-21),
and apparently by Mr. Round (*E.H.R.* xiv. 423).

[90] *D.N.B. s.n.* " Albini, William de ".

[91] *Red Book*, i. 397 ; Farrer, *Yorkshire Charters*, i. 461. As to his shrievalty,
see *Pipe Roll*, pp. 133, 134.

[92] *Ibid.* p. 115.   [93] *Ibid.* p. 35.

[94] Odard in Northumberland ; Hugh fitz Baldric, Northamptonshire ; Geoffrey
de Furnell, Cornwall and Devon ; Ruallo de Valognes, Kent ; Hildret, Carlisle ;
Reiner of Bath, Lincolnshire ; and a Vernon in Westmorland.

to hold no position at the *curia*, and of these two are otherwise known to have had the king's special confidence.[95]

This year shows the highest centralization in local government since the period following Tinchebrai. The situation was now very exceptional, inasmuch as two curials, Aubrey de Vere and Richard Basset, jointly held eleven shires. This famous arrangement is known to have been a sudden creation.[96] It was formed in the main at Michaelmas 1129, when Robert fitz Walter and Maenfinin each surrendered his two shires, and Fulk the three which had been for some twenty years in his family.[97] Probably at Easter it was completed by superseding Hugh of Warelville in the shires of Leicester and Northampton, which he had recently obtained for a period of five years, and William of Eynesford in Essex and Hertfordshire under exactly the same circumstances.[98] As Mr. Round has observed,[99] Aubrey and Richard did not farm these counties according to customary usage, but were in the position of the later *custodes*. The king was thus free to dispose of all the profits arising within a considerable portion of the realm, and the fact affords the one plausible explanation[100] of this very remarkable innovation. Moreover, in still other directions this is a year of innovation in exchequer accounting.[101] It is certain that several of the retiring sheriffs were sadly in arrears and that the two special administrators placed in control of this strong fiscal unit were able to advance a large sum, a *superplusagium*, over and above their receipts, a part of which went to supply the king's needs in Normandy.[102] Aubrey

[95] Odard and Reiner, the latter classed by Ordericus along with Ralph Basset and Hugh of Buckland among the persons raised by the king from the dust.

[96] Round, *Geoffrey de Mandeville*, p. 298.

[97] *Pipe Roll* 1130, pp. 44, 90, 100.

[98] *Ibid.* pp. 53, 85. Maenfinin had held his shires for four years of what was probably a longer term (*Pipe Roll*, p. 100). Fulk had probably served but part of a five-year term.

[99] Round, *Geoffrey de Mandeville*, pp. 297-8.

[100] The accusation of treason brought against Geoffrey de Clinton at the Easter court, 1130 (*Henry of Huntingdon*, R.S. p. 252), came rather too late to give the explanation. Moreover, there is no intimation that Geoffrey's associates, either sheriffs or treasury officials, were involved. At the following Michaelmas he was still sheriff. William of Pont de l'Arche was in custody of land which he had held in Oxfordshire, and Richard Basset of his land in Leicestershire (*Pipe Roll*, pp. 6, 81).

[101] The blanched farms of the previous year in the eleven counties give way to farms paid by weight, as one might expect. But the same change occurs also in the counties of Berkshire, Hampshire, Wiltshire, Kent, Warwickshire and Lincolnshire, and for half of the year in those of Nottinghamshire and Derbyshire.

[102] Six hundred out of a thousand marks (*Pipe Roll*, p. 63).

de Vere, the king's chamberlain,[103] had long been a special agent of the administrative *curia* and had had much experience as a sheriff as well.   The same may be said of Richard's father, Ralph Basset. All three held the highest judicial position [104] and did wide itinerant service.   Richard by marriage also [105] was identified with the same circle at court.   The placing of two officials of this type over so wide a region, and the dismissal of tried and experienced sheriffs, are sufficient indications that the step has administrative rather than political significance, and that the matter lay very close to important interests of the king.

The height of the administrative revolution was apparently soon passed.   How long this group of counties held together is unknown.   But Aubrey de Vere before the next Michaelmas agreed to give a hundred marks to be permitted to withdraw from the shrievalties of Essex and Hertfordshire.[106]   Fulk seems again to have been sheriff of Huntingdonshire after 1133,[107] and Robert fitz Walter, as Mr. Round has shown, was certainly sheriff in East Anglia until a decidedly later period.[108]   The new curial control over sheriffs, which largely repaired the chief defect in the Norman plan of local government in England, of course broke down in Stephen's reign.   A few baronial heads of shires who survived from King Henry's time were destined sometimes to aid, more often to plague, his successor.   The changes in the personnel of the office during the first third of the twelfth century none the less show convincingly a steady approach towards conditions of the Angevin period, and, as individual cases prove, follow a corresponding trend in the selection of the king's central administrative staff.[109]

It has already become clear that groups of shires, usually pairs, were often governed by a single sheriff.   Several of the familiar pairs

---

[103] He was probably not great chamberlain until 1133 (Farrer, *Itinerary*, no. 698). Concerning him and his family, see Round, *Geoffrey de Mandeville*, pp. 388-96.

[104] Aubrey is designated *iustitiarius totius Angliae* (Round, *Geoffrey de Mandeville*, p. 80), a title also given the two Bassets by Henry of Huntingdon R.S. p. 318 ; cf. *Chron. Abingdon*, ii. 170).

[105] With the daughter of Geoffrey Ridel (Sloane MS. xxxi. 4 (47)), another justiciary of all England.

[106] *Pipe Roll*, p. 53.

[107] *Ramsey Chartul.* i. 152 ; cf. iii. 176.   [108] *E.H.R.* xxxv. 483-6.

[109] Through the kindness of Dr. Curtis H. Walker, the writer, after this was written, was enabled to read in proof the article " Sheriffs in the Pipe Roll of 31 Henry I " (*E.H.R.* xxxvii. 67).   Dr. Walker's chronology of the shrievalties was in several instances more exact than that of the writer, and acknowledgement of the debt is gratefully made.

of latter days already appear. Moreover, the joint sheriffs, familiar in the usage of the Angevin kings, are occasionally mentioned. The shrievalty of London and Middlesex even now presents unusual features. King Henry's charter, probably dating between 1131 and 1133, confers upon the citizens of London the special privilege of electing.[110] Moreover, Fulcred fitz Walter, who during the fiscal year 1128–9 held this position by the payment of a *gersoma*,[111] gave place in the succeeding year to four sheriffs.[112] Before 1130 an undersheriff[113] and the sheriff's *ministri*[114] are mentioned at London as well as elsewhere, although it is clear that in the former place there was also a reeve. Here the sheriff in the earlier years of King Stephen is known to have employed a clerk.[115]

There is some allusion to the duties of this official in levying distress and making arrests. If he made distraint unjustly, according to the *Leges Henrici*[116] he rendered double amends to the injured person. If the case proceeded higher, he was not at once to sell the goods seized.[117] His power and that of his assistants to arrest criminals was limited by the rights of sanctuary possessed by churches.[118] As early as 1130 the sheriff from whose custody a prisoner escaped was heavily liable.[119]

Reference to his judicial functions is occasionally made in the scanty records of the period. King Henry apparently righted a grievance as old as his brother's reign when, a few years after Tinchebrai, he directed that the courts of shire and hundred should sit only at the same times and places as in the reign of King Edward.[120] The king reserved the right to summon special sessions when his

---

[110] Liebermann, *Gesetze*, i. 525-6.

[111] *Pipe Roll* 31 Henry I., p. 144.

[112] *Ibid.* pp. 143, 144, 149. In 1130 they paid 100 marks to have a sheriff *ad electionem suam* (p. 148).

[113] William de Eynesford, sheriff of London at some time between 1114 and 1130, had such a subordinate (*Chron. Ramsey*, i. 249), as had Robert fitz Walter in Norfolk and Suffolk (*ibid.* 266) in the same period. Hugh of Buckland may have had such an official in Albericus de Berchescira (*Chron. Abingdon*, ii. 91, 93).

[114] *Ibid.* ii. 76, 83 ; cf. note 162. In 1130 are mentioned in Berkshire (*Pipe Roll* 31 Henry I., p. 126) *collectores*, who apparently handle the Danegeld under the sheriff (*E.H.R.* xxxiii. 169, n. 227).

[115] Below, note 268. It is difficult to say how far the salutation in such a writ form as *Henricus rex Anglorum vicecomitibus et ministris suis de Dorsete* (*E.H.R.* xxxv. 390, no. 18) is simply euphonious, and how far intended to recognize the existence both of high sheriffs and undersheriffs.

[116] 51. 4. This appears to be a transaction in the hundred court.

[117] *Ibid.* 51. 6.

[118] *Leges Edw. Conf.* 5.

[119] *Pipe Roll* 1130, p. 53.   [120] Liebermann, *Gesetze*, i. 524.

necessity required, but the sheriff was to have no such right, a provision from which has been drawn the inference that the latter had made the local courts instruments of fiscal oppression.[121]   The sheriff issued summons to the county court seven days in advance of its session, and in doing so took witnesses to the house of the person summoned.[122]   This official seems occasionally to assume jurisdiction [123] or rights claimed by the holder of a liberty.   Upon complaint of the archbishop of York that the customs of his church were much infringed by Osbert the sheriff, the king in 1106 commissioned certain justices to go to York and inquire what these customs were.[124]

An occasional glimpse is afforded of a sheriff presiding over one of the sessions of the county.   Hugh of Buckland presided in a full county court in Berkshire in which men of two other sheriffs assisted, and before him the famous abbot, Faritius, established his right to certain lands and immunities in regard to which he had suffered violence.[125]   A royal writ in favour of the bishop of Norwich directed that he be not disseised of certain lands except by just judgement of the county.[126]   At Oxford the two sheriffs of the shire, Thomas of St. John and Richard de Monte, held a plea in the house of Harding, a priest,[127] and to the church of Abingdon was adjudged the right to toll on shipping at Oxford.   Again, William, son of Hugh of Buckland, in 1119 held court at Sutton on the Monday after Martinmas, and before him the whole county received the pledge of a man of the abbey of Abingdon that certain lands of his demesne were quit of Danegeld.[128]   This court held by the sheriff was regarded by royal officials as important also to the maintenance of the peace.[129]   The view of frankpledge, already made in the hundred court twice a year,[130] must have been held by the sheriff or his agent now,[131] as it was a half-century later.   The sheriff might direct a hundred

---

[121] Stubbs, *Const. Hist.* 6th ed. i. 429.

[122] *Leges Henrici*, 41. 2, 41. 5; cf. 46. 1.   Thus the *iusticia regis, ibid.* 53. 1, is probably the sheriff.

[123] The abbot of Ramsey (*Chron. Ramsey*, 218) procures a writ directing the sheriff that his men are to plead only where they pleaded in the time of King Edward.

[124] Farrer, *Itinerary*, no. 166.   [125] *Chron. Abingdon*, ii. 117.

[126] Brit. Mus. Cotton MS. ii. 6 (prior to 1107).

[127] *Chron. Abingdon*, ii. 119.   [128] *Ibid.* ii. 160.

[129] Liebermann, *Gesetze*, i. 524 (sect. 4).

[130] *Leges Henrici*, 8. 1.

[131] A writ to various sheriffs (*E.H.R.* xxiv. 427) directs that with certain exceptions the men of the abbot of St. Edmunds shall not leave the soke *pro plegiis suis et treingis renovandis*.

to do justice in a given case,[132] and in the reign of Stephen at least one sheriff is found holding a recognition in a hundred.[133]

Evidence of the importance to the king's administrative system of these sessions of the shire in which the sheriff presided is becoming clearer than in the earlier days of Norman rule. The king's grants and many of his administrative orders were directed to the county court, that the sheriff might have them read aloud and thus published. An ordinary grant to an abbot was said to be " ventilated " by being read in the county court ; [134] and the great charter of liberties issued at the king's coronation was sent to be read in every shire.[135] Some of the royal writs thus published by the sheriff extended the king's peace,[136] breach of which after it is thus proclaimed was amended by a payment of a hundred shillings ; [137] others made grants of the king's property and of peculiar privileges [138] as well as exemption from the usual fiscal obligations, such as Danegeld, scot, toll and castlework.[139] Furthermore, they might acquit the men of the grantees of the duty of attending shire and hundred courts,[140] or confer certain jurisdictional rights like that of dealing with thieves,[141] even granting away the great pleas of

[132] *Chron. Abingdon*, ii. 117.
[133] Below, note 274.
[134] *E.H.R.* xxxiii. 164, n. 163.
[135] Liebermann in *Trans. R.H.S.* N.S. viii. 22.
[136] Thus the monks of Tavistock are to have their possessions with a firm peace (*Monasticon*, ii. 501) ; the monks of St. Augustine a fair seven days each year when all coming are to have a firm peace and the monks are to have toll (*Hist. Monast. St. Aug.* 358) ; the same monks a market in the Isle of Thanet, all coming and going to have peace and the abbot all customs, forfeitures and pleas (*ibid.* 365-6). Both the sheriff and the *ministri regis* may proclaim the king's peace. This may also be done in the hundred (*Leges Henrici*, 79. 4 ; *Instituta Cnuti*, iii. 50).
[137] *Leges Henrici, ibid.*
[138] The bishop of Chichester and the abbot of Ramsey were granted warren (*Monasticon*, vi. part iii., 1168 ; *Chron. Ramsey*, 229) ; the church of St. Mary of Sarum, tithes of the New Forest (*Register of St. Osmund*, R.S. i. 206) ; the abbot of Westminster, stalls (Robinson, *Gilbert Crispin*, 157).
[139] *Ramsey Chartul.* i. 242-3 ; *Monasticon*, i. 44 ; vi. 53 ; *Chron. Abingdon*, ii. 88, 95 ; Robinson, *Gilbert Crispin*, 149, 150 (toll and passage) ; *Monasticon*, vi. part iii., 1168, no. 25. William II. acquitted the men of Ramsey of labour dues (*operatio*, Davis, *Regesta*, no. 354 ; cf. *Pipe Roll*, 31 Henry I., p. 33) ; cf. *de conductu thesauri et omni operatione*, *Monasticon*, v. 13, no. 5.
[140] *Monasticon*, i. 44, vi. part ii., 1043 ; *Ramsey Chartul.* i. 249 ; Rymer, *Foedera* (1819), i. 12 ; *Two Charts. of Bath* (Somerset Record Soc.), 49 ; Robinson, *Gilbert Crispin*, 144.
[141] The king gave to abbot Faritius justice upon the *latro presbyterus qui in captione sua in Abbendona est et de aliis latronibus suis faciat justitiam suam similiter vidente comitatu* (*Chron. Abingdon*, ii. 90).

the crown,[142] and in rare instances recognizing a judicial immunity from which the sheriff was partially or wholly excluded.[143] The ministerial functions of the sheriff are occasionally mentioned. As at an earlier time, the king depended upon him to put in effect judgement given by the royal justices,[144] and to see that the king's right was done.[145] The sheriff of Northumberland was directed to place Richard, the king's chaplain, in seisin of four churches which had been granted him.[146] The sheriffs of Berkshire and Oxfordshire were informed by the king's writ that Faritius, abbot of Abingdon, was to have his rightful customs in all things on the waters of the Thames ; [147] the sheriff and barons of Nottinghamshire were notified that the bishop was authorized to divert the royal highway at Newark.[148] Sheriffs and *ministri* were directed to see that abbot Faritius had his fugitive serfs with their chattels.[149] For failure to carry out these royal commands it was usually specified that the *forisfactura* should be ten pounds, a penalty often provided also for failure of officials to observe the immunities granted to individuals by charter.[150] Already there occurred a form of royal writ, also found in Normandy,[151] which was in effect a *praecipe* commanding one person to do justice to another in a given case, the alternative being that the sheriff was to do it.[152]

[142] To the archbishop of York pleas of thieves and false moneyers (*Monasticon*, vi. part iii., 1180) ; cf. Rymer, *Foedera*, i. 9.

[143] *Chron. Abingdon*, ii. 164, 165 ; *Chron. Ramsey*, ii. 83 ; *Monasticon*, v. 13, no. 5. Henry I. conferred on the abbey of Evesham the hundred of Blakenhurst so that no sheriff or other *minister placitet vel exigat* (Brit. Mus. Harleian MS. 3763, fo. 82 d, formerly lxxix d). His charter of liberties to St. Edmunds specifies that no *minister regis* in any way intervene in the borough, confers soke of the eight and a·half hundreds held by this church with the liberties and forfeitures pertaining to the crown, with quittance of all scots, geld and aids, and also with quittance of suits at the county court of Suffolk, so that all *libere tenentes* are to come to the *magna placita* of the abbot, who may distrain to enforce this right (Brit. Mus. Add. MS. 14847, m. 39 *dorse*).

[144] *Chron. Ramsey*, 280. The *Ramsey Chartul*. i. 239 contains notification to a county court of recovery of land before the king's justice.

[145] Hugh of Buckland is to see that the abbot of Ramsey has full right in a mill recently constructed, *et ita ne rectum inde amittam* (*Ramsey Chartul*. i. 247).

[146] *Historians of Hexham* (Surtees Soc.), i. app. p. ix.

[147] *Chron. Abingdon*, ii. 95.

[148] *Monasticon*, vi. part iii., 1273, no. 24.

[149] *Chron. Abingdon*, ii. 81.

[150] As in *Monasticon*, v. 13, vi. part ii., 1043 ; Round, *Calendar*, no. 479 ; *Ramsey Chartul*. i. 249.

[151] Haskins, *Norman Institutions*, 93.

[152] *Chron. Ramsey*, 280 ; Bigelow, *Placita Anglo-Normannica*, 117. In one instance the king commands Jordan de Sackville to do justice for land taken ; if he fails, Walter Giffard is to do it ; if he fails, Hugh of Buckland, who is sheriff

Legal process of the reign of Henry II. was also foreshadowed by a writ of his grandfather to the sheriff of Lincoln commanding the latter to reseize the abbot of Ramsey of his land at Trickingham, since the king was unwilling that he lose unjustly in any way what pertained to the food and clothing of the monks.[153]   The same is to be said concerning the continuation of the Anglo-Norman usage by which the king directed the sheriff to summon a jury of recognition to settle a disputed legal point.[154]   Such a precept was often tantamount to an order to the sheriff to institute proceedings in a case in the county court.   But a sheriff might be directed to summon a hundred to try a cause [155] or to hold a recognition.[156]   On one occasion nine hundreds came before the sheriff of Norfolk as inquisition witnesses.[157]   The sheriff might be authorized to employ this primitive jury procedure to determine either facts essential to his execution of a special administrative order of the king or points involved in ordinary administrative routine.[158]   The *juratores comitatus*, who in 1130 were being amerced by itinerant justices,[159] were probably inquest jurors.

The position of the sheriff with reference to judicial affairs already showed signs of the decline which was to continue for centuries.   It was being assailed both by feudal and by royal influence.   Baronial assumption of jurisdictional powers over feudal tenants after the Norman Conquest must have affected the

---

(*Chron. Abingdon*, ii. 85).   Likewise the king directs William of Houghton, his chamberlain, to perambulate the bounds in Cranfield between the land of the abbot of Ramsey and the king of Scotland (*Ramsey Chartul.* i. 247-8), and unless William does this openly the sheriff shall cause it to be done.   See the king's writ to Odo, sheriff of Pembroke, in Haskins, *Norman Institutions*, 305.

[153] *Ramsey Chartul.* i. 237.   For the corresponding writ of Henry II. *ibid.* i. 251.

[154] The king's writ to the sheriffs of Berkshire and Oxfordshire at some time in the period 1100-1103 bids these officials direct the men of their counties to speak the truth concerning land which Ruallo de Avranches claims against the abbey of Abingdon (*Chron. Abingdon*, ii. 84-5).   Similarly Hugh of Buckland is commanded to cause the county court of Bedfordshire to sit and cause recognition to be made of the *divisiones* of the abbot of Ramsey (*Ramsey Chartul.* i. 246).

[155] *E.H.R.* xxxv. 392-3, no. 27.

[156] Stubbs, *Const. Hist.* i. 427, n. 1.

[157] *Chron. Ramsey*, 267.

[158] As in the perambulation of bounds (*Ramsey Chartul.* i. 247), and probably in such a writ as that to the county court of Yorkshire (*Appendix to Deputy Keeper's 30th Rep.* p. 204) prescribing that the monks of St. John of Beverly are exempt from paying geld if they did not pay in the reigns of King William and King Edward.

[159] Stubbs, *Const. Hist.* i. 428-9.

hundred much more than the shire court.[160]   Yet the prelates and magnates, who by the king's grants and the statement of the *Leges Henrici* [161] had sake and soke and some other franchises, evidently excluded part of the sheriff's older jurisdiction.   Ordinary soke in the *Leges Henrici* was exercised under but three kinds of officials, manorial reeves, or reeves of hundreds and boroughs or sheriffs.[162]   Regarding matters which still belonged to the latter in the county court, few particulars other than those already noted are given.   The importance of his civil jurisdiction is clear.[163] The occasional consent of the king to a grant of immunity which excluded the sheriff [164] and his regular notification to sheriffs and county courts when the king alienated crown pleas, convincingly demonstrate the importance of this side of its jurisdiction.   Cases involving outlawry had been disposed of here even in the Domesday period.   Obstructions upon the king's highways other than the king's four great ways, and which led from one town to another or to a market, or upon lesser waters, were said to be *sub lege comitatus* ; [165] the same was true of the boundaries (*divisae*) of hundreds and wapentakes.   Theft, in some cases at least, was a *placitum comitatus*,[166] and by special direction of Henry I. jurisdiction over land cases between mesne tenants [167] of different lords was conferred upon the county with the requirement that trial be by battle.

The theory that the Norman king was the fountain of justice naturally made for the enlargement of the sheriff's sphere as justice of crown pleas.   These pleas were said to be above ordinary law and everywhere to preserve immobility.[168]   Although no complete list has come down to us, it is clear that the items included were fairly numerous.[169]   The king, moreover, might command the

[160] It may be doubted whether the shire court declined so much in power as Stubbs (*Sel. Charters*, 103) and Davis (*England under Normans*, 139) hold.   Cf. *Am.H.R.* viii. 486-7, and p. 94 below.

[161] 20. 2, 30, 55, Liebermann, *Gesetze*, i. 560, 563, 575.

[162] *Leges Henrici*, 20 ; Liebermann, *Gesetze*, i. 560.

[163] Concerning litigation between parties in the shiremote see *Leges Henrici*, 7. 3, 7. 3 a, and pp. 55, 102, n. 240.

[164] See p. 89.

[165] *Leges Edw. Conf.* 12.9-13.1 ; Liebermann, *Gesetze*, i. 639-40.   These obstructions are said to be amended *erga comitem et vicecomitem* (*ibid.* 12. 10) ; the judicial income from the *divisiones* of hundreds and wapentakes to go to the same officials *cum iudicio comitatus* (*ibid.* 13. 1).

[166] *Furtum probatum et morte dignum* (*Leges Henrici*, 13) was a plea of the crown.

[167] Liebermann, *Gesetze*, i. 524, sect. 3. 2.   [168] *Leges Henrici*, 9, 10 a.

[169] *Omnia placita ad coronam meam pertinentia* (*Ramsey Chartul.* i. 241).   The long list of *dominica placita regis*, which the king has of all men (*Leges Henrici*, 10)

county court to do justice in a specific case not ordinarily justiciable there, even directing the sheriff, like an itinerant justice, to summon a special session.[170]   He might also extend or alter the list of pleas which belonged to his courts, and which the sheriff had been accustomed to try as his agent.[171]   This particular jurisdiction, as the reign advanced, was evidently conferred more and more upon royal justices.   It had grown at the expense of the ordinary county pleas in the sheriff's hands ; its loss, therefore, meant a marked curtailment of his former judicial province.

The reign of Henry I. brings into view the legal and judicial arrangements by which the sheriff's fiscal responsibility to the crown was enforced.   By the middle of the reign the exchequer was in existence, employing its characteristic methods of accounting,[172] and about this period occurs an allusion to the fact that the sheriffs appeared there to give an accounting.[173]   Moreover, in a royal grant of 1131 is mentioned a well-known course of justice followed by the exchequer to collect the king's *ferm*.[174]   This is excellently illustrated by the record of the sheriff's debts and payments under this head in the Pipe Roll of 1130.   A former sheriff was declared *in misericordia regis* for failure to account for the *ferm* of lands which he had held in custody.[175]   The exchequer in all probability had by this time devised means to assure the sheriff's payment also of a large revenue derived from miscellaneous sources, the amount of which had not been known in advance to the officials.[176]   It was the pride of the exchequer in later days that it was able to discern

---

and which are essential to the peace, as well as the special list which places a man in *misericordia regis* (*Leges Henrici*, 12. 3), may be supplemented from Domesday Book.   See above, p. 55, note 110.   Neither source mentions some of the cases listed above as triable in the county court.

[170] As in *Chron. Abingdon*, ii. 93 ; *Ramsey Chartul.* i. 246.   Professor Adams differentiates between "king's county court" and the "people's" or "sheriff's county court" (*Pol. History of England*, 152).   Cf. *Am.H.R.* viii. 489-90.

[171] In 1108 he made theft and robbery punishable by hanging (*Florence of Worcester*, Eng. Hist. Soc. ii. 57), and in the same period took measures to enforce an earlier decree which declared that for counterfeiting or debasing the king's currency the king's justice was to be done (Liebermann, *Gesetze*, i. 523).   The idea of a king's peace promotes constant changes of this kind.

[172] Poole, *Exchequer*, 37-40, 49-50.   The *iudicis fiscalis ius* is mentioned in the *Leges Henrici*, 24.

[173] Round, *Commune*, 123.

[174] Round, *Calendar*, nos. 1387, 1389.

[175] *Pipe Roll* 31 Henry I., p. 2.   Compare the *forisfactura scaccarii* (*ibid.* p. 91).

[176] According to Sir James Ramsay's computation more than half the king's revenue for the year 1129-30 was derived from sources not included at a fixed figure in the sheriff's *ferms* (*Foundations of England*, ii. 328).

in just such difficult matters whether the sheriff had acted wrong-fully ; [177] moreover, this seems implied by the form of the Pipe Roll of 1130 and by the apparent assumption in the *Dialogus* that the usual form of writ of summons to the sheriff, whereby specific sums from unfarmed sources were specifically demanded, dates from the reign of Henry I.[178]

The principal audit of the year was already at Michaelmas, and by 1113, possibly as early as 1108, the trial of a case involving revenue was held at the treasury in the castle of Winchester.[179] The king's confirmation, at some time in the period 1118–23, of Queen Matilda's grant to the canons of St. Trinity, London, gives significant information. The grant had conferred two parts of the revenue of Exeter. The king now specified that the sheriff at Exeter, whoever he should be, was to send the money to London and make payment to the canons at Easter and Michaelmas.[180] Thus the semi-annual periods of accounting to be known for cen-turies were already being observed,[181] but a distant sheriff's fiscal transactions at London appear as yet to be quite exceptional.

Some detail illustrative of this fiscal custody of the shire may be added. As at an earlier time, the sheriff occasionally owed an additional *ferm* as special guardian of a royal manor or other estates in the king's possession.[182] The *ferm* of Canterbury and Dover in 1130 clearly passed through the sheriff's hands ; [183] but those who rendered account of the *firma burgi* of Northampton and Colchester [184] accounted directly to the exchequer. The burghers of Lincoln paid to the king a sum that they might hold in chief,[185] one of the earliest known instances of a borough's farming directly of the crown. The Pipe Roll of Henry I. shows that the sheriff as

---

[177] This was the *excellens scaccarii scientia* (*Dialogus*, 67-8).

[178] *Additum fuit in summonitionibus hoc subscriptum ex novella constitutione hoc est post tempora regis Henrici primi* (*Dialogus*, 114).

[179] Round, *Commune*, 79-81, and *Calendar*, xliii-xlv ; *Dialogus*, 43. As to the case concerning the dues of the abbot of Abingdon before the justices in the treasury see Round, *Commune*, 94, and Poole, *Exchequer*, 34, n. 2, 36.

[180] Rymer, *Foedera* (1819), i. 12.

[181] As at a later time the sheriff's annual account could not always be closed at Michaelmas, and debts were carried over sometimes for several years. William of Buckland even owed upward of 13 pounds of the old *ferm* of the county of the time of his father (*P.R.* 31 Henry I., p. 127), deceased some fifteen years earlier.

[182] As in *P.R.* 31 Henry I., p. 40.

[183] As in *P.R.* 31 Henry I., p. 63.

[184] *Ibid.* 135, 138. The reeve of Malmesbury appears to pay its *ferm* independ-ently of the sheriff's action (*ibid.* 16).

[185] *Ibid.* 114 ; Ballard, *Domesday Boroughs*, 93.

a disbursing official was even now paying from the *ferm* of the shire the king's fixed alms and other sums expended on his account or by his writ,[186] and, as in the days of Angevin rule, receiving credit therefor at the exchequer. An outgoing sheriff was held rigidly to account for depreciation of the property in his care. Serlo de Burg, late sheriff of Northampton and Derby, owed sixty pounds for the restoration of these counties, Baldwin de Clare the same amount for the restoration of Berkshire,[187] Aiulf the chamberlain fifteen pounds for Dorsetshire.[188] Restold, sheriff of Oxfordshire 1128–29, was recorded as debtor to the amount of some two hundred and forty pounds for *defectus* of the county in grain, hay, houses, granges, mills, fisheries, land bearing no profit, *bordarii, buri* and oxen ; also twenty-eight pounds for loss of cattle, swine and sheep, and seven pounds for woods so far destroyed that no sheriff could render *ferm* [189] for them.

Of the sheriff's payments to the exchequer other than those related to the *ferm* of the shire may be identified sums arising from fairs,[190] treasure trove,[191] the *census* of the king's forests,[192] the Danegeld,[193] by 1130 an annual tax, the *donum* or *auxilium* [194] from the boroughs, which supplemented the Danegeld, and the pleas of the crown. Since the income from the ordinary pleas of the county was still included in the *ferm* of the shire,[195] it is evident that amounts from judicial sources entered individually upon the Pipe Roll were

[186] For a good example see *P.R.* 12-13. Cattle and oxen for the restoration of manors under the sheriff's supervision are purchased at the king's expense by William de Pont de l'Arche (*ibid.* 122). Work on the castle of Arundel is paid for in the same way (*ibid.* 42). Expenditures duly witnessed after the fashion familiar to the writer of the *Dialogus* are also recorded.

[187] *Ibid.* 31, 122.

[188] *Ibid.* 14.

[189] *P.R.* 31 Henry I., p. 2.

[190] *Ibid.* p. 153 (*feria Exoniae*).

[191] *Ibid.* p. 68 (*inventio*). According to *Leges Henrici*, 10. 1, *naufragium* as well as *thesaurus inventus* is one of the king's rights ; cf. *Ramsey Chartul.* i. 241.

[192] Farrer (*Itinerary*, no. 681, and note 7) calls attention to a writ of the king to the exchequer officials in the period 1131-33 authorizing the release to the monks of Rievaulx of two shillings in Danegeld and serving as the necessary quittance to the sheriff (*Chartul. of Rievaulx*, Surtees Soc. no. 196). In the Chartulary of Hyde abbey (Brit. Mus. Cotton MS. Domitian A xiv. m. 34) a quittance from gelds is addressed to William de Pont de l'Arche and the collectors of Winchester.

[193] Instances will be found in Madox, *Exchequer*, i. 600-601.

[194] *P.R.* 31 Henry I., p. 17. There were also sums *de minutis forestariis* (*ibid.* 39).

[195] *Leges Henrici*, 8. 11, 10. 4. In *P.R.* 1130 (p. 142) Odoard renders account of 10 pounds *ferm pro placitis Caerloil quae ad vicecomitatum pertinent.*

derived from the king's special pleas. Among them were included the older *forisfacturae* which went to the king, for the sheriff is specifically mentioned as turning over sums collected for *hamfare* [196] and breach of the peace.[197] It is obvious, furthermore, that he handled the income from some miscellaneous pleas.[198] The same is true of amercements laid upon communities, including *murdra*,[199] unsatisfactory judgements given in the hundred,[200] and unsatisfactory conduct on the part of judgement finders in the borough [201] and the county.[202] The *forisfactura comitatus*,[203] for which the sheriff occasionally accounted, is thus explained. A sum paid in by the sheriff of Huntingdonshire for the men of the county, on account of the hedges at Brampton,[204] suggests trespass against the king. Various manors or vills appear to pay collective emercements in the same way.[205]

The accounts of judicial origin thus collected by the head of the shire often, perhaps usually, arose in the later years of Henry I. from pleas held before the itinerant justices. It is reasonably clear that sums imposed upon communities by the king's justices were brought to the exchequer by the sheriff; [206] there is question concerning amounts for which individuals have incurred the same liability. For these, Sir James Ramsay believed [207] responsibility rested with others. In some instances individuals who, according

---

[196] *P.R.* 112.

[197] *Ibid.* 97, 113.

[198] *Minuta placita* (*ibid.* 160) ; *placita de minutis hominibus* (*ibid.* 96).

[199] The amount paid by the sheriff at the exchequer on account of *murdra* is curiously variable ; see *P.R.* 1130, p. 21, and *Leges Edw. Conf.* 15, 4. According to *Leges Henrici*, 92. 16, the burden of collection rested upon the hundred.

[200] The sheriff accounts for money due from a wapentake on account of a certain judgement (*P.R.* 1130, 10) ; cf. note 202.

[201] The sheriff renders 40s. *de judicibus burgi de Buckingham* (*ibid.* 101) arising from the pleas held by Geoffrey de Clinton.

[202] In Suffolk the sheriff collected 25 pounds *de judicibus comitatus et hundredorum* (*ibid.* 97) ; in Kent over 18 pounds *de placitis G. de Clinton de juratoribus comitatus* (p. 65).

[203] *Ibid.* 2, 13, 24. Of much the same nature appear to be the *placita communis comitatus* for which the sheriff in one instance makes an accounting quite separate from his *ferm* (*ibid.* 123).

[204] *Ibid.* p. 48.

[205] Robert owes 40s. *pro hominibus suis de placitis Ricardi Basset* (*ibid.* p. 93). The sheriff of Lincolnshire renders account of 10 marks silver for breach of peace in Traho wapentake by the men of Hervey Belet, and 20 marks silver for the same offence in Lovenden wapentake by the men of Ralph de Albini and Jordan fitz Alan (p. 113).

[206] This is the rule laid down later in the *Dialogus*, 142-3.

[207] Ramsay, *Foundations of England*, ii. 325.

to the Pipe Roll of 1130, owed amounts on account of pleas were persons who had been amerced.[208]   But there is also mention of persons who can hardly be other than custodians of sums locally collected.[209]   Although at first sight it might seem that the sheriff was relieved of all responsibility when these accounts at the exchequer were charged against the names of individuals, several facts tend to show that this was not the case.   In the first place, it is clear that one who was entered as owing money for the pleas held by royal justices might make payment through the sheriff.[210]   The latter, moreover, was occasionally recorded as making collective returns for *placita* as well as returns for individual sums received from such sources.[211]   This not only coincides with usage set forth a few decades later in the *Dialogus de Scaccario,* but also with the sheriff's earlier function as collector of the king's judicial income.[212]

The sheriff was still reckoned an oppressor, intent upon his own financial advantage as well as insistent upon the exacting claims of the exchequer.   The second charter of liberties issued by King Stephen in 1136 promised that exactions and injustice and mis-

---

[208] Various persons in Wiltshire who owe sums *de placitis Radulfi Basset* are pardoned *pro paupertate sua* (pp. 18-19).   A certain *Nicholas scutiger Wandregesili nichil habet nec potest inveniri* (p. 19).   A Suffolk man *non est inveniri comitatu nec quicquam habet* (p. 97).   Ulric Bulehals pays 2 mk. silver for loss of a duel (p. 48).

[209] *Beniamin reddit compotum de iiii. li. et v. s. ut custodiat placita que Coronae Regis pertinent.   In thesauro lvi. s. et viii. d.* (p. 91).   *Et idem vicecomes reddit compotum de c. s. pro placitis pecuniae Beniamin in hundrede de Clavering.   In thesauro liberavit* (p. 93).   As to this Benjamin's official status see *Sel. Coroners' Rolls,* ed. Gross, Selden Soc. p. xvii.   Baldwin de Redvers renders account of 500 mk. of the pleas of the forest and other sums for the old aid of the knights of Devonshire (p. 153).   Roger accounts for 40s. of the pleas of Thetford (p. 93).   Ralph and Godric, reeves of Grimesby, owe 2 mk. gold of the pleas of W. de Albini (p. 118).   Moreover, it is clear that the lord or his representative is sometimes the person against whom is charged what is due from peasants through pleas of the justices.   See the sums due from Peisson, *homo Patricii de Cadurcis* (*ibid.* 124), Allan de Creon (p. 112) and William fitz Rannulf (p. 48).   Cf. note 114 above.

[210] *Ulwinus de Bedewinda debet xliii. s. et iiii. d. de placitis Robertorum.   Sed liberavit Willelmo vicecomiti ut dicit* (p. 20).

[211] The sheriff renders account of 108s. 8d. of the pleas of *Robert Arundel de burgensibus Exon* (p. 154).   Hait, sheriff of Pembroke, accounts for the pleas of Miles of Gloucester and Payn fitz John (p. 136).   The sheriff accounts for both pleas and pleas of the forest held in Devonshire and Cornwall by Robert Arundel (p. 155).   Geoffrey de Clinton as sheriff accounts for 100s. in old pleas and *murdra* (p. 105).   The sheriff also accounts for moneys derived from *minuti homines* on account of pleas (*P.R.* 31 Henry I., pp. 96, 142, 143, 160).

[212] *Dialogus,* 142, 215 ; above, p. 67.   The *Dialogus* holds, however (ii. xii. A), that when debts arising from *placita* and *conventiones* have not been satisfied, individuals answer for themselves.

kennings,[213] whether due to the sheriff or any one else, should be utterly destroyed.[214] Mr. Round has shown that the *gersoma* paid by sheriffs at this time for the privilege of farming their shires points to a wide margin of profit in the undertaking. He also finds indications that the sheriff might not arbitrarily increase the sums due from the demesne lands of the crown and that these were stationary in amount.[215] His deduction is that the sheriff so manipulated the ordinary pleas of the shire and hundred, the remaining source of his *ferm*, as to make them produce enough to assure a good profit.[216] The mode of manipulating pleas in this period makes this highly probable.

But other devices were evidently profitable. The sheriff of Suffolk, commanded by King Henry to desist from his collection of toll on the stone being brought to build the church of Bury St. Edmunds, and to release the ships seized on this account,[217] seems to have attempted one. Moreover, Westminster Abbey, immediately after King Henry's coronation, was acquitted of the obligation to pay sheriffs' aids [218] which, therefore, must have been demanded in the reign of William Rufus. In the third decade of King Henry's reign there is mention of aids both of sheriffs and of their *ministeriales*.[219] Restold, late sheriff of Oxfordshire in 1130, owed the crown over a hundred and sixteen pounds which he unjustly took from the *villani* and burghers of the king's manors.[220] He had been caught in a practice of which he was hardly the sole exponent. His offence in the eyes of the exchequer was hardly his injustice but rather his unwarranted appropriation of revenues from sources which were peculiarly the king's. The towns were apparently made also to contribute to the sheriff's pocket. It is well known that

[213] Miskenning is defined (*Leges Henrici*, 22) as *mislocutio*. It apparently occasioned exaction upon the person who had blundered over the technical phraseology of pleading.

[214] Stubbs, *Sel. Charters*, 121.

[215] Round, *Geoffrey de Mandeville*, 191.

[216] *Ibid.* 359-61.

[217] Brit. Mus. Add. MS. 14847, m. 33 d. The name of the sheriff is given as William de Chames (probably Chaines, one of the many forms of *Caisneto*).

[218] Brit. Mus. Cotton MS. vii. 8 ; Farrer, *Itinerary*, no. 2.

[219] *Foedera* (1819), i. 12. A charter of Stephen (Round, *Calendar*, no. 295) remits aid to justices and sheriffs as has been done formerly by King Henry's charter. Cf. *ibid.* no. 1261. As to its character see Stubbs, *Const. Hist.* i. 382, 499 ; Round, *Feudal England*, 500-501 ; Liebermann, *Gesetze*, ii. 422 (161). Powicke (*Loss of Normandy*, 64) shows that in Normandy the sheriff's aid or *gravaria* was included in the *ferm* and that it was raised by a land tax.

[220] *P.R.* 31 Henry I., p. 2.

Lincoln was willing to give to the crown two hundred marks of silver and four marks of gold for the privilege of paying its *ferm* directly instead of through the sheriff.[221] Furthermore, the men of London gave Henry I. a hundred marks of silver that they might elect their own sheriff.[222] It was probably this circumstance which led a writer in the decade 1140–59 to declare that the sheriff of the shire ought to be elected annually in the full folkmote of the shire.[223]

But if the sheriff of the year 1130 was in a remunerative position, he was subject to the hard rule of the exchequer, and his post was no easy one. Not infrequently he had to collect sums which the agents of former sheriffs failed to obtain.[224] His burdens had been increased through the fisco-judicial policy of the exchequer and the king's justices, which seems to impose the utmost farthing.[225] Sheriffs when they came before the exchequer were so filled with fear that Gilbert of Surrey made a reputation because of his happy and joyful mien.[226] The four sheriffs of London in 1130 paid a hundred marks *ut exeant de vicecomitatu*.[227] Aubrey de Vere, one of the king's favoured officials, was required to give five hundred pounds and four *dextrarii* for the escape of a prisoner from his custody and for *forisfactura comitatuum*.[228] It occasions little surprise to find him also paying a hundred marks that he might leave the shrievalty [229] at the end of the year.

The subordination of the office to the king's justices stood, aside from the close control of its personnel and its strict legal responsibility for accounts, as the third great phase of its development in the reign of King Henry. The justices of William the Conqueror could make little progress in this direction.[230] The sheriff

---

[221] *P.R.* 31 Henry I., p. 114.

[222] Round, *Geoffrey de Mandeville*, 361-3.

[223] *Leges Edw. Conf.* 32 B 1, 32 B 8, Liebermann, *Gesetze*, i. 655-7.

[224] See the obligations resting upon Osbert Sylvanus (*P.R.* 31 Henry I., p. 7).

[225] See below, p. 101. Round has declared (*Geoffrey de Mandeville*, 105) that under this regime " the very name of ' plea ' became a terror to all men ".

[226] Round, *Commune*, 123.

[227] *Ibid*. 363.

[228] *P.R.* 31 Henry I., p. 53. The *forisfactura comitatus* is sometimes heavy at this time. Maenfinin, late sheriff in the shires of Buckingham and Bedford, owed 115 marks under this head (*ibid*. 100).

[229] *Ibid*. 53. Besides the hundred marks he owed at the exchequer £515 : 6 : 8 when he left office.

[230] They were sent out by special commission and seem usually to deal with but one plea (*Yale Law Journal*, xxiii. 490-510). In only one known instance are the justices of the Conqueror said to try *placita* ; and in the reign of Rufus occurs but one statement to that effect, although there is allusion to what may be another

was not only the important judicial official of the shire, but apparently in nearly every case the justice who handled the pleas of the crown. The Pipe Roll of the year 1130, on the other hand, shows that *placita*,[231] sometimes including pleas of the forest, were being held by the king's justices in nearly every county in England. It is clear, moreover, that this eyre system has been employed for several years. A writ of this king later than July, 1123, refers to a usage by which four men from each wapentake are being summoned to the king's pleas and counties.[232] The king's *curia* has for some time assumed supervision over important criminal pleas. The fact is witnessed by the punishment of false coiners from all over the realm at Winchester in 1125 under the direction of the chief justiciar,[233] and also by the famous assize the preceding year in Leicestershire at which Ralph Basset surpassed all records for the punishment in one day of thieves and other criminals.[234] A concession to the canons of the church of St. Trinity, London, reserving cases concerning their tenements to a hearing before the king or his capital justice,[235] was not without precedent, and exemplifies the same tendency in civil causes.

The eyre made inroads into the sheriff's former jurisdiction by taking over the pleas of the crown; it gave to the justices a precedence also in some matters of administration; and it served to reduce the fiscal independence of the head of the shire. It is well recognized that the itinerant justices were financial as well as judicial agents of the king.[236] In the reign of Henry II. the eyre rolls furnished the exchequer officials with information upon which was

---

case of the same usage (Davis, *England under Normans*, 523; cf. *Memorials of St. Edmund's Abbey*, R.S. i. 156). The *justicia* or *justiciarius* from the reign of Rufus and the earlier years of Henry I. precedes *vicecomes* in the salutation in the king's administrative writs (as in Davis, *Regesta*, no. 389; *Ramsey Chartul.* i. 241), but this may be a local justice.

[231] Only occasionally is there a hint as to their character; but Ralph Basset dealt with a case concerning treasure trove (*P.R.* 31 Henry I., p. 18; cf. p. 32) and Geoffrey de Clinton with breach of the peace (*ibid.* 78). Pleas of false money (*ibid.* 8), pleas relating to *minutis hominibus* (*ibid.* 56, 96), *minuta placita* (p. 142) are mentioned as well as pleas of the forest (*ibid.* 13, 32, 47; cf. *Leges Henrici*, 17).

[232] *Monasticon*, vi. part iii., 1272 (no. 20).

[233] *A.-S. Chron.*, 1125; *Florence of Worcester*, Eng. Hist. Soc. ii. 79.

[234] *A.-S. Chron.*, 1124. In 1106 Ralph served as one of the justices sitting for a special purpose at York (Farrer, *Itinerary*, no. 166). As early as 1116 along with others he held pleas at Huntingdon, among them one involving concealment of treasure trove (Farrer, *Itinerary*, no. 370).

[235] Rymer, *Foedera* (1819), i. 12.

[236] Stubbs, *Const. Hist.*, 418.

based the demand that the sheriff pay certain specified sums.[237] There is reason to believe that this held for the year 1130. Specific amounts due from *placita* and agreements before the itinerant justices were according to the Pipe Roll of this year being demanded at the exchequer.[238] The sheriff in the case of *murdra* and various crown pleas as well as the aids of towns and counties, all possibly levied by the justices [239] even now, could be called to pay the exact amount due.

There is no reason for holding that in 1130 the justices in eyre could long have occupied so strong a position. The handling of regular routine presupposes a frequency and regularity in these visits to the various shires which even the organizing genius of Bishop Roger could hardly have attained except after much experience. By the earlier years of Henry I. justices along with sheriffs were being instructed to observe liberties granted by the king.[240] But the *Leges Edwardi Confessoris* and the *Leges Henrici*, the latter written by a lawyer of the *curia regis* in or a little before 1118, both assume the existence of a king's justice who is locally resident and ordinarily within reach when crime is committed.[241] It has been suggested that the sheriff was sometimes the official who in these treatises is called the justice.[242] There appears no escape from this conclusion, for in the early Norman period the sheriff held exactly this position [243] so far as crown pleas were concerned. But it is nevertheless certain that in the reign of Henry I. there was a justice of this description quite distinct from the sheriff.[244] This king

[237] See *Dialogus*, pp. 113, 142. *Murdra* levied upon hundreds, as well as other sums derived through pleas of the crown, might be traced in this way. The justices of Henry II. clearly had knowledge of other fiscal matters, since they levied aids upon towns and the common assize upon the counties (*Dialogus*, ii. xiii C ; Poole, *Exchequer*, 125).

[238] This seems clear when the collector has failed to turn in the sums due. The use of a summons to the sheriff as an order for collecting these also proves the point (below, p. 95).   [239] See above, note 237.

[240] Thus in a Ramsey charter of 1107 (*Cartae Antiquae*, P.R.O. 42, no. 2) : *Prohibes ne aliquis de justiciis aut vicecomitibus vel ministris meis pro aliqua occasione de praedictis libertatibus iniuriam aut vexationem eis faciat vel facere permittat nec se intromittat.*

[241] *Leges Henrici*, 75. 6 a, 91, 92. 4 a ; and *Leges Edw. Conf.* 15, 20. 1 a, 20. 6.

[242] Stubbs, *Const. Hist.* 6th ed. i. 420, n. 1 ; Liebermann, *Über die Leges Edwardi Confessoris*, 73 ; *Am.H.R.* viii. 490.

[243] A trial held before the sheriffs of Oxfordshire (*Chron. Abingdon*, ii. 119) is authorized by royal writ directed to the justiciars and sheriffs.

[244] Round, *Geoffrey de Mandeville*, 110, 373 ; Stubbs, introd. to *Benedict*, ii. , p. lviii.

employed similar officials in Normandy,[245] and they are mentioned often enough in England to show that their appointment was no unusual occurrence.[246]   According to the *Leges Henrici* [247] either the *iusticia* or the sheriff has jurisdiction involving theft, robbery, arson and other crimes.   The justice, moreover, had functions strikingly similar to those of the later coroner.[248]   Among the well-known privileges of London conferred by Henry I. toward the close of his reign were the right of the citizens to elect their own *iusticiarius,* and his exclusion of all other persons, including sheriffs, from the privilege of holding the pleas of the crown.[249]   The local justice, then, seems to have prepared the ground for the precocious eyre system of 1130.   The office is to be traced back as far as the ministry of Flambard.[250]   So far as evidence suggests, this was the royal agent who first permanently weakened the position of the powerful Norman sheriff.

   Regarded as a whole, the history of the sheriff's office in the reign of Henry Beauclerc well illustrates his favourite methods.   Unobtrusive, yet persistent and powerful, measures were adopted to bring it under control.   The great problem which it presented in the Conqueror's time was all but solved.   More than half the sheriffs in 1130 were members of the newer nobility, and a great baron's occupancy of the office was a very rare occurrence.   On the other hand, it was in many cases conferred upon members of the king's *curia*, the leading exponents of the newer centralizing movements.   An even greater revolution took place through the development of curial institutions.   Special machinery and process were devised to make the sheriff rigidly responsible at fixed periods for the render of accounts.   A means had probably been devised for checking the return from crown pleas and other unfarmed revenues.   The influence of the king's justices constantly relegated the sheriff to a secondary position.   His freedom in general administration was assailed by the loss of his judicial headship of the shire which had been its pivotal support.

[245] Haskins, *Norman Institutions*, 100.
[246] Davis, *England under Normans*, 523.   One case (*Ramsey Chartul.* i. 149) which discloses the sheriff acting with the justice of his shire in a matter involving the king's rights, may date from the reign of William Rufus.   [247] 66. 9.
[248] *California Law Review*, vii. 239 ; *Sel. Coroners' Rolls*, pp. xv-xvii ; *Pol. Sci. Quarterly*, vii. 656-72.
[249] Liebermann, *Gesetze*, i. 524-5 ; Stubbs, *Sel. Charters*, 108.
[250] Davis, *England under Normans,* 520, 523 ; *E.H.R.* xxxiii. 159, n. 112.   Page, *London*, 204-8, traces the succession of London justiciars.

In a few instances the sheriffs of the period escaped in a measure the effect of these centralizing forces. The great officers of the king, who in Henry's later years often served as sheriffs, were too powerful to fear the influence of a local *iusticia regis*.[251] The king frequently employed them as itinerant justices, and they sometimes held eyres even in their own shires.[252] Such lack of supervision seems to account for the extortion practised by Miles of Gloucester and Payn fitz John.[253] Bishop Stubbs has made the observation that sheriffs of the curial circle as barons of the exchequer would audit their own accounts.[254] But it is apparent that in some instances at least no special tenderness was shown on this account.[255] Far more dangerous was the perpetuation in a small number of instances of the powerful hereditary shrievalty of earlier days and its combination with the office of justice.

The first four years of Stephen's reign are naturally to be regarded as continuing the administrative system of Henry I. In his charter of liberties of 1136 the king sought popularity and support by promising relief from the exactions and injustice of sheriffs.[256] There is every reason to believe that in this, as in other matters which implied strength and vigour, Stephen's concession was a dead letter. The few powerful sheriffs, hard to control under Henry I., were of the type which persisted ; and we soon hear of others. Upon Miles, constable and hereditary sheriff, *sicut baroni et justiciario* Stephen conferred anew the castle and the shrievalty of Gloucester.[257] The strength which he brought to Matilda after his

---

[251] This is illustrated by diplomatic evidence which is hardly accidental. Despite the precedence of the justice over the sheriff as evidenced by the king's writ formulary (Round, *Geoffrey de Mandeville*, 110), Hugh of Buckland, sheriff of Berkshire, but also curial and justice, takes precedence over the king's local justices (*Chron. Abingdon*, ii. 115). Richard fitz Baldwin, hereditary sheriff of Devon and possibly justice in eyre (*E.H.R.* xiv. 421-2), takes a similar precedence (*Monasticon*, ii. 501) over the justice of Devonshire and Cornwall. Another Berkshire writ in which *vicecomes* precedes *justitiarii* (*Chron. Abingdon*, ii. 164) is later than Hugh's time, but may belong to that of his son William.

[252] Geoffrey de Clinton had done this when sheriff of Warwick (*P.R.* 31 Henry I., 106).

[253] They held the king's pleas in 1130 in Gloucestershire and Staffordshire (*P.R.* 74, 78) as well as in Pembroke (*ibid.* 136). Miles also held pleas in Hants (*ibid.* 38). Payn seems to be sheriff of Hereford and Shrewsbury. The *Gesta Stephani*, 16, says that he had *dominatum* over these *provinciae* as Miles had of Gloucester, and that from the Severn to the sea no man was free from their extortion and litigation. Payn, like Miles, seems to have been employed at the exchequer. He is among those pardoned Danegeld (*P.R.* 31 Henry I., 6).

[254] *Const. Hist.* i. 423-4.   [255] See above, p. 100.

[256] Stubbs, *Sel. Charters*, 121.   [257] Madox, *Exchequer*, 134, note (i).

defection from the king in 1139 induced her to create him an earl two years later. William de Beauchamp turned against Stephen in 1138. Baldwin de Redvers, or more properly de Reviers, who in a chronicle of 1142 is called Baldwin the sheriff, was one of the earliest and most determined rebels of the new reign, in 1136 fortifying the castle of Exeter against the king.[258] William fitz Alan, *municeps et vicecomes Scrobsburiae*, husband of a niece of Earl Robert of Gloucester, also rebelled prior to August 1138, when he was defeated and put to flight.[259]

But however great the weakness of the king's position on account of local feudal revolts, his writs to the county courts [260] show that formally at least the sheriff was still in his old administrative position. The same may be said of writs issued by Stephen's rival, the Empress Matilda.[261] The impression that the office had no history in the troubled years of the reign has sometimes been conveyed by the assumption of Bishop Stubbs that the arrest of Bishop Roger and the officials of his family in 1139 completely disorganized the work of the exchequer,[262] and by the statement of Ralph of Diceto that one of the reforms agreed upon by Stephen and Henry Plantagenet in 1153 was the restoration of the sheriffs to their places.[263] But the continuity of the exchequer during these fifteen years is well established despite the statement of Bishop Richard in the *Dialogus* to the effect that his father, Nigel of Ely, the treasurer of Henry I., was called by Henry II. to restore the form of the exchequer, the *scientia* of which had been almost lost in the long period of warfare.[264] Aside from the strong case presented by Mr. Howlett,[265] Mr. Round has called attention to exchequer operations involving even accounting for the sheriff's *ferm* in four shires at the height of the anarchy,[266] and Mr. Turner has found that the *ferm* of Wiltshire was collected the fiscal year prior to the first year

[258] Liebermann, *Ungedruckte anglo-normannische Geschichtsquellen*, 29. Baldwin had been employed in 1130 upon the fiscal business of his county (*P.R.* p. 153). His rebellion is described in *Gesta Stephani*, 21-9, and in *Florence of Worcester*, ii. 96.

[259] Ordericus Vitalis, v. 113. He appears again as sheriff of Shropshire, 1155-59.

[260] As in *Chron. Abingdon*, ii. 178, 179, 181 ; *Chartul. Monast. Sancti Johannis Baptiste de Colecestrie*, ed. S. A. Moore, 31-4, 51-3.

[261] *Monasticon*, i. 44, v. 107 ; Round, *Calendar*, no. 1056.

[262] *Const. Hist.* i. 351-3.

[263] *Diceto*, R.S. i. 297 ; Stubbs, *Const. Hist.* i. 361.

[264] *Dialogus*, i. viii G.

[265] *Gesta Stephani*, introd. i. pp. xxviii-xxxi.   [266] Below, p. 107.

of Henry II.[267] King Henry doubtless restored sheriffs in the regions which did not recognize his predecessor, but in some shires no restoration was necessary.

This conclusion is sustained by evidence of a very specific nature. It is well known that, at the crisis of the struggle between the two claimants for the throne in the years 1141 and 1142, Geoffrey de Mandeville gained the office of sheriff in three shires including London and Middlesex. In this later sheriffdom he was preceded by Osbert Huitdeniers.[268] There is mentioned a sheriff of Essex in this period [269] who probably succeeded Earl Geoffrey in the position; and Payn, sheriff of Huntingdonshire at the beginning of the reign of Henry II., appears in this office before the death of Abbot Walter of Ramsey in 1140.[270] Robert fitz Walter, who was a follower of Stephen,[271] probably held Norfolk and Suffolk in this reign; and it is certain that he was succeeded in turn by his two sons, John and William de Chesney.[272] Matilda's recognition in 1141 of William de Beauchamp in Worcestershire may also be cited.[273] Mr. Howlett has discovered a sheriff in Wilts, a theatre of warfare, and also other sheriffs in the period 1146–50, one of whom performs judicial functions at the king's direction.[274] The names of their shires give some important information concerning the extent of King Stephen's power in this period. The form of his writs to officials in Kent, moreover, is so definite as to leave no doubt as to the existence of a sheriff there.[275] The Abingdon Chronicle relates

[267] *Trans. R.H.S.*, N.S. xii. 127.

[268] Sheriff about 1134–42. His young kinsman, Thomas Becket, was employed by him at this time as notary (Round, *Geoffrey de Mandeville*, 374). Fitzstephen says (*Vita S. Thomae*, R.S. 14) that after his return from Paris, Thomas was *et vicecomitum clericus et rationalis effectus.* It is added that his father, Gilbert, had once held the same shrievalty.

[269] *Stephanus rex Angl. Ricardo de Lucy justiciario, et vicecomite de Essex* (Madox, *Exchequer*, i. 33, n. 3; *Chronicles of Stephen*, ii. p. xliii).

[270] *Ramsey Chartul.* i. 253; *baronis regni praepotentis* (*Chron. Mon. de Bello,* 65; cf. p. 112, n. 3).

[271] He witnessed a charter of the king at Guildford in company with various dignitaries (*Red Book of Canterbury*, Claudius, DX. fo. 99.

[272] Cf. Round, *E.H.R.* xxxv. 481-9.   [273] Below, p. 108.

[274] *Gesta Stephani*, p. xxxv, n. 1. About 1146 the king addresses the sheriff of Warwickshire (*ibid.* p. xxx), and at some time in the period 1148–53 sends an agent to Norwich to dispose of a judicial matter (*ibid.* p. xxvi) at the settlement of which two sheriffs were present. About 1147 Maurice, sheriff of Essex, held an assize of novel disseisin in the hundred court at Maldon (*ibid.* p. xxxviii). William de Chesney is mentioned as sheriff of Norfolk in the period 1146-9 (*ibid.* p. l), and in 1150 William, sheriff of Yorkshire, witnesses a deed (*ibid.* p. xxx).

[275] He addresses the sheriffs of Kent and *ministri* of Faversham; again, the sheriffs and reeves of Canterbury (*Chron. Abingdon*, ii. 383, 384).

that in 1152, the year when Stephen and Henry were reconciled, the former directed Henry, the sheriff of Oxford, that certain lands of the abbey occupied by one Thurstan be restored.[276]   The additional statement, that Henry the sheriff yielded to bribery and continued to permit the church of Abingdon to be deprived of its possessions, is a sad commentary upon the state of Stephen's power in this region.

The fundamental difficulty was not the lack of sheriffs in the districts which remained loyal to Stephen, but rather his inability to achieve that control over them which had marked the rule of his family.   Feudal reaction meant the irresponsibility of local magnates and local representatives of the king's government.   The conduct of the sheriffs of the period was bad.   One of the reforms agreed upon in 1153 between Stephen and his successor was the appointment to the office of men who would not use it to gratify private revenge or hatred, nor extend indulgence to crime, and who would give every man his own.[277]   Beside Henry of Oxford who received bribes, may be placed William de Chesney of Norfolk, who levied blackmail upon his own nephew, Gilbert Foliot, the abbot of Gloucester.[278]

The powerful baronial sheriff was the harder to manage because of his weight in casting the balance between the two claimants to the throne.   Baldwin de Redvers became one of Matilda's earls,[279] and Miles of Gloucester by turning from Stephen contrived to gain the earldom of Hereford.[280]   Mr. Round has revealed the astounding series of grants made by the king and his rival to Geoffrey de Mandeville from 1140 to 1142 by which Geoffrey gained successively the earldom of Essex, the hereditary custodianship of the tower of London, the shrievalty and hereditary justiciarship of Essex, the same two offices in London and Middlesex for the *ferm* which his grandfather had paid, and also in Hertfordshire as well as Essex under the same condition.[281]   This wholesale appropriation of the office of sheriff by an earl whose conduct rivalled the most unruly great vassals on the Continent was fortunately of brief duration.[282]

[276] *Chron. Abingdon*, ii. 184.
[277] *Diceto*, i. 297.
[278] *E.H.R.* xviii. 638.   This may have been before William was sheriff (*ibid.* n. 18).
[279] *Gesta Stephani*, 79 ; *Monasticon*, v. 106, 107.
[280] Rymer, *Foedera* (1816), i. 14 ; Round, *Geoffrey de Mandeville*, 123.
[281] *Ibid.* 51-2, 89, 92, 140-43, 166 ff.
[282] Geoffrey was killed in 1144.

Matilda in July 1141 granted to William de Beauchamp, the grandson of Urse d'Abetot and heir to the extensive possessions of that house, the family shrievalty and the forest of Worcestershire, to be held by hereditary right for the same *ferm* which William's father had paid for this shire.[283]   Whether or not William became reconciled to Stephen, he certainly held his shrievalty in the earlier years of Henry II.

In some instances the sheriff of this period was probably the dependent of a great baron as he was at an earlier time in palatinates. Mr. Round construes a certain grant of Roger, earl of Warwick, as indicating that he held the shrievalty of that county from the king and regranted it to his son-in-law, Geoffrey de Clinton.[284]   It is to be noted, however, that although the writs of William of Aumale, earl of York in Stephen's reign, more than once mention his *vicecomes*,[285] this is not the sheriff of Yorkshire but an official of the count of Aumale who appears in the reign of Henry I.,[286] and again in the time of Henry II.[287]   Under whatever influence the sheriff here may have fallen, it is clear that legally he was still the king's official.[288]

Great feudatories of the Norman period of English history often had sheriffs to whom they addressed their writs resembling those of the king.   In this the count of Aumale was not alone.   There appear similar well-known officials in various rapes of Sussex, notably in that of Hastings held of the count of Eu and in that of Arundel.[289] Moreover, William of Warenne in the reign of Stephen had a sheriff for his rape of Lewes.[290]   Earl Robert of Gloucester and his son William likewise employed a line of sheriffs in Glamorgan.[291]   The

[283] Round, *Geoffrey de Mandeville*, 313.
[284] Round, "A Great Marriage Settlement", *The Ancestor*, xi. 153-4.   Geoffrey is to hold *comitatum de Warr' de me et meis heredibus eodem modo quo de rege habeo et habere potero*.   The statement (*Chron. Monast. de Bello*, 66) concerning William of Ypres *qui Cantiae comitatum possidebat* may imply that he was sheriff, as he certainly was not earl.
[285] Farrer, *Yorkshire Charters*, iii. nos. 1306, 1314.
[286] *Ibid.* nos. 1326, 1327.
[287] *Ibid.* nos. 1373, 1374.
[288] See *ibid.*, no. 1448, the writ issued in Jan. 1154 by Stephen addressed *justiciariis et vicecomitibus et ministris et omnibus fidelibus suis de Eboraco et Eboraciscira*.   See also note 274.
[289] *V.C.H. Sussex*, 380, 384.
[290] Book of Lewes, P.R.O., *Anc. Deeds*, A, box 265, fo. 45.   The *Guido vicecomes* of fo. 27 seems to be the person who witnesses in *Monasticon*, v. 14, no. 9. *Ibid.* no. 10 has a sheriff named Adelulf.
[291] *Cartae et munimenta de Glamorgan* (1910), i. 35, 48, 51, 54, 55, 99, 104, 106, 108, ii. 202, 235.

office of *vicecomes* was well adapted to the requirements of feudal administration even before 1066.   The noble could do no better than adopt the plan of the king.[292]   That the latter was sure to have a sheriff in the region which he controlled the restoration of the office by Henry II. well attests.

[292] The bishop of Durham also had a sheriff in his palatinate in this period (Lapsley, *Durham*, 80).

V

THE ADMINISTRATIVE SHRIEVALTY OF THE FIRST
TWO ANGEVIN REIGNS, 1154–99

THE reigns of Henry II. and his sons mark a rapid development of institutional absolutism which of necessity brought into prominence the sheriff, who was its principal local agent. From the time when King Henry restored sheriffs in regions where they had disappeared during the feudal anarchy, the activity of the office is again written large in administrative and constitutional history. Outstanding functions identify it with the action of the central government in the administration of justice, the levy of taxation, the collection of revenue, the enforcement of feudal military service, and the summoning of local representatives to meet the justices of the king. It was an integral portion of a centralized governmental organization which constitutes the permanent groundwork of the English constitution. As this grew stronger in the reigns of Henry and Richard the office of sheriff became more important. As Angevin absolutism reaches the culmination in the reign of John, it demands a separate treatment.

Feudal spirit and opposition, which in 1154 seriously threatened the king's control of the sheriff, were soon overcome. It is obvious that by 1170 the king completely dominated the office, and this was undoubtedly true some years earlier. Henry II., of course, at his accession set out to appoint sheriffs who were his own adherents. In London a new sheriff was at once installed.[1] The excerpts from the lost Pipe Roll of the year 1154–5, printed in the Red Book of the Exchequer [2] and preserving the names of the sheriffs of some-

---

[1] John fitz Ralph, also joint-sheriff with Gervase for the fiscal years 1155–7, in 1161 (*P.R.* 7 Henry II., p. 17) owed £100 *ferm* of London *de termino quo Rex Stephanus mortuus fuit usque ad Pasca proxime sequens.* Gregory (*Red Book*, ii. 658) was thus the sheriff for the second half-year.

[2] ii. 648–58.

111

what more than half of the shires for this first year of the reign, indicate that King Henry found difficulty in filling the office with loyal adherents. A number of the new sheriffs were persons of consequence.[3] Henry's need of strong support stands out in his employment of four earls. Hugh Bigod, despite his title of dignity conferred by Stephen, held Norfolk and Suffolk, as did his father and brother before him. Earl Patrick[4] likewise retained for six years the shrievalty of Wiltshire, and Earl Richard de Redvers in Devon held for a shorter period the post filled by his father before his revolt in 1142. To gain the support of Roger of Hereford[5] the king conceded him the lands and dignities held by his father, Miles of Gloucester, including the earldom and castle of Hereford, the custody of the castle of Gloucester and the shrievalty of Gloucestershire.[6] After Roger's death in 1155, his brother Walter was for a time sheriff of Herefordshire and also of the county so long held in the family. William de Beauchamp until 1170 continued the rule of the line of Urse d'Abetot in Worcester, and for some years succeeded Walter also in his two counties. The fact that in the first fiscal year of the reign Abbot Geoffrey served as sheriff in Leicestershire, and Robert, bishop of Chichester, in Sussex,[7] points to a scarcity of strong partisans of the Angevin cause in those regions. On the other hand, the long-continued terms in office of some of the less prominent sheriffs[8] of this year prove their trustworthiness. The displacement at Michaelmas 1155 of a full dozen, including Earl Hugh and William Martel, seems to designate these as men who had been available rather than satisfactory.

[3] Robert of Stafford was a son and a grandson of Norman baronial sheriffs ; Bertram de Bulemer, a sheriff of Henry I. and the head of an honour ; William fitz Alan, appointed in Shropshire after the year was well advanced, a sheriff and an old feudal opponent of Stephen ; Richard de Lucy, who was to rise high under the new regime, a baron who had served the late king as justice, though not yet as sheriff (Madox, *Exchequer*, i. 33, note; Madox, *Formulare*, 40) ; and Payn, sheriff of Cambridgeshire and Huntingdonshire, had held the latter position also at an earlier time.

[4] But not for life ; he was slain in 1168 by Guy of Lusignan, while returning from a pilgrimage to the shrine of St. James of Compostella (Hoveden, R.S. i. 273). His successor in the shrievalty, Richard the clerk, or Richard de Wilton, was probably a personal adherent of the earl. Except for the year 1162–63, Richard was in office continuously from 1160 to 1179.

[5] For his attempt at rebellion, *Gervase of Canterbury*, R.S. i. 161-2.

[6] *Rotuli Chartarum*, Record Commission, i. part i. pp. 53, 61.

[7] His term was but three-fourths of the year ; Robert died Mar. 28, 1155.

[8] Ralph Picot, sheriff of Kent, was in office six years, for five of which he also held Surrey and for three of which he held Sussex. Payn, appointed sheriff of Cambridge and Huntingdon and the next year of Surrey, was in office nine years.

Subsequent changes in sheriffs were not as a rule rapid.[9]   In 1162 new sheriffs were appointed in about half of the counties, and the names of several persons well known as faithful servants of the king begin to appear.[10]   The famous Inquest of Sheriffs in 1170 involved an investigation into exactions, fiscal administration and various other matters of complaint,[11] after which nearly all the sheriffs were removed in the middle of the fiscal year.   In the majority of cases men already connected with the king's curia were substituted.[12]   For the remainder of the reign changes in a given year were as a rule comparatively few.[13]

The title of the sheriff in popular speech already has much of its modern sound, as witnessed by the appearance of the word *schirreue*.[14]   The writer of the *Dialogus de Scaccario* states that a knight or other discreet person may be created sheriff even though he hold nothing of the king.[15]   Pope Alexander III., in a decretal addressed to Roger, bishop of Worcester, repudiates a practice of Henry's early years in declaring that priests may hold neither the shrievalty nor the office of secular reeve, since no man in sacred orders is permitted to take part in a judgement involving the shedding of blood.   This was enforced for England by a canon of Archbishop Richard's council held in 1175.[16]

The sheriffs of Henry II. who filled vacancies occurring during the year were apparently reckoned at the exchequer as entering office on one of the quarterly term days.[17]   In London prior to 1191

[9] The Pipe Rolls show an unusual number in 1157, 1159 and 1162.   At the end of 1159 Maurice of Tilney, Ralph Picot and William de Chayney, prominent sheriffs, went out after but three months of the fiscal year had passed, and three clerks, Alexander (Stafford), Richard (Wilts) and the bishop of Chester (Surrey), held the office for the whole of it.   Some half-dozen other changes also testify to dissatisfaction with those in office prior to the preceding Michaelmas.   One of the removed sheriffs, Robert of Stafford, seems to have been heavily amerced for misconduct in office (*P.R.* 7 Henry II., p. 41).

[10] Gervase of Cornhill (Surrey), William Pipard (Gloucester), Oger the *dapifer* (Norfolk and Suffolk), Thomas Basset (Oxford), Ranulf Glanville (York).   Ralph Basset (Warwick) had already been in office three years and Hugh Gubiun (Northampton) one.

[11] Stubbs, *Const. Hist.* i. 510-11 ; introd. to Benedict ; Stubbs, *Sel. Charters*, 147-50.

[12] According to the conclusions of Stubbs (introd. to Benedict) five of the seven who were retained were members of the king's household, and sixteen of the new sheriffs were employed at the exchequer.

[13] In 1175 and in 1185 sheriffs retired in about a dozen counties.

[14] *P.R.* 14 Henry II., p. 104.   [15] *Dialogus*, 125.

[16] Benedict of Peterborough, *Gesta Henrici Secundi*, R.S. i. 85.

[17] They account for the *ferm* of the county for either one-fourth (as in *P.R.* 7 Henry II., 49), one-half or three-fourths of a year (as in *ibid.* 13 Henry II., 170).

there were from two to four sheriffs at a time, except in the period 1181–88, when there was only one.[18] In other counties also there were occasionally two, more rarely three, acting jointly, apparently in most cases as special custodians. In the reign of Henry II. the sheriffs were not chosen from the high feudal baronage, but after 1170, as in the past, they often belonged to the class of curial officials, and in Richard's reign the highest officials of state were also among the sheriffs. It was not unusual for them to serve among the itinerant justices. In both reigns sheriffs often accompanied the justices who assessed tallages,[19] frequently levying these in their own counties.[20]

That the farming of the shire was still highly profitable is an inference drawn from the sheriff's willingness to pay heavy sums to obtain his office.[21] His other emoluments included, first, the sheriff's aid, a sum varying in amount in different localities, but given to sheriffs and bailiffs in recognition of their goodwill in matters involving claims and inconvenience, and which the king, probably in 1163, appropriated to the treasury.[22] It seems always to have been levied on land, as it certainly was in the reign of John.[23] The chattels of the manifest thief, according to the *Dialogus*, went to the sheriff [24] if the latter dealt summarily with him and his case was not brought into court ; [25] and the chattels of an outlaw fell to whomsoever seized him. The same authority states that the demesne lands of sheriffs are quit of Danegeld [26] because of their labour in collecting this tax. But the concession could have been worth little in this period, for Danegeld was not levied after 1163. A grant of Henry II. to a monastic establishment prescribes as a penalty for disturbing the monks, in addition to the usual forfeit to the king, twenty shillings *ad opus vicecomitis Eboracensis*.[27] The

[18] See Norgate, *England under Angevin Kings*, ii. 471.

[19] As in *P.R.* 33 Henry II., p. xxxv.

[20] The *assisa burgi de Oxineford* for which the sheriff accounted in 1175 was levied *per ipsum vicecomitem* and the constable of Oxford (*ibid.* 21 Henry II., p. 13). There is an instance in the preceding year of the collection by a sheriff of an *assisa* laid by himself (*ibid.* 20 Henry II., p. 143).

[21] Poole, *Exchequer*, 129, and note 1.

[22] *Red Book*, ii. 768 ; *Materials for Life of Becket*, R.S. ii. 374, iv. 23 ; Round, *Feudal England*, pp. 500-501 ; Davis, *England under Normans*, 211, and note 2. In Normandy it was part of the *ferm*.

[23] Inspeximus charter of 1236, based on a grant of 10 John (*C.C.R.* i. 216) : *sive capitur per hidatas terre sive per carucatas.*

[24] Cf. Glanville, *Tractatus de legibus*, vii. 17.

[25] *Dialogus*, 140.   [26] *Ibid.* 102.

[27] Farrer, *Yorkshire Charters*, iii. no. 1390.

fact of a general inquiry in 1170 concerning sheriffs' exactions by judicial and non-judicial means from hundreds, vills and individuals [28] and concerning the return by them or the bailiffs of hush money, points to a tendency on the part of sheriffs to find perquisites by oppressive and irregular methods.

The sheriff now clearly had an organized office force. Under-sheriffs are frequently mentioned. In the reign of Richard high baronial personages and state officials who acted as sheriffs some-times depended upon such officials to discharge their duties. Sheriffs also had *servientes* [29] and clerks, the former acting in matters of police and execution of judgement,[30] the latter by the reign of Richard regularly accounting for such financial items as payments of hidage and aid.[31] The sheriff appointed individuals to act as summoners.[32] It was assumed that he had a seal.[33] By the days of King Richard a sheriff sent information required by the king's court at Westminster, either *per breve suum* or *litteris suis sigillatis*.[34]

The sheriff's guardianship of the peace and his custodianship of criminals are traceable in the records of the exchequer, and from 1194 in those of the central courts. The Assize of Clarendon in 1166 seems to have brought a marked extension of his powers in regard to criminals and their arrest. No one, even within a city, borough or castle, was to deny the sheriff admission to his land or soke to apprehend criminals accused by jury ; upon notice from a sheriff of another county he was to seize and place under suretyship fugitives from that county ; and he was also to enroll the names of fugitives from his own county so that the justices might seek them throughout the realm.[35] Moreover, the same enactment provided that robbers, murderers, thieves, and their abettors should be handed over to the sheriffs, who were to receive them without delay ; also that jails be built in counties where there were none, and placed either in a borough or a castle so that sheriffs might in

---

[28] Inquest of sheriffs, Stubbs, *Sel. Charters*, 148-9.

[29] As in *P.R.* 33 Henry II., 58 ; *P.R.* 1 Richard I., 109.

[30] *Select Pleas*, i. 2, 20.

[31] *Placitorum Abbreviatio*, Record Com., 13, 13 b, 15, 15 b ; *Rot. Cur. Regis*, Pipe Roll Soc. no. 14, p. 90 ; cf. *Dialogus*, 123 : the sheriff's *generalis procurator*.

[32] See p. 120.

[33] *Dialogus*, 86, 1, 35. In the time of Henry II. it is declared that even a minor knight now uses one (Bigelow, *Placita Anglo-Normannica*, 175 ; *Chronicon Monasterii de Bello*, 106).

[34] *Plac. Abbreviatio*, 22. *Vicecomes significavit per litteras suas sigillatas patentes* : *Curia Regis Roll*, 21, m. 10 d, 2 John (P.R.O.).

[35] Assize of Clarendon, chs. 11, 17, 18.

these guard *per ministros qui hos facere solent et per servientes suos* the criminals who were taken.[36]   Ten years later it was directed [37] that the thief be delivered into the custody of the sheriff and, in the absence of the latter, to the nearest castellan, who was temporarily to keep him in custody.   The well-known writ of the justiciar, which in 1195 gave directions for the enforcement of the king's peace,[38] practically repeats the main provisions of the Assize of Clarendon on the subject.   Criminals when taken were to be handed over to assigned knights of the bailiwick, who were to deliver them to the sheriff.   An allowance to the sheriff for conducting prisoners from place to place is often mentioned,[39] as also his presence at the execution of criminals,[40] although special executioners [41] as well as keepers of jails [42] are mentioned.   Restoration of an outlawed man to his original status was made by the king's act of pardon, addressed to the sheriff and read in the county court.[43]

The rule of the law which made the sheriff or reeve responsible for the escape of his prisoner was well established by the reign of Henry II.[44]   Robert de Vaux, dismissed from the shrievalty of Cumberland in 1186, had been heavily amerced for this and other shortcomings.[45]   The next year William fitz Isabel was dismissed from office in London and was in debt to the huge amount of £566 for many amercements levied against him, but the particular offence which is named is that of taking from Robert the Lame, arrested on an appeal of counterfeiting, twenty shillings as a preliminary to placing the prisoner under poor pledges who permitted him to abscond.[46]   In the Assize of Clarendon [47] the sheriff was charged

---

[36] Chs. 6, 7.

[37] Assize of Northampton, ch. 12.

[38] Hoveden, *Chronica*, R.S., iii. 299 ; Stubbs, *Sel. Charters*, 264.

[39] As in *P.R.* 9 Henry II., 26 ; *in conducendis prisonibus de Tanton ad Windsorum,* 21s. 4d.   *P.R.* 32 Henry II. (p. 175) records an amercement for taking a thief out of the hundred in which he was caught into another *sine visu servientis regis.*

[40] *P.R.* no. 38, 4 Richard I., Oxfordshire : the son of Alan de Furnell, a former sheriff, owes a sum which Alan received from Robert de Estropa *pro homine suspenso sine visu justiciae et vicecomitis.*

[41] *Dialogus*, ii. vii. B.   The porter of Canterbury received a fixed payment (as in *P.R.* 28 Henry II., 149) *qui facit justitiam civitatis.*

[42] According to *P.R.* 33 Henry II., 39, the *custos* of the jail of London was paid 76s. for a half-year.   By a grant issued in the king's name in November 1189, William Longchamp made over this same post to his brother Osbert (*Cartae Antiquae*, P.R.O. 17, no. 9).

[43] *Select Pleas*, i. 22.

[44] Thus Robert Pukerel, former sheriff of Somerset, in 1171 owed 15 marks *pro prisone qui ab eo evasit* (*P.R.* 17 Henry II., 14).

[45] *Ibid.* 32 Henry II., 98.   [46] *Ibid.* 33 Henry II., 43.   [47] Ch. 4.

with the duty of leading certain prisoners before the justices and of sending with them men to bear the record of proceedings in the hundred or county where they were indicted. In the reign of Henry II. the sheriff of York was once amerced for causing accused persons, saved from the ordeal of cold water, to abjure the realm, as the assizes of Clarendon [48] and Northampton [49] required, but without assent of the justices.[50]

The military importance of the sheriff is now clearer. Aside from his services in providing castle work on royal castles,[51] equipment and stores for their defenders, the sheriff sometimes accounted for the income from the castle estate [52] or acted as custodian of the castle itself. Again he paid the wages of knights on garrison duty [53] in the king's strongholds.[54] In 1166 a tenant-in-chief states that he has rendered the knight service which he owes to the king's sheriffs of the shire [55] as if this were quite the rule. During the revolt of 1173 the castles of Worcester and Northampton were held in person by the sheriffs of these counties.[56] In the preceding year the sheriff of Shropshire was in the army operating in Leicestershire along with sergeants of his county.[57] This military activity of a sheriff is mentioned after the great feudal revolt was suppressed.[58] In 1182 Ralph Puer the sheriff of Gloucestershire, ordered by the king's justices to aid a certain magnate who had fortified a castle in the Welsh march, was killed in an onslaught made upon it by the Welsh.[59]

---

[48] Ch. 14.

[49] Ch. 1.

[50] Madox, *Exchequer*, 547, note (d); *P.R.* 31 Henry II., 70.

[51] Cf. Bridgnorth and Shrewsbury castles, *P.R.* 19 Henry II., 108.

[52] Robert de.Vaux, sheriff of Cumberland in 33 Henry II., owed over £50 *de redditu* from the castle of Sowerby for ten years (*P.R.* 95).

[53] William Bendenges, sheriff of Dorset and Somerset, accounts for certain sums *ad·custodiam castelli de Bristou per breve regis* (*P.R.* 29 Henry II., 27).

[54] See the payments made by the sheriff of Northampton (*P.R.* 20 Henry II., 51, 54-5). In the same year ten knights were paid £36 by writ of Richard de Lucy for thirty-six days' service in Worcester Castle, and ten.sergeants were paid sixty shillings for the same term after Robert de Lucy had.custody of the castle (*ibid.* p. 26).

[55] *Red Book*, i. 400.

[56] *Et militibus et servientibus quos idem vicecomes retinuit in castello Wigornense xl. mk. per breve regis* (*P.R.* 20 Henry II., 26). *Et in liberatione x. militum residentium in Castello de Norh. cum ipso vicecomite a crastino S. Michaelis anni praeteriti usque ad Vincula S. Petri sequentis anni, scilicet de ccc. et vi. diebus c. et liii. li. per breve regis* (*ibid.* 51).

[57] *P.R.* 19 Henry II., 107-8.

[58] Mr. Round calls especial attention to the £60 paid in 33 Henry II. for the support of the sheriff *et suos in custodia turris de Doura* for the past year and a half (*P.R.* 33 Henry II., p. xl).

[59] *Benedict of Peterborough*, R.S., i. 288.

The king sometimes employed the sheriff in administering the details of knight service. The record concerning knights' fees, embodied in the famous *cartae* of 1166, originated in a writ of the crown issued to the sheriff [60] of each county, the terms of which were communicated to the several tenants-in-chief whose *capita baroniae* lay within the jurisdiction.[61] A similar investigation made by John concerning military tenants-in-chief was also managed by the sheriffs.[62]

In executing the orders of the crown and the law of the land the sheriff of this period had other functions, necessarily varied. It was his duty to see that the royal regulations regarding circulation of the coinage were enforced. Robert de Vaux was amerced because he kept the old money in circulation after its disuse had been ordered,[63] in 1180.[64] At one time fifty-five carts of lead from a mine at Carlisle were delivered to the sheriff for transport to Caen.[65] His authority, or that of his agent, seems essential in the reign of Henry II. to the disposition of the body of a person found dead.[66] As the Assize of Clarendon testifies, it was his duty to apprehend law-breakers in general and to enforce the king's orders concerning them. The King's directions in 1166 for the expulsion of the heretics excommunicated at Oxford required that sheriffs and their *ministri*, as well as the *dapiferi* of barons, should take oath to do these things.[67] According to the Constitutions of Clarendon,[68] when no one dared accuse offenders against canon law the sheriff, at the request of the bishop, was to swear in before the latter twelve law-worthy men of the vicinage or the vill to make presentment. Again, the sheriff as in the past was the official primarily responsible for the return of escaped villains.[69] In the event of a clash between secular and ecclesiastical authority he was peculiarly likely to be

[60] Round, *Feudal England*, 239-40.

[61] Cf. *Red Book*, i. 263, 412, 416.

[62] *Ibid.* ii. p. cclxxxv ; cf. *Benedict*, i. 138, for a similar inquest ordered in 1177.

[63] *P.R.* 31 Henry II., 187.

[64] Benedict of Peterborough, *Gesta Henrici Secundi*, Rolls Ser., i. 263.

[65] *P.R.* 14 Henry II., 109.

[66] A Yorkshire vill is amerced *quia sepelivit mortuum inventum sine visu servientum vicecomitis* (Madox, *Exchequer*, i. 547, note (g) ; *P.R.* 31 Henry II., 71). Even in the reign of John, after the coroner appears, it is shown that the burial of the body of a certain woman (*Select Pleas*, no. 108) is supposed to be by view of coroner and sheriff.

[67] Assize of Clarendon, chap. 21, Stubbs, *Sel. Charters*, 145-6.

[68] Chap. 6.

[69] See the king's writs in *Chron. Abingdon*, ii. 235 ; *E.H.R.* xxxv. 399, no. 45.

involved. Gerald de Barry, with a commission from the arch-bishop to act as his legate in the diocese of St. Davids, about 1175 excommunicated the sheriff of Pembroke for refusing to restore to the priory of Pembroke cattle which he had seized.[70] William Longchamp, after losing control of the government in 1191, freely used his authority as bishop to direct the excommunication of his political enemies,[71] half a dozen of them being sheriffs for the year. At the same time he excommunicated Hugh de Nunant, bishop of Coventry, for serving as sheriff contrary to the provisions of canon law, on the ground that he had taken an oath not to do this.[72]

The functions of the sheriff pertaining to the judicial system are of unusual interest in these days of the beginnings of the common law system. The logical starting-point here is the customary local work of the sheriff, for the keynote of Angevin government is local action through central guidance. In 1166 occurs mention of the sheriff's view of frankpledge which is not confined to the hundred court ; for placing men under such pledge the sheriff may claim of any man admittance to his land and court.[73] The tourn is men-tioned in 1205,[74] and a record of 1214 permits the actual tracing of the sheriff of Cornwall from hundred to hundred as he holds such sessions.[75]

Of the crimes punished by loss of life and limb, according to Glanville, theft alone was tried before the sheriff in the county court. This and some other pleas, so it is said,[76] were handled in various counties according to diverse customs. Some cases in-volving trial by duel came before the county.[77] In the event of failure of manorial courts, scuffles, blows and wounds were also matter for trial before the sheriff, unless the plaintiff alleged breach of the king's peace.[78] These in John's time were to be classed as breaches of the sheriff's peace. The judging of fugitives *per legem comitatus*,[79] which is the old process of outlawry, is mentioned. Before the sheriff in this county assembly as well as before the

---

[70] Giraldus Cambrensis, *Opera*, R.S., i. 25.
[71] *Benedict*, ii. 223.
[72] *Ibid.* ii. 224.
[73] Assize of Clarendon, chap. 9.
[74] *Rotuli Chartarum*, i. 132.
[75] *Select Pleas*, i. 68-9.
[76] i. 2.
[77] A county court is amerced 100 marks for sending a duel to a hundred (*P.R.* 23 Henry II., 20).
[78] Glanville, *Tractatus de legibus*, i. 2. In twelfth-century London the alderman decided whether battery or affray with bloodshed was a plea for the sheriff or for the king (*Borough Customs*, Selden Soc. ii. p. cxlvii).
[79] *Plac. Abbreviatio*, Record Com. 18 b, 19.

reeve in the hundred court, wounds were displayed and crimes of violence denounced by the injured person.[80] Already there occurs mention of the *plenus comitatus*.[81] The amercement of a county *pro falso iudicio* [82] shows that in theory judgements given in its assembly are not those of the sheriff.

This body as of old was a useful forum for bringing official acts to public attention. In its presence and hearing a private grant of land might be made.[83] The sheriff was liable to amercement if the summoners whom he appointed failed in executing the king's writ unless he had publicly enjoined them in the county court.[84] In order to escape financial risk he was likewise required to receive in the county assembly the pledge of a tenant-in-chief that he would make due satisfaction for the debt he owed the king.[85] From John's reign comes the information [86] that he who wishes to free his serf is to hand him by the right-hand to the sheriff in *pleno comitatu*.

In relation to the justices and the common law the sheriff was attaining a permanent but subordinate position highly character-istic of the developments of the period. In the first place, between 1154 and 1216 he all but lost original jurisdiction in crown pleas. Even before 1166 these were hardly tried exclusively by the sheriff.[87] The rapidity with which from that year on they were added to the regular routine of the itinerant justices is obvious.[88] The king by special as well as general acts made over to the justices certain civil pleas.[89] Moreover, the appointment of sheriffs to serve as justices

[80] Glanville, xiv. 6 ; *Select Pleas*, i. 3, 18. By the time of John this must be done as a preliminary to further appeal (*ibid.* p. 19).

[81] See below, note 94.

[82] Cf. county of Stafford, *P.R.* 25 Henry II., p. 99.

[83] *Ancient Charters*, Pipe Roll Soc. x. 73.

[84] Glanville, *Tractatus de legibus*, i. 30. In case he has done so they are liable, and they cannot dispute the county court in this matter (*P.R.* 22 Henry II., 210 ; 23 Henry II., 205).

[85] *Dialogus*, p. 151.

[86] Liebermann, *Gesetze*, i. 491 (15, 1).

[87] See page 101 above.

[88] Compare Assize of Clarendon, chs. 1, 4-6 ; Assize of Northampton, chs. 1, 7, 9, 11 ; Glanville, i. 2, 3 ; Form of Proceeding on the Judicial Visitation of 1194 in Roger of Hoveden, iii. 262-7 and Stubbs, *Sel. Charters*, 258-63 : *Select Pleas*, i. 1-37.

[89] Glanville (i. 3) gives a partial list of civil pleas heard only in the king's court. A charter of Henry II. to the abbey of Eynsham (*Chart. Abbey of Eynsham*, Oxford Histor. Soc., i. 40) specifies that the abbot be placed in no plea regarding a tenement which he holds in demesne *nisi coram me*. John conceded to the prioresses of Appleton *quod non ponatur in placitum de aliquo dominico suo nisi coram rege vel capitali justiciario* (Madox, *Exchequer*, i. 408, note p).

was in 1194 limited to counties which they had not held since King Richard's coronation.[90]

But the sheriff's judicial province might be greatly extended by royal act.   To him belonged all pleas whence he had a writ of the king or the capital justiciar.   Among those assigned him in regular course were the plea of right when a lord's court had failed to do right and the plea *de nativis*.[91]   The former of these, with some other pleas, was often transferred to the county court as a step toward bringing it into the king's court.[92]   The plea upon the writ *Justicies*, when the lord sought his relief or services of his tenant, was before the sheriff in the county court, and the defendant who lost not only rendered a reasonable relief, but was also *in misericordia vicecomitis*.[93] In the reign of Henry II., just as in Norman days, the sheriff was directed by royal precept to assemble his county to hold special recognitions.[94]   Payment was made to the king by interested persons to expedite the sheriff's execution of such writs.[95]

The procedure which Glanville describes in the last decade of the reign of Henry II. already illustrates the functions of the sheriff with reference to the common law and the courts which he has never lost.   Throughout this treatise one learns of summoners to be sent by the sheriff to direct the appearance of parties before the justices, of inquests to be held by him to ascertain facts material to a decision in causes at issue, of record of a case to be made by the sheriff in the county court and brought by discreet persons before the justices, of men sent by him to measure, divide or make view of land, of mandates directing him to place a successful claimant in possession.   When, for instance, the writ of *mort d'ancestor* was received by the sheriff, security for prosecuting the claim had to be given in the county court; twelve free and law-worthy

---

[90] Stubbs, *Sel. Charters*, p. 260 (sect. 21).

[91] Glanville, xii. 9 ; i. 4 ; for details cf. ix. 10, and v. 1.   As to a local usage in Cornwall involving in 1201 an independent inquest by the sheriff regarding villains see *Select Pleas*, i. 2.

[92] *Ibid.* xii. 1.   So also cases concerning dower, vi. 6-8.

[93] Glanville, ix. 10.

[94] The sheriff of Berkshire is to convene a *comitatum plenum* at *Ferneberg* and to submit to the oath of twenty-four older men the question whether the Abbot of Abingdon had a full market in the time of Henry I. ; he is also to submit to the witness of the whole county the question whether in the same reign this monastery had the tithe of *Mercha* for the lighting of the church and whether it was unjustly withheld (*Chron. Abingdon*, ii. 226, 228).   A delay in carrying out a writ *per consilium comitatus* : *Rot. Curia Regis*, no. 14, p. 34.

[95] As, for example, the payment by R. de Cruce, *P.R.* 19 Henry II., 93.

men of the vicinity were chosen as inquest jurors and their names
enrolled by the sheriff ; and the latter then summoned the occupant
to be before the king's justices on the day set.[96] Finally, Glanville
supplies writ forms directing the sheriff to put the successful claimant
in possession.[97] Procedure to determine ownership of land involved
a still longer series of acts on the part of the sheriff under direction
of the justices, each act enjoined by writ.[98] Advowson cases,[99]
pleas of debt,[100] essoins,[101] and the restoration of fugitive serfs [102]
are some of the many other matters which occasion writ mandates
to sheriffs. The justices might also intervene to direct him when
fiscal matters were involved.[103]

The sheriff of the time of Henry II. also assumed new duties in
relation to the criminal law which he performed as an aid to the
king's justices. By the Assize of Clarendon he, as well as the
justices, was commanded to have criminals presented before him
by juries representing vills and hundreds. Mr. Maitland held that
henceforth this was done in his tourn of the hundred as well as
in the county court,[104] and in John's time the fact is proved by
records.[105] Yet in the main such cases were triable only in the
king's court. Failure of a hundred to present a plea of the crown
in the county court [106] or to make known to a sheriff a violent

[96] Glanville, xiii. 7.
[97] Ibid. xiii. 7, 8, 14, 16.
[98] In the initial stage of proceedings he sent law-worthy men to make view of
the land in question, four of whom he must have before the justices at the appointed
time to testify to this view. When a duel has taken place in the king's court, the
sheriff places the victor in seisin. When one party places himself on the grand
assize, the sheriff orders the lord in whose court the plea was originally brought
to stay proceedings. The sheriff is next to summon four law-worthy knights to
come before the justice at Westminster or elsewhere to choose upon oath the
twelve knights who best know the truth in the matter. The sheriff must also
summon the defendant to be present at this election; also the twelve to be before
justices at a fixed time and place prepared to make recognition as to which party
has the better right; also the defendant to be present. At last the sheriff places
in seisin the party to whom the land is adjudged (ibid. ii. 2, 4, 8, 9, 11, 15, 20).
[99] Ibid. iv. 4.
[100] Ibid. x. 2, 4, 7, 9, 16.
[101] i. 14, 19, 21.
[102] xii. 11.
[103] De predictis debitis monetariorum habet vicecomes plegios et plures debet
accipere in comitatu precepto justiciarum (P.R. 23 Henry II., 208).
[104] Cf. Maitland, Select Pleas in Manorial Courts, Selden Soc. p. xxi ; Stubbs,
Sel. Charters, 143 (sect. 1).
[105] Select Pleas, i. 68-9 ; Rot. Char. i. 132.
[106] For this the hundred of Dorchester pays 5 marks (P.R. 30 Henry II., 72. Cf.
ibid. 32 Henry II. introd. xxx). In 1201 a presentment of a hundred which goes
to the justices without having reached the county is called silly (Select Pleas, i. 6).

death [107] were offences punishable by amercement.   In anticipation
of the coming of the justices the sheriff was by the king's order of
1166 required to assemble the county court and make inquiry
concerning those newly come into the county.   These persons were
to be placed in pledge or in custody against their appearance at the
eyre.   As a supplement to this process individual sheriffs prepared
coram comitatibus lists of fugitives so the justices might seek them
in other counties.[108]   Again, the sheriff was directed without
summons to lead criminals before the justices either in his own
county or in the nearest county to which they came.[109]   The
justices in eyre are already to be found ordering the sheriff to place
a complainant under pledges [110] or a suspected person [111] under
plevin, or to make a summons.[112]

The reigns of Henry of Anjou and his sons witnessed a development of absolutist devices for raising revenue which were very
advanced for the age and which brought the sheriff to the pinnacle
of his importance in fiscal affairs.   The *Dialogus de Scaccario* and
the annual Pipe Rolls from the year 1156 provide for finance such
detailed and consecutive information as no other branch of Angevin
administration may claim.   The recovery of the royal income was
evidently a prime consideration leading to the prompt re-establish-
ment of some shrievalties at the accession of Henry II.   The
wastage of war upon the king's manors had to be repaired, and lands
bestowed lavishly by Stephen reclaimed.[113]   The Pipe Rolls, however, show no sheriff again accounting for the revenues of Carlisle [114]
or Northumberland until 1158, of Westmoreland until 1177, nor
of Cornwall until 1176, after the death of the king's uncle, Earl
Reginald, who held many of the royal estates.[115]   The honour of
Lancaster appears with its separate account in 1167.[116]   Rutland,
apparently under the guardianship of a court official in 1130, from

---

[107] Madox, *Exchequer*, i. 553, note (u).   As early as 1168 a bailiff was amerced
for not making known to the sheriff proof of Englishry that had been offered him
(*ibid.*).

[108] Assize of Clarendon, chs. 18, 19.   [109] *Ibid.* chs. 4, 6.

[110] *Rot. Curiae Regis*, Pipe Roll Soc., vol. 14, p. 98.

[111] *Placitorum Abbreviatio*, Record Com., 17 b.

[112] *Rot. Curiae Regis*, Pipe Roll Soc., vol. 14, 104.

[113] Poole, *Exchequer*, 133-5.

[114] First recorded under the name of Cumberland in 1178, clearly regarded as a
county with a *firma comitatus* earlier.

[115] See below, note 145.

[116] After 1182 it is clear that there is a shire of Lancaster (Farrer, *Lancashire
Pipe Rolls and Early Charters*, pp. xvi, 49).

1155 to 1179, with one brief interval,[117] accounted for its small *ferm* of £10 through Richard de Humez, the king's constable. After due allowance is made for some minor portion of the king's revenue handled by special royal agents [118] and sometimes unrecorded in the Pipe Rolls, and even for a peculiar class of income handled by one person for the whole realm,[119] the sheriff was nevertheless the great collector. Indeed, with the exception of special custodians of manors and towns and tenants-in-chief who might, if they chose, answer directly in person at the exchequer, he and his bailiffs were practically the sole collectors of the king's revenues.

No side of the shrievalty of these reigns is so well known as its accountability to the exchequer. But the treatment of its fiscal functions has necessarily been partial and incidental.[120] A few well-known details may be repeated. Each year he received from the exchequer clerks summons directing him to be present at Westminster, Winchester or elsewhere on the morrow of the close of Easter and again on the morrow of Michaelmas, and to have with him what was due the king from certain sources.[121] Only after he had paid the sums due and had receipt tallies for them was he supposed to sit at the exchequer. Failure to appear at the place and time specified was contempt of the king's mandate, punishable by heavy amercement. The number of recognized excuses aside from illness and absence in the king's service was small, but even these acquitted the sheriff only in case he sent a representative to meet his obligations.[122] No excuse sufficed to confer immunity

[117] Apparently 1156-63. The counties for which the sheriff paid the whole of his *ferm* by tale are these five together with Salop and Sussex (G. J. Turner, *Trans. R.H.S.*, N.S. xii. p. 118).

[118] Hilary Jenkinson in *E.H.R.* xxviii. 218-20, 732.

[119] Thomas Fitz-Barnard from 1178 answered for the pannage of all England (*Pipe Roll*, 25 Henry II., pp. xxiv, 60). This marks a new development. As early as 1200 the profits of the coinage of the realm were farmed by one person (*Chancellor's Roll*, 3 John, 111). King Richard in 1189 for a short time employed John Marshal as custodian and receiver of all his escheats in England (*Benedict of Peterborough*, ii. 91).

[120] Such accounts are given in Madox, *Exchequer*; Publications of Pipe Roll Soc. iii. 42-57; Hall, *Antiquities of the Exchequer*, chaps. 5-6; Poole, *Exchequer*, chaps. 6-7; *Dialogus* and the introduction to the Oxford edition, pp. 47-53, as well as the notes to this edition, pp. 163-240.

[121] There are some sources of income that must necessarily come through the hand of the sheriff though no summons concerning them be issued, but these are casual rather than fixed (*Dialogus*, 113).

[122] *Ibid.* 121-3. The *Dialogus* holds that on the third day of his absence the sheriff becomes liable in all his goods, and on the fourth day in body as well as goods. The recorded cases, however (*ibid.* 212) seem to show for the reigns of Henry II. and Richard a uniform fine of one hundred shillings a day.

upon the person who withheld from the king the money collected for him in the county.[123] According to the text of the *Dialogus de Scaccario* [124] no one, not even the sheriff's eldest son, might take his place at the completion of the annual account. But in the Pipe Rolls of Henry II. his brother or his steward, as well as his son, is seen to appear in his place.[125]

The sheriff of a given county after appearing awaited his turn until he was called by the voice of a herald. The debtors of the county were notified before he rendered his account.[126] Other sheriffs were excluded, and he sat with his attendants.[127] Some of the debtors of the county were present to acquit their obligations, but the attendance of those who had paid amounts to the sheriff was probably encouraged as a means of checking his report.[128]

The barons of the exchequer in auditing the sheriff's account gave attention also to disbursements which he had made on the king's behalf. Not until these were approved was it possible to compute the balance of the *ferm* which was due. The three-fold classification of them is a familiar one. First, the treasurer in beginning proceedings made inquiry concerning the fixed payments, met from the *ferm* of the county, and consisting of alms and tithes and also of allowance to be made the sheriff for loss of income from lands the king had granted away.[129] With these should be grouped the earl's third penny in some counties and occasionally fixed payments to cover the wages of minor officials and guards.[130] Payments specifically ordered by the king's writ formed a second category. The personal requirements of the king and those of his family, his

[123] *Dialogus*, 124.

[124] Poole, *Exchequer*, 123.

[125] The statements of the *Dialogus*, as its latest editors show (pp. 213-14), probably mean that the sheriff may send an attorney with his *profer*, but that at the accounting for the whole year he must have special permission to appoint some one acceptable to act for him.

[126] *Ibid.* ii. iv. E ; ii. xx.

[127] *Ibid.* pp. 70, 125.

[128] See *Dialogus*, 215.

[129] Parow, *Compotus* (Berlin, 1906), pp. 27-8, gives tables showing for three years of Henry II. that only about one-third of the amount collected was actually paid in by sheriffs, the remainder being disbursement. But there were in addition considerable debts uncollected.

[130] Parow (*ibid.* p. 14) shows that the 30*s.* 5*d.* allowed a *falconarius* is equivalent to a penny a day for a year of 365 days, that similarly the *parcarius* receives 1½*d.* per day, the *elemosynarius* 2*d.*, and the *clericus* 3*d.* It is well known that the wages of a knight were at the rate of 8*d.* per day (Round, *Feudal England*, 271-2). *In liberationibus constitutis portariis et vigilibus castelli Lareste*, £9 : 2 : 6 (*P.R. 2 Henry II.*, 57).

court,[131] or his service were usually supplied by these. The fortification of his castles and the erection and repair of his buildings were authorized by a special form of writ which designated two or three persons by whose view the work was to be done,[132] and who at the time of accounting made public oath that the operations involved were to the king's advantage.[133] Inability of sheriffs to produce authorization of military expenditures during the hostilities of 1173–4 [134] was for some time a source of confused accounts. A third set of disbursements was sanctioned only by the custom of the exchequer, and these were attested by the sheriff's oath. They included the expense of fulfilling justice and judgement,[135] of transporting the king's treasure, caring for his vines and his vintage, and supplying ships and other necessities.[136] Sums were sometimes paid out by the sheriff at the order of the chief justiciar or justices,[137] especially for restoring manors that had been wasted, an object of no small importance in the earliest years of Henry II. and again after the revolt of 1173. The expending of funds for these various objects frequently required that the sheriff keep on hand a surplus. If he failed to retain a sufficient sum from the last payment at the exchequer he himself had to advance it. This amount advanced is the sheriff's *superplusagium* of the Pipe Rolls,[138] which is treated at the exchequer of the ensuing Michaelmas as a credit toward his *ferm* of the shire.[139]

[131] Poole, *Exchequer*, 162, gives interesting examples. Hundreds might be cited. Parow, p. 16, classifies.

[132] Cf. sect. 7 of the Assize of Clarendon concerning jails. The Pipe Roll of this year shows work on a number of jails, among them those at Canterbury and Rochester (*P.R.* 12 Henry II., 111), Oxford (117), Cambridge, Huntingdon (84), Salisbury, Malmesbury (72), and Newcastle-on-Tyne (75). In 1178 31s. was expended *per breve regis* for making a jail at Ipswich (*ibid.* 24 Henry II., 19).

[133] *Dialogus*, 130 ; Poole, *Exchequer*, 161.

[134] See *P.R.* 20 Henry II., 107 ; 30 Henry II., p. xxviii.

[135] Such as the cost of holding ordeals, the wages and travelling expenses of the king's approvers, the cost of conveying criminals, and sums paid executioners in some places. The porter of the city of Rochester *faciebat justitiam comitatus* (*P.R.* 2-4 Henry II., 101).

[136] *Dialogus*, 127-9 ; Poole, *Exchequer*, 162-3. The cost of pickling the great fish which was claimed for the king was also included.

[137] Madox, *Exchequer*, i. 201. In 1185 the justices authorized expenditure from the revenues of the honour of Peverell, then in the king's hands, for the stocking of two *vaccariae* for forty cows (*P.R.* 31 Henry II., 116). The expense of restocking Evesham in 1178 was incurred by the king's writ, testimony of the itinerant justices and view of the men of the vill (*ibid.* 24 Henry II., 19).

[138] Madox, *Exchequer*, i. 285-6 ; ii. 231 ; *Pipe Roll Society*, iii. 95 ; Poole, *Exchequer*, 132.

[139] *Et habet de suppl.* £10 : 16 : 6 *blanch quod computatum erit ei in firma Huntendonscire* (*P.R.* 2-4 Henry II., 11). In 1172 the sheriff of Hants was ;credited with 61s. 7d. bl. *in suo superplusagio de firma civitatis* (*P.R.* 18 Henry II., 79).

The largest item for which the sheriff was regularly responsible was of course the *ferm* of the shire. Its amount was not recorded on the Pipe Roll until the last year of Richard I, and was known through the *Rotulus Exactorius* now lost. The careful labour of Mr. G. J. Turner, however, has given the totals for the various shires for ten years in the earlier period of the reign of Henry II.[140] The argument is advanced in the *Dialogus* that the sheriff ought to be summoned not only for the *ferm* of the year, the *nova firma*, but also for any unpaid balance of the old *ferm* due even from his predecessors.[141] It is clear, however, that in this latter case responsibility rested alone upon the former sheriff. In the time of Henry II., when sheriffs were changed in the course of the fiscal year they were reckoned as dividing the annual obligation for the *ferm* either by halves or by fourths.[142] When there were two or more sheriffs they bore proportionate shares.[143] The *firma comitatus* was a composite affair. When pairs of counties were farmed together there was sometimes but one *firma*. It still consisted in part of the judicial income of local courts.[144] Just how detailed was the allotment of its parts to localities within a given county is an unsolved question, but the fact that certain lands of the county bore a fixed portion is frequently mentioned.[145] Until a much later time the king's manors,

[140] *Trans. R.H.S.*, N.S., xii. pp. 142-9. Parow (*Compotus*, 25) gives a useful table. The highest *ferm* was that of Lincolnshire, £836 : 1 : 8 *blanch* plus £140 tale. The three lowest were : Staffordshire, £140; Sussex, £40; Rutland, £10. Parow makes the total of *ferms* of the counties, 10 Henry II., £8817 : 15 : 5 *blanch* plus £1326 : 17 : 0 tale.

[141] *Non de parte eius set de universo quia firma est cuius solutio differri non debet* (*Dialogus*, 131-2).

[142] Thus in 1165 (*P.R.* 11 Henry II., 63) Richard de Raddon, who retired from the shrievalty of Somerset in the course of the fiscal year 1160–61, owed £90 *blanch de firma Somerset de quarta parte anni*. His successor, Walter de Lisores, paid the remaining three-fourths. The *ferm* paid by the sheriff of London from the death of King Stephen (October 25, 1154) to the next Easter (*P.R.* 7 Henry II., 17) was undoubtedly that of a half-year. But at Southampton one *fermer* accounts for a third of the year, another two-thirds (*P.R.* 2-4 Henry II., 53).

[143] In 1174 four sheriffs of London owed three-fourths of the *ferm* of the current year and all that of the year preceding. Each accounts for £114 : 0 : 6 *numero de quarta parte superioris debiti* (*P.R.* 20 Henry II., 8-9). In 1170 Hugh de La Lega owed £39 : 12 : 4 *blanch* for his part of the old *ferm* of Beds. and Bucks., Richard fitz Osbert, £53 : 14 : 0 *blanch* for his part (*P.R.* 16 Henry II., p. 26).

[144] *Dialogus*, 109 : *ex magna parte de placitis provenit*.

[145] G. J. Turner, *Trans. R.H.S.*, N.S., xii. 131-2 ; Poole, *Exchequer*, 131. Thus Geoffrey Hose owes £79 : 0 : 2 *blanch* of the old *ferm* of Oxfordshire for three years when he held the county *quae remanserunt super terram monachorum de Tama* (*P.R.* 31 Henry II., 106). In 1170 there was a difference of opinion as to whether certain lands in Clakelose hundred, Norfolk, belonged to the *ferm* of the shire (*P.R.* 16 Henry II., 3). In 21 Henry II. certain lands which the sheriff of

K

hundreds or towns were often let to bailiffs who responded to the sheriffs for their revenues.[146]   The Inquest of Sheriffs of 1170 [147] represents, so it has been shown,[148] the only general attempt of the period to look into the operation of the system and to ascertain the amounts of *ferm* exacted both from the king's lands and through the courts of shire and hundred.

The income from the king's lands was still sometimes enhanced by the imposition of an increment (*crementum*).   That the burden of this was shifted from the sheriff to the county is shown by the provision in Magna Charta [149] to the effect that, except upon the king's demesne manors, the ancient *ferms* of counties, hundreds, wapentakes and trithings are to be collected without increment.   In the reign of Henry II. a *crementum comitatus* was accounted for by many sheriffs.[150]   Occasionally it was but a nominal due sometimes rendered in hawks.[151]   More often it was a heavy sum of money.[152] In John's time one sheriff charged with an increment was acquitted of the debt by making fine with the justiciar.[153]   It is difficult to say how the burden was borne, but parts of the county increment were probably distributed among the manors of the *corpus comitatus*. Certain it is that royal towns belonging to the *ferm* of the county and certain estates accounting separately made such payments.[154]

York had *in manu* belonged to the *ferm* of Northumberland (*P.R.* 21 Henry II., 173).   In 28 Henry II. (*P.R.* p. 27) manors of Cornwall which Earl Reginald had held and which now belonged to the *ferm* of Devon brought in £122 : 10 : 0 *blanch*. Henry II. granted the abbot of Ramsey the hundred of Hurstingstone *ad firmam*, the latter to render for it four marks of silver, so that whoever was sheriff he might require no more (*Ramsey Chartul.* ii. 85).

[146] Assize of Northampton, sect. 10, Stubbs, *Sel. Charters*, 152.

[147] *Ibid.* 148-50.

[148] Poole, *Exchequer*, 129.

[149] Magna Charta, sect. 25 ; Articles of the Barons, sect. 14 ; Stubbs, *Sel. Charters*, 291.

[150] Mr. G. J. Turner shows reason to hold that it had existed in all those counties for which the sheriff accounted partly in blanched money and partly by tale.   These were fourteen in number (*Trans. R.H.S.*, N.S. xii. pp. 118, 124).

[151] This occurs in the Pipe Rolls for Buckinghamshire and Bedfordshire (*P.R.* 16 Henry II., p. 25 ; *P.R.* 20 Henry II., 82).

[152] For Warwickshire and Leicestershire together £60 *numero* (*P.R.* 14 Henry II., 56) ; for Lincolnshire £80 are due for the *crementum* of two years (*P.R.* 13 Henry II., 50 ; 16 Henry II. 22) ; for Norfolk and Suffolk together £100 a year (*P.R.* 13 Henry II., 18 ; 16 Henry II., 2 ; 1 Richard I. 40).

[153] Madox, *Exchequer*, i. 202.

[154] Gilbert Basset, *fermer* of a part of the Honor of Wallingford, accounts for 56s. *de cremento de Achelay per justicias et Gilbertum Basset* (*Pipe Roll* 32 Henry II., 115).   In Northamptonshire and Derbyshire in 1173-4 (*P.R.* 20 Henry II., 57) the *crementum mancriorum* amounted to over £20.

The king's cities and boroughs still as a rule contributed their annual render to the *firma comitatus*.[155] Sometimes the sheriff held the town as *custos* and answered for its issues independently of his *ferm*.[156] In some instances this was done by a *custos* other than the sheriff.[157] Such an official is found at Northampton and at Lincoln from 1154.[158] A further step in some towns took the whole matter of the collection and payment of the *ferm* from all royal agents and placed it in the hands of the burghers. Thus at Northampton, where either the sheriff [159] or a special *custos* [160] had been accounting for the annual *ferm* of £100 tale, in 1186 the burgesses agreed to give 100 marks *pro habenda villa sua ad firmam in capite de rege*.[161] In 1170 the burgesses of Shrewsbury gave two marks of gold, those of Bridgnorth twenty marks of gold, to have their respective vills at farm,[162] and were even willing to increase these sums by a *crementum* in order to retain the privilege.[163] King Henry's charter made over

---

[155] See Madox, *Firma Burgi*, 18. In 1170 the burgesses of Bridgnorth owed two and a half marks *de cremento ville sue per annum per Hugo de Bello Campo praeter illam firmam ville quae est infra firmam comitatus (P.R.* 16 Henry II., 133). At the beginning of the reign *(P.R.* 2 Henry II., 105) Winchester owed money *in firma comitatus.* In 1197 (Hoveden, iv. 33) it seems to be regarded as exceptional if cities and boroughs are not *in manu vicecomitis.*

[156] In 24 Henry II. *(P.R.* p. 83) the sheriff accounts for £37 : 11 : 8 *firma burgi de Legercestr' de termino Pasche et termino St. Michaelis dum fuit in manu regis.* At Gloucester, which was let to a special custodian, the sheriff at one time answered only for the *crementum burgi*, amounting annually to 100s., and for 7s. 6d. *firma* of the purprestures of the borough *(P.R.* 27 Henry II., 118). The sheriff's render of the *ferm* of Launceston is separate from that of the *ferm* of the county *(P.R.* 23 Henry II., 9), and the same is true at Winchester, in 1172. Richard fitz Thurston, former sheriff of the shire in this year, still owes over £173 *blanch* of the old *ferm* of the city as well as a lesser amount of the old *ferm* of the county *(P.R.* 18 Henry II., 80, 84).

[157] *Red Book*, ii. 655, 657.

[158] Thus at Gloucester *(P.R.* 12 Henry II., 80 ; 17 Henry II., 87) by Osmund, and a little later by William *nepos* of Warner *(P.R.* Richard I., 73) ; at Winchester early in the reign of Henry II. by one Stigand *(P.R.* 2-4 Henry II., 52), although later by the sheriff *(P.R.* 12 Henry II., 104 ; 16 Henry II., 125 ; 18 Henry II., 80, 84 ; 29 Henry II., 147) ; at Southampton in 1156 *(P.R.* 2-4 Henry II., 53) by a separate *fermer*, in 1169 *(P.R.* 13 Henry II., 193, 194) by three reeves, four years later by Richard de Limesia *(ibid.* 17 Henry II., 42), and at the close of the reign and the beginning of the next *(ibid.* 29 Henry II., 148 ; 31 Henry II., 215 ; 1 Richard I., 206) by Gervase the reeve.

[159] *P.R.* 17 Henry II., 47 ; 24 Henry II., 49.

[160] *Ibid.* 2-4 Henry II., 42 ; 15 Henry II., 77.

[161] *Ibid.* 32 Henry II., 3.   [162] *Ibid.* 16 Henry II., 133.

[163] In 1170 the *crementum* of Bridgnorth, then contributing to the *firma comitatus*, was 2½ marks *(ibid.)* ; in 1176 the render is 10 marks *firma ville cum cremento (P.R.* 22 Henry II., 56, 57), that of Shrewsbury £20 *de firma burgi.* The next year Shrewsbury contributes (23 Henry II., 37) 11 *fugatores de cremento firme de Salopsberie de hoc anno.*

the vill of Cambridge to his burgesses to be held *in capite* by the same *ferm* they had been accustomed to pay to the sheriff, but for which they were now to answer nowhere except at the exchequer.[164]   The reign of King Richard has usually been regarded as marking a new era in the history of English municipal enfranchisement.   It is clear that by his time many towns were accounting directly to the exchequer.   This liberty often prevented the sheriff from holding pleas as well as from making collections.[165]   Mr. Ballard has shown that King John conferred this privilege of the *firma burgi* upon no less than twenty-three boroughs and four vills.[166]   From the third year of Richard I. this fixed sum due from the citizens of London was paid at the exchequer on their behalf by the sheriffs.[167]

The sheriff was sometimes, though by no means always, *custos* of certain royal estates,[168] as well as towns, which were not included in the *ferm* of the shire.   Such a *custos* was summoned to the exchequer for the king's debts in the same way as the sheriff of the shire.   In some places he also made fixed annual disbursements or even paid out sums by custom of the exchequer, as well as by special writ.[169] When a feudal estate fell into the king's hands in escheat or temporarily in wardship it was not farmed, but let to a custodian, who rendered to the exchequer annually a sum assumed to cover everything he received except provisions,[170] and fixed in advance by the king's agents,[171] who in some cases were the justices.[172]   The

[164] In King John's charter of 1207 the annual *ferm* was £40 *blanch*, and a *crementum* of £20 tale was also paid (Maitland and Bateson, *Charters of the Borough of Cambridge*, 8).

[165] See Richard's charter to Hereford, 13th *Rept. Hist. MSS. Commission*, Appendix, part iv. 284.

[166] *E.H.R.* xiv. 97, 102-3.   John's charter to Huntingdon in 1205 (*Rot. Cart.* i. 157) expressly forbids the sheriff to interfere concerning its *ferm*.   The charter to Derby (*ibid.* i. 138) in 1204 provides that the reeve is to answer for the *ferm*; that of Appleby (*ibid.* i. 41) in 120. that the burghers are to render it to the sheriff.

[1-7] Madox, *Firma Burgi*, 165 ff.

[168] For a list of these and their income together with *alia regalia constituta*, 16 Henry II., see Parow, *Compotus*, pp. 36-7.

[169] Thomas Becket when in charge of Berkhampstead paid sums for castle work and restoring manors (as in *P.R.* 2-4 Henry II., 21).   Expenditures from the revenues of Winchester (as *ibid.* 52) illustrate these features.

[170] *Dialogus*, 158, 239.   Provisions might not be taken by way of procuration. For a list of honours and great estates *in manu regis*, 9-21 Henry II., together with the amount of income, 16 Henry II., see Parow, *Compotus*, 33.

[171] *Dialogus*, 158.

[172] Thus in 1186 the sheriff of Warwickshire and Leicestershire accounts for the issues of two manors *per rotulum justiciarum* (*P.R.* 32 Henry II., 134); and in 1190 the sheriff of Northumberland answers for £42 *de firma terre Walteri de Bolebec per rotulum justiciarum* (*P.R.* 1 Richard I., 244).

Rotuli de Dominabus [173] of the reign of Henry II. were founded upon an eyre and used to determine these very questions. It is assumed in the *Dialogus* [174] that a sheriff is likely to be appointed custodian even of a barony or honour.

Along with parts of the king's feudal income sheriffs also accounted at the exchequer for sums derived through purprestures. Escheats first appear upon the Pipe Roll under a separate head in 1165.[175] They were probably removed for a short time from the sheriff's control when a general custodian was appointed by King Richard [176] in and after 1189. When the sheriff is recorded as turning over assart fines,[177] or in some cases the income from the king's mines,[178] he is acquitting himself also of his regular obligations.

Certain profits from escheats [179] were linked with income arising through judicial sources, and both were collected by the sheriff in the same way. When a tenant-in-chief of the king guilty of felony was according to feudal law sentenced to forfeiture, the sheriff took over the revenues from his lands and turned them in at the exchequer. The proceeds of the sale of his moveables also fell to the crown. Similarly the chattels of offenders against the assizes of Clarendon and Northampton were sold, the accounts rendered by the sheriff and the amounts involved entered on the Pipe Roll for the year.[180] Possibly no summons was issued, for ordinarily exchequer officials hardly knew the exact amounts involved.

[173] *Pipe Roll Soc.*, vol. 35. In 1177 the king ordered a similar inquiry to be made by sheriffs and bailiffs (*Gesta Regis*, i. 138).

[174] Pp. 133-4.

[175] Concerning these payments see Poole, *Exchequer*, 167-8. The income from the former seems to be regarded as a kind of rent. In 1182 Gervase of Cornhill, former sheriff of Kent, owes £4 of the old *ferm* of purprestures of the twelfth year of the reign (*P.R.* 28 Henry II., 151).

[176] *Benedict of Peterborough*, ii. 91.

[177] *Dialogus*, 103. For making clearances in the king's forests the fine levied was 1s. for an acre of wheat, 6d. for an acre of oats. This entry, however, shows him merely acting as collector when another had failed to render what he owes. *Vic. debet 63s. de wastis et essartis in Staffordscr. qui remanserunt super Gillebertum forestarium qui fuit occisus in servicio regis* (*P.R.* 25 Henry II., 98).

[178] Thus the sheriff of Devon (*P.R.* 2-4 Henry II., 74) *r.c. de £16 : 13 : 4 de veteri minaria stagni.*

[179] *Tertium genus excidentium vel escaetarum quod fisco provenit iure perpetuo* : *Dialogus*, 135.

[180] *Ille vicecomes reddit compotum de catallis fugitivorum vel mutiliatorum per assisam de loco illo N. scilicet de hoc V. de illo X.* (*Dialogus*, 135-6). In accounting in 1176 for the chattels of fugitives and those who perished by the ordeal of water *per assisam de Norhanton*, a sheriff explained (*P.R.* 22 Henry II., 45) that these had been disposed of according to the precept of the justices by four knights of the county, and that for this reason he did not have particulars in single cases.

It is significant that this is one of the special items concerning which investigation was made in the Inquest of Sheriffs.[181]   A like process accounted for treasure trove,[182] wreckage of ships [183] and chattels of deceased Christians proved to have been usurers ; and it absorbed some further sums, arising *pro enormibus culpis*, which were also casual in origin and uncertain in amount.[184]   But the collection of the ordinary income from *placita* not included in the sheriff's *ferm* was enforced by more specific process.   Such amounts were copied from the rolls of the itinerant justices [185] and listed upon the sheriff's writ of summons.   At the exchequer the chancellor's clerk was thus enabled to make demand upon him *seriatim* for individual sums due from fines and amercements.[186]   Here the function of the sheriff was purely that of collector, the *onus* of these debts resting upon individuals except when a whole community was involved, as in case of the *murdrum* levied upon hundreds and the common *misericordia* imposed upon counties.[187]

These usages regulating the collection of the profits of justice through the king's courts also held for other sources of the royal income which bore in germ the principles of a system of national taxation.   Along with the profits of *placita* were collected and recorded sums paid through *conventiones* or agreements, which, technically speaking, were fines.[188]   The *Dialogus* classes feudal reliefs among these, but recognizes that reliefs resemble pecuniary penalties rather than voluntary oblates.   Debts which in theory arose by covenant were often compulsory in fact.   This was especially true of the forms of arbitrary levy to which the barons in 1215 objected under the name of *auxilium*.   One variety, the tallage, though often assessed in the various communities of the county by

---

[181] Sect. v., Stubbs, *Sel. Charters*, 149.

[182] See Madox, *Exchequer*, i. 342.

[183] *Vic. r.c. de ij. s. de navicula fracta vendita* (*P.R.* 24 Henry II., 44, Dorset and Somerset) ; *Vic. r.c. de 6s. 8d. de navicula vendita* (*ibid.* 62, Northumberland). Madox (*Exchequer*, i. 343) shows that the men of the communities accounted for this.

[184] *Dialogus*, 138-9.   [185] *Ibid.* 119.

[186] *Ibid.* 142.   *Vicecomes reddit compotum de £63 de minutis debitis et perquisitionibus per Rand. de Glanvill et Hug. de Cressi factis quarum particule sunt in brevi quod est in thesauro et de quibus id vicecomes habet rescriptum* (*P.R.* 21 Henry II., 179).

[187] *Dialogus*, 95, 143.   Thus the sheriff renders account of a *misericordia* imposed upon a town because its jurors did not report a plea ; also a sum promised the king that the taking of assizes within the city might be respited (*P.R.* 31 Henry II., 67-8).

[188] See Poole, *Exchequer*, 169-70.

the justices, was styled a gift (*donum*). But the *Dialogus* shows that the procedure enforced in collecting this in towns left but one option, namely, either leaving this levy to be parcelled out to individuals by the justices or offering collectively a sum which would be accepted as sufficient. In the former case the sheriff, ultimately the exchequer, enforced the process usual for collecting the king's debts ; in the latter, as in the usual case of the communal aid or *assisa*,[189] the sheriff was personally bound to produce the amount at the exchequer, though, according to the *Dialogus*, if he found the community unable to pay because of declining prosperity, he might grant some indulgence.[190] The sheriff is also mentioned as bringing to the exchequer sums arising from the *assisa* of bishops [191] and religious houses as well as the *dona* of the Jews.[192]

Of the other forms of taxation, the land taxes and the scutage, the sheriff was usually the collector. The *Dialogus* treats *Danegeld* as a communal burden and therefore holds the sheriff alone responsible for turning it in.[193] As chief collector of the carucage levied in the reign of Richard I., he will be seen to have been under a new form of control.[194] The scutage and the scutage fine were also paid by the sheriff at the exchequer,[195] except when tenants-in-chief accounted directly.

The sheriff's settlement for the *ferm* was delayed until he gave satisfaction concerning his various other obligations. In the possibility of appropriating the fixed sum of money the sheriff brought to cover his *ferm*, or, what amounted to the same thing, refusing him credit for outlays,[196] the exchequer found a ready

---

[189] *Dialogus*, 142. This levy was also called *assisa* of the king's demesne, *auxilium burgorum et villarum* (as in *P.R.* 23 Henry II., 134-5), and *assisa* or *donum comitatus*. The last form was sometimes levied to cover common debts incurred through judicial sentence (as in *P.R.* 22 Henry II., 29 ; 24 Henry II., 118).

[190] *Ibid.* 145. The exchequer officials may inquire *per fidem vicecomitis* to ascertain whether citizens were originally able to pay what they promised. If they are not, others may be found to make good the amount, or what remains may be redistributed *per commune*. As to the collection of tallage cf. Madox, *Exchequer*, i. 736.

[191] *P.R.* 2-4 Henry II., 94.

[192] Mitchell, 61.

[193] *Dialogus*, 95, 143.

[194] Below, p. 139.

[195] Letters close of the king to about twenty sheriffs (*Rot. Lit. Claus.* i. 43) in 1205 direct them to permit that the scutage of William de Albini from the knights' fees which he holds in chief be collected by his bailiffs, since he responds thence to us *et nobis scire facias summon. scutagii*. It is stated concerning certain sums due from the first scutage of Richard I. that those who are charged with them are acquitted, and the sheriff has not rendered anything (*Red Book*, i. 126 ; cf. *ibid.* i. 48).

[196] See Poole, *Exchequer*, 140.

means of disciplining him. For negligence in collecting the king's
debts, to borrow the exchequer phrase, he might be coerced *per
firmam*.[197] He was warned in his summons to have at the exchequer
the king's debts *vel capientur de firma tua*.[198] The same penalty was
sometimes incurred if he failed to maintain the king's estates [199]
or came late to his accounting,[200] or failed to take pledge publicly
of a tenant-in-chief for his debt to the king. It was only after his
accounts were surveyed that in a full session of the exchequer his
sum was cast. Due credit was given for the countertallies representing payments made at Easter, and the balance due was ascertained. A part of his money was sent to be blanched,[201] and
deduction made from the sum paid equivalent to the loss by
combustion. The *ferm* of Shropshire, Sussex, Northumberland and
Cumberland was paid by tale alone.[202] Before a sheriff or *custos*
was formally acquitted the marshal administered an oath upon his
conscience that he had made a rightful accounting.[203] The *Dialogus*
says that, if he were still held to the payment of any debt at this
session, he also bound himself not to go without special leave more
than a league from where the exchequer sat except to return the
same day. This, of course, did not hold for accounts upon which
he was allowed more time. In scores of cases sheriff's debts are
carried over from year to year. A shortage in the payment of the
last year's *ferm* is a contingency taken for granted in the *Dialogus*
itself.

In his oath, taken before he was acquitted, the sheriff was
required to swear that he had diligently inquired concerning uncollected debts due the king, and had found no chattels upon which
to levy.[204] Concerning the process of distraint, there are mentioned
some rigid exchequer rules. The sheriff's agents first took moveable
goods, the plough oxen being spared as much as possible. When

---

[197] *Dialogus*, 142-3.   [198] *Ibid*. 114.

[199] *Vic. debet* £4 : 4s. *de veteri firma de Oreford qui remanserunt pro defalta
instauramenti perditi tempore werre* (*P.R.* 24 Henry II., 20). This damage was
actually done by Roger Bigot, who made it good.

[200] Alexander, sheriff of Staffordshire, was mulcted 100s. *qui capti fuerunt de
firma sua pro mora compoti* (*P.R.* 7 Henry II., 42).

[201] For details see *Dialogus*, 28-32 ; Poole, *Exchequer*, 75-8, 173.

[202] *Dialogus*, 63, 91 ; in Rutland also (Parow, *Compotus*, p. 26).

[203] *Dialogus*, 73, 159.

[204] *Ibid*. 143-4. Of a former sheriff who owed some £25 it is said, *mortuus est
et de suo nihil invenerunt in hoc comitatu* (*P.R.* 21 Henry II., 42). Adam de
Cattemere owed £22 : 12d. old *ferm* of Oxfordshire, *sed nihil habet in hoc comitatu*
(*P.R.* 20 Henry II., 78).

the debtor's goods failed, the chattels of his bondsmen were taken in the same order. The sheriff also had to ascertain whether any debt were due a king's debtor whom he declared to have no chattels, to be sure that the latter had paid nothing due the sheriff so long as he failed to pay the king's debts, and also to inquire whether by marriage or otherwise the person in question had again become solvent.[205]   The sheriff's power of collection extended both to the moveables of the debtor and to the issues of the land which he held.[206]   The legal test of the standing of a debtor of the king was the opinion of the sheriff and of the community.[207]   A tenant-in-chief holding ·a barony escaped dealings with the sheriff by tendering before him in the county court either his own oath or that of his seneschal, undertaking·to give security for his obligations to the barons of the exchequer.   Only if he incurred judgement of the exchequer in default of this engagement did the sheriff act.[208] His servants then went through the debtor's estates selling his chattels.   Similar usage certainly prevailed in many another case when the king's debtor failed to pay at the exchequer.   The sheriff was evidently expected in some instances to see that suit was brought to collect what was due.   The author of the *Dialogus* holds that he was bound to prevent the king from being defrauded through lack of legal measures by the person who had promised him an oblate.[209]   According to the same writer, the new sheriff received from his predecessor a rescript of the debts of the king in his county.[210]   If a sheriff were summoned concerning the debt of a person without chattels in the county, ·it was expected from the time of Henry I. that he notify a sheriff or bailiff of a county wherein the debtor did have such property, when possible presenting the notice in the county court or, at any rate, *coram pluribus*.[211]

The period of King Richard's crusade and of his absence from England until 1194 witnessed partisan struggles for control, which tended to involve the sheriffs of the time, and which led in turn to some new measures to reduce their independence and power.   The relations of Richard to his late father were such that his accession

---

[205] *Dialogus*, 148-50.
[207] *Ibid*. 153.
[209] *Ibid*. 155.   The queen's gold, an additional 1 per cent upon oblates promised the king, was collected by the sheriff according to rules followed in case of the king's debtor (*ibid*. 157).
[210] *Ibid*. 142.

[206] *Ibid*. 144, 148.
[208] *Ibid*. 151, 153.

[211] *Ibid*. 114.

at all events probably would have meant a change of officials. But, as it was, his desire to take prominent men to enhance the importance of his crusade, and his willingness to let offices for money to finance the enterprise, brought in a set of sheriffs who were nearly all new.[212] Of the few sheriffs of Henry II. who were retained [213] half were transferred to other counties. The slate was all but swept clean. The sheriffs were in general politically insignificant, and the way was prepared for the rule of the king's chancellor, William Longchamp.

Longchamp's supremacy shows a full recognition of the political as well as the administrative importance of the shrievalty. William of Newburg stigmatizes the sheriffs whom the chancellor set over the counties as most wicked instruments of his own avarice, sparing neither clergy nor laity,[214] and retaining in their service bands of armed men who perpetrated enormities. His attempts to dominate some of the shrievalties have an important part in the history of the period. In 1190 he deposed John Marshal, a friend of the king and sheriff of Yorkshire, because of his part in precipitating a massacre of the Jews.[215] Marshal had become infuriated with the Jews of York, who through their fear of violence refused to give up the tower of the castle assigned them temporarily as an asylum. He and the custodian of this castle then called upon the knights of the county and the men of the city to retake the tower, and a scene of carnage ensued. Longchamp now conferred this county on his own brother Osbert, who already held another northern county.[216] His onslaughts on two sheriffs, adherents of Earl John, the king's brother, had much to do with his downfall. Against one of these, Hugh de Nunant, bishop of Coventry, who had obtained two counties by fining for them, he proceeded in the ecclesiastical court, obtaining from the archbishop of Canterbury an injunction against Hugh's holding the office of sheriff,[217] and actually forcing him for a time

---

[212] Hoveden, iii., introd. pp. xxviii–xxix.

[213] The fiscal year 1190–91 found only five of these among twenty-five sheriffs, most of them persons of importance. These were Hugh Bardolf, William fitz Henry, William fitz Audelin, William Rufus and Henry of Cornhill (*P.R.O. Lists and Indexes*, no. 9).

[214] *Chronicles of Stephen*, i. 334.

[215] *Gesta Regis*, R.S. ii. 108.

[216] Westmoreland. Another brother, Henry, held Herefordshire at this time.

[217] Hoveden, iii., introd. p. xxxi. Hugh held Warwickshire and Leicestershire from about Easter 1190. He seems to have given up these shires while the archbishop lived, but acquired Staffordshire, and in the fiscal year 1191–2, after Longchamp's downfall, accounted for all three.

to give up the position. Longchamp's struggle against Gerald de Camville, sheriff of Lincolnshire,[218] is better known.

The justiciarship of Walter of Coutances, from the autumn of 1191 to Christmas 1193, differed radically from that of his predecessor in its attitude toward powerful sheriffs. The four strong men who were appointed Walter's coadjutors, William Marshal, Geoffrey fitz Peter, Hugh Bardolf and William Brewer,[219] were already sheriffs. The two sheriffs deposed by Longchamp returned to office, and de Camville was retained. Oger fitz Oger, *dapifer* of Henry II., regained his shrievalty, temporarily lost, as did also William Earl of Salisbury, the king's half brother. These with two or three prominent officials of Henry II.,[220] and an equal number of local feudal nobles,[221] practically completed the list of sheriffs of the period. Among them are to be found the men who ruled England under Walter of Coutances. Some held high judicial office and went on circuit. The control of the counties by so small a group was made possible by the fact that the king's bestowal upon Earl John of seven counties, held as palatinates,[222] considerably reduced the number of sheriffs. To these, furthermore, must be added an eighth county, Northumberland, which the bishop of Durham was at this time holding in much the same way.[223]

With King Richard's return to England, in the spring of 1194, begins a new and final period in the history of the shrievalty of the reign. The king's dissatisfaction with affairs in England called for some sweeping changes. The boldness of the group who had deposed Longchamp was to be rebuked. That revolution had in a sense marked a victory for the king's brother, John, who was now for disloyalty to suffer the forfeiture of his numerous estates, including

[218] He obtained by fine the shrievalty of Lincolnshire, uniting it with the castellanship of Lincoln, which his wife, Nicholaa de Hai, had inherited (Stubbs, introd. to Hoveden, iii. p. lvi). Longchamp's struggle with de Camville is related by Newburg in *Chronicles of Stephen*, ii. 337-8.

[219] Stubbs, *Const. Hist.* i. 538.

[220] William Rufus and William fitz Audelin had been sheriffs continuously since Richard's accession.

[221] William fitz Alan (Shropshire), William de Braiose (Herefordshire) and William de Beauchamp, who since Richard's accession had obtained the ancestral office in Worcestershire.

[222] These counties, according to the Pipe Rolls, did not account at the exchequer for the years from 1189 to 1193. They were Cornwall, Devon, Dorset, Somerset, Nottingham, Derby and Lancaster. According to *Benedict of Peterborough* (ii. 99), the four in the south-west were conferred in Dec. 1189.

[223] This county also disappears from the Pipe Rolls from 1190 until the second half of the year 1193-4.

the whole counties he held.   His special adherent, de Camville, was at once removed from the shrievalty of Lincolnshire, and Hugh de Nunant relieved of office and placed on trial.   As a means of providing much needed funds, counties were again let to the highest bidders, York going to Archbishop Geoffrey for the tremendous sum of three thousand marks in addition to a hundred marks of increment.[224]   At Easter there was a change of sheriffs almost as general as that of 1189.[225]   A few of the ruling group remained in the office, but these were practically all transferred to other counties,[226] thus being assigned far less important posts.   William and Osbert Longchamp reappear for a time on the list, but the great majority of sheriffs for the remaining five years of the reign were obscure and insignificant persons.   Because of his hesitancy in carrying out a precept of the king, Bishop Hugh de Puiset was in July 1194 disseised of the county of Northumberland and the castles of Newcastle-on-Tyne and Bamburg, for which he had agreed to pay two thousand marks.[227]   It was obviously a part of the new policy to restrain sheriffs within proper bounds.

Various measures adopted by the new justiciar, Hubert Walter, which reduced the sheriff's independence, may be readily explained in the light of the circumstances just set forth.   In this relation three provisions of the form of eyre, issued in Sept. 1194,[228] are of especial interest.   The best known of these declared that no sheriff was to be justice in his own county, nor in any county which he had held since 1189.[229]   This seems to show dissatisfaction with the conduct of sheriffs since the beginning of the reign, especially the powerful group lately in office, and indicates a determination to make every sheriff subject to a higher judicial authority which

---

[224] Stubbs, *Const. Hist.* i. 541.

[225] Only eight sheriffs were retained in the counties they held at Richard's return.   These were William fitz Audelin (Cumberland), Reginald de Argentine (Cambridge and Huntingdon), Reginald of Cornhill (Kent), William fitz Alan (Shropshire), Hugh Bardolf (Westmoreland), William de Braioso (Hereford), William Marshal (Sussex) and William Earl of Salisbury (Wiltshire).   It is possible, however, that some of the sheriffs of the counties now taken over from John had been his agents in administering these.   It is obvious that in Devon (*P.R.* no. 40, P.R.O., 6 Richard I.—Devon) the accounts of debts which were due John are now being collected by the sheriff.

[226] William Brewer was given the counties of Nottingham and Derby instead of Oxford and Berks, Hugh Bardolf continued to hold Westmoreland, but received Northumberland instead of Yorkshire.   The county which William Marshal continued to hold was Sussex.

[227] Hoveden, iii. 260-61.

[228] *Ibid.* iii. 262-7 ; Stubbs, *Sel. Charters*, 259-63.   [229] Sect. 21.

may investigate his acts. The accompanying regulation, prescribing a process for electing local juries to make presentments concerning pleas of the crown,[230] has been interpreted with every degree of
reason [231] as aimed to take from the sheriff the power of nominating
these jurors. A third section of the same document, requiring the
election in each county of four coroners to act as keepers of the
pleas of the crown,[232] established a regular procedure for preserving
a record of many matters touching the preservation of the peace and
the king's fiscal interests by which the sheriff was bound before the
king's justices. Still another limitation upon the powers of this
official is to be noted in the regulations of 1195 for the preservation
of the peace. The men of the realm in general, assigned knights in
particular, are to arrest malefactors and deliver them to the sheriff
for safe keeping, and they may not then be liberated except through
the precept of the king or his capital justice.[233]

Usage of the period 1194–99 in fiscal matters imposes some
further limitations upon the sheriff. The carucage of 1195 was
levied under special regulations, which in some respects anticipate
methods of the next century. A clerk and a knight acted with the
sheriff of the county in fixing the amount of the levy, each retaining
a copy of the assessment roll according to which the sheriff responded
at the exchequer. The seneschal of each baron also had a copy of
that portion of the roll which pertained to the lands of this baron,
so there could be no possibility of over-exaction. The money was
actually received and turned over to the sheriff, not by his bailiffs,
but by two knights of each hundred assisted by the bailiff of the
hundred.[234] King Richard at times suspected that his difficulty
in raising funds was due to the dishonesty of sheriffs. This feeling
possibly dated from 1193, when the collection of the various taxes
to meet the king's ransom was in the hands of a body of men strongly
intrenched in control. In 1194 a general inquest into all collections and exactions (*de prisis et tenseriis*), made since the king's first
coronation by all bailiffs of the king, including justices and sheriffs,
was ordered, though subsequently suspended.[235] The project, which

---

[230] The opening section.
[231] Stubbs, introd. to Hoveden, iv. pp. xcviii-xcix. The Constitutions of
Clarendon (sect. 6) show that the sheriff was the official who empanelled at least
one kind of jury.
[232] Sect. 20.
[233] Hoveden, iii. 299 ; Stubbs, *Sel. Charters*, 264.
[234] Hoveden, iv. 46 ; Mitchell, 7-8.
[235] The form of eyre, 1194, sect. 25.

would have amounted to a repetition of the Inquest of Sheriffs on a larger scale, was revived in 1196, when the king sent two confidential agents from Normandy at Easter to make the inquest *de prisis.*[236] According to William of Newburg, the sheriffs were actually summoned to appear at London on a fixed day to explain their methods,[237] but the death of the abbot of Caen, who was in charge, seems to have terminated proceedings. It is noteworthy that the exchequer this year employed extraordinary measures to facilitate a view of the sheriffs' current obligations.[238] Their accounts had been in poor condition in 1193, but the Pipe Rolls of the closing years of the reign make a remarkable showing for them.[239] It is extremely probable that the exchequer reforms which are in evidence in John's first year [240] had their origin in these later years of his brother's reign, when Archbishop Hubert was attempting to vindicate his fiscal administration. With the year 1197 begins the practice of placing at the head of the Pipe Roll the total amount of the sheriffs' *ferm* of the shire for the year.

It is obvious that during these five years substantial progress was made toward curbing the power of the great sheriff. But the prime object as yet was protection of the royal rather than of popular interests. In one particular it is clear that the new regulations were not followed to the letter. In the reign of Richard [241] as well as in that of John [242] a person employed at court was some-

---

[236] Hoveden, iv. 5 ; Stubbs, *Const. Hist.* i. 546.

[237] *Chronicles of Stephen,* ii. 464-5.

[238] The Pipe Roll of the preceding year (7 Richard I., *P.R.* no. 41, P.R.O.) shows the formation of a separate escheat list to simplify the statement of current debts. The Roll of 1196 (*P.R.* no. 42, P.R.O.) is in a new script, as if a new system were being installed. Sewed in with its first membrane is a schedule of county and borough *ferms* paid in, an evidence of careful supervision which appears twice in the rolls of John's reign.

[239] The schedule mentioned in the above note shows no less than twelve sheriffs marked *quietus,* an unusually good showing. In the Pipe Roll for the year arrears of *ferm* are cleared up and nearly all sheriffs' accounts under this head are acquitted in full, something which rarely happened.

[240] Hilary Jenkinson, in *Magna Charta Commemoration Essays,* 259, 273.

[241] According to the Pipe Rolls of 1195, 1197 and 1198 (*P.R.* nos. 41, 43, 44, P.R.O.), William Brewer, sheriff of Nottingham and Derby, is among the justices who have held pleas in these counties. In the Roll of 1197 as well as in that of 1198, Hugh Bardolf, who is sheriff of Northumberland, appears as one of the justices who hold pleas of the forest in that county. Sheriffs were also among those who tallaged their own counties (William fitz Aldelin did so in 1197, *P.R.* no. 43, P.R.O. —Cumberland), and Stephen de Turnham in 1198 (*ibid.* no. 44—Berkshire). But this has probably come to be regarded as the work of an assessor of taxes rather than a justice.

[242] Engelard de Cigogné was one of John's justices in Gloucestershire while

times sent on judicial circuit, even in a county which he held as sheriff. It was unavoidable that a justiciar of the realm like Geoffrey fitz Peter should hold pleas anywhere. The possibility of rewarding a useful official with a shrievalty was more important to the king than that of his making local exactions which might go undetected.

---

sheriff of that county (see p. 160, below). The Pipe Roll of 1200 (*P.R.* no. 46, P.R.O.) records that Hugh Bardolf was one of the justices who levied amercements in Nottingham and Derby. He was sheriff of these counties, however, only the second half of the preceding year. Exactly the same statement is made concerning William Brewer, one of the judges who is recorded the same year as among the justices in the county of Southampton. The Rolls for both 1200 and 1203, however, show that the justiciar, Geoffrey fitz Peter, held pleas in Yorkshire, of which county he was sheriff both years (although accounting at the exchequer by an under-sheriff).

VI

THE OFFICE AT THE CULMINATION OF ANGEVIN ABSOLUTISM, 1199–1216

THE sheriff of the Angevin period of English history was still the personal agent as well as the official of the king. Of the many duties, fiscal, judicial, police, military, miscellaneous, which fell to his predecessor in the days of Norman rule, he had lost only those assumed by itinerant justices.[1] The chancery records which show King John's acts in detail, especially the Close Rolls, create the impression that at times the greater part of executive action was initiated through letters to various sheriffs.[2] The process of municipal enfranchisement and the prevalence of private franchises had, to be sure, somewhat restricted the territorial limits of their power. Their work had become much more ministerial in the course of a century, and in matters of detail they were more directly responsible to the king and his agents. Up to 1215 the coroner was not so effective a foil as he became later. In John's time baronial sheriffs were not unknown, but it is clear that these, like all the rest, owed their places to royal favour rather than feudal rank or local power. As an instrument of centralized absolutism the shrievalty attained its maximum importance in this reign.

The mode of designating the incumbent of the office well illustrates the king's control over it. Notwithstanding the repetition in law tracts of the legend that sheriffs were elected in every county

---

[1] The escheator mentioned in *Benedict of Peterborough*, ii. 91, functioned but a short time. The custodians of escheats of John's time (*Rot. Lit. Claus.* i. 127, 128, 133, 146, 157, 160, 183, 195; *Rot. Lit. Pat.* i. 103) were apparently only the usual temporary custodians. In 1212 a sheriff acts by their view and counsel (*Rot. Lit. Claus.* i. 125).

[2] Sheriffs of Pembroke are mentioned in this reign (Giraldus Cambrensis, *Opera*, R.S. iii. 214, 217), and sheriffs in at least three regions of Ireland (Davis, *England under Normans*, 362).

L

in full shiremote,[3] they were chosen directly by the crown except in the palatinates, in London, where annual election of the two sheriffs already occurred,[4] and for a time under a peculiar arrangement in Cornwall. The form, by which the appointment of a sheriff was announced to the men of the county and they were told to heed and obey him in that capacity, seems to be a new one when it appears on the Charter Roll in the year 1200. Neither the wording of the address nor of the body agrees in the two earliest extant documents of this type.[5] With the rise of the Patent Roll a little later it became the depository of this record, which was substantially the same as that familiar a few decades later on the pages of the Memoranda and Fine Rolls. A Liberate writ of Oct. 26, 1200,[6] illustrates the earliest known example of another well-known form. It ordered William de Braiose to deliver to his successor, Geoffrey fitz Peter, the county of Hereford with his writs and rolls, and also the castle of Hereford.

The term of office was regularly said to be during the king's pleasure. Appointees of the first quarter of the year assumed fiscal accountability from the preceding Michaelmas.[7] In partitioning responsibility between an old and a new sheriff the exchequer still recognized no term less than a quarter. The duties of the shrievalty, especially when held by a prominent person, were often discharged by an under-sheriff, who, in rare instances, was also named by the king.[8] In designating a sheriff it was not unusual to specify that he was to render an increment above the usual *ferm* of the shire, which thus increased the king's revenue. On the other hand, a payment made by a sheriff to be relieved of office still occurs.[9] Beginning with the year 1204, many a sheriff was appointed as *custos*, and thus accounted for all the revenues of the county collected, which in this way fell more directly under the king's control than when they were farmed.

[3] *Leges Angliae London. saec. xiii. collectae*, 32 B 1, Liebermann, *Gesetze*, i. 656. It is said (32 B 8) that the election is to be held on the calends of October.
[4] *Liber Albus*, ed. Riley, ii. 250-51.
[5] *Rotuli Chartarum*, Record Com. i. 61, 100.
[6] *Liberate, Misae, Praestita Rolls*, Record Com., 7 ; cf. *Rot. Chart.* i. 100 (Jan. 1201).
[7] The appointment of Thomas de Multon, Dec. 22, 1205, is effective from the preceding Michaelmas, and his predecessor delivers to him the rolls of the first quarter of the year (*Rotuli Selecti*, Record Com., p. 7).
[8] In committing Sussex to William Brewer in 1208 (*Rot. Lit. Pat.* 8, 84), the king specified that Robert Camerarius was to be *subvicecomes*. See note 122.
[9] By Jocelin de Stiveclea, *P.R.* 53, Cambridge and Huntingdon (9 John).

Perhaps no feature of John's administrative system so clearly shows its peculiarities and defects as the conditions attached to some of these appointments. For a good and sufficient monetary consideration the king twice designated a county *custos* for a term of seven years,[10] one of these on the condition that he be not disseised within that term except by judgement of the king's court.[11] In 1204 he gave the custody of the counties of Essex and Hertford to Mathew Mantell and his heirs *in perpetuo* under the terms by which the preceding sheriff held them.[12] Mantell must have found the bargain a bad one, for he held the office but three and a half years. The men of a county also negotiated with the king in regard to the appointment. The county of Lancaster engaged to pay a hundred marks to have Richard de Vernon as sheriff.[13] The men of Cornwall entered into a more remarkable engagement. They agreed to give a large sum for various concessions, among them the appointment to this office of persons from among their own number.[14] They were to choose names from among the better men of the county for presentation to the monarch, who was to select a sheriff from this list. If no suitable person were found in the county, it was to have a sheriff who would treat its men well and not hate them. The abuses here disclosed are clearly due to the misgovernment of one of John's foreign followers, Richard the Fleming,[15] who was thus removed from office. The men of Dorset and Somerset in concluding, apparently at the end of the year 1210, a similar bargain for reform and a resident sheriff specified that he was to be neither William Brewer, their oppressor at that time, nor any of his aids.[16]

[10] Thomas of Multon in 1205 received the county of Lincoln for such a term, promising a bonus of five hundred marks and five palfreys, and also the ancient *ferm* plus three hundred marks increment (Madox, *Exchequer*, i. 461, n. (g); *Rot. Lit. Pat.* i. 57). In 1207 Fulk fitz Theobald received the counties of Cambridge and Huntingdon for seven years from Easter of John's eighth year (*ibid.* i. 70) at the ancient *ferm* plus a hundred and twenty marks and three palfreys. For the privilege he also gave a hundred and twenty marks and two palfrey (*P.R.* 53, Cambridge and Hunts; Madox, *Exchequer*, i. 462, n. (i)). He actually served as sheriff five and a half years. [11] *Rot. Selecti*, 7.

[12] The grant included the king's manors and the pleas of the Jews. The amount to be paid annually was £640 : 2 : 4 *alb.* and 50 marks increment (*Rot. Chart.* i. 125).

[13] Sheriff 1200–05. In 1207 £52 were still due on this account (*P.R.* 53, Lancaster; Madox, *Exchequer*, i. 461, n. (f)).

[14] *Ibid.* i. 405, n. (t); cf. 410, n. (l).

[15] Sheriff of this county in 1201 and until Easter 1202. Concerning John's further favour to him see *Rot. Chart.* i. 165. It is doubtful if the change improved conditions. The sheriff of the next year and a half was William Brewer.

[16] Madox, *Exchequer*, i. 508, n. (k). Cf. *P.R.O. Lists and Indexes*, no. 9; or Pike's List of sheriffs in *Deputy Keeper's 31st Report*; *Lit. Claus.* ii. 169.

Definite information concerning the sheriff's headship over the reeves of the county, an arrangement which must have subsisted for ages, is now appearing in its earliest form. It is stated that reeves of hundreds and wapentakes have their dignity *sub vicecomitibus regis per universum regnum*.[17] Already in Richard's reign the bailiffs of Norwich had been called to answer why they had not executed a mandate of the sheriff in conformity with a precept of the justices.[18] The bailiffs of Portsmouth presented a letter (*breve*) of the sheriff as sufficient authority for taking pontage and other customs of the water at Southampton.[19] The sheriff was the intermediary between the courts and the holders of franchises, as at a later time passing on the writ of the justices to be executed by seignorial bailiffs.[20] The sheriff of Herefordshire declared that he dared not lay hand within the liberty of William de Braiose even in response to the king's order.[21] To carry out the directions of the justiciar of the realm for the removal of a market, wrongfully established by the Abbot of Ely within the liberty of the Abbot of St. Edmunds, whither the sheriff might not enter, the latter sent notice *per breve suum*, and later the hundred reeve, coming, privately as it seems, on the market day, issued formal prohibition, exhibiting the letters both of the king and the sheriff.[22] The official last named had bailiffs who had much to do with arrests [23] and attachments. The under-sheriff of Rutland testified that he seized a certain thief after the bailiffs of the franchise refused to act.[24] The clerk of the sheriff of York declared that he was unable to attach the defendants in a suit because they were clerks.[25] The sheriff sometimes sent a bailiff (*serviens*) to represent him at a private view of frankpledge.[26]

The function of the sheriff's office in serving and executing a great bulk of judicial writs, so abundantly attested by Glanville in the days of John's father, now finds mention in numerous acts enjoined upon the sheriff in furtherance of judicial process. Royal

[17] *Leges Angliae Lond. saec. xiii. collectae*, 32 A 1, in Liebermann, *Gesetze*, i. 655.
[18] *Rot. Curiae Regis*, ed. Palgrave, i. 413.
[19] *Abbreviatio Placitorum*, Record Com. p. 94.
[20] *Rot. Curiae Regis*, ii. 86 ; i. 426. The sheriff's sealed letters to the bailiffs of Wye are mentioned in the reign of Henry II. (*Chron. of Battle Abbey*, trans. Tower, 1851, p. 33). Cf. John's grant to the archdeacon of Wells (*Rot. Chart.* i. 129).
[21] *Rot. Curiae Regis*, ed. Palgrave, i. 426.
[22] *Chron. Jocelini de Brakelonda*, Camden Soc. 98.
[23] *Select Pleas*, i. nos. 106, 111.
[24] *Rot. Curiae Regis*, ii. 160.
[25] *Ibid.* ii. 32-3.   [26] *Rot. Chart.* i. 134.

letters occur ordering him to enforce judicial decree,[27] to make special inquest,[28] to bring the assizes of novel disseisin of the county before the justices of jail delivery.[29] Court rolls relate that he is ordered to attach certain persons,[30] to seize land or chattels,[31] to record by law-worthy men proceedings formerly held in the county court,[32] to make inquests [33] or partition of lands,[34] to answer why writs or lists of pledges had not been sent to Westminster as directed.[35] Eyre rolls already disclose a responsibility of the sheriff or undersheriff to account for criminals' chattels entrusted to their care.[36] At the Lincolnshire eyre of 1202 the court proceeded to judgement against the sheriff for not imprisoning persons accused of homicide and for not producing a writ.[37]

Besides the collection and payment at the exchequer of the annual county *ferm* with or without increment, a similar obligation for *proficua* or perquisites, probably based largely on sheriffs' aid, was now being laid upon the head of the shire.[38] Charged in only a few counties when it began to appear on the Pipe Roll of 1205, this soon tended to become general.[39] The exchequer was by 1208 keeping the record of the view of the sheriff's account which is so familiar later in the Memoranda Rolls, and it is clear that he was expected at Easter to pay a definite sum toward his *ferm*. A new rule of this reign required him to exercise due discretion in accepting a lord's steward as surety for the prospective obligations of the latter.[40]

[27] As in *Lit. Claus.* i. 150.   [28] *Ibid.* i. 69.   [29] *Ibid.* i. 83.

[30] *Rot. Curiae Regis*, ii. 113 ; *Salt Arch. Soc.* iii. 135, 142 ; *Placitorum Abbreviatio*, 25. The sheriff attaches (*ibid.* 46) a defendant in a suit to appear *coram rege*.

[31] *Ibid.* 99. The land and chattels of one who has married an heiress in wardship are seized that satisfaction be made to the guardian.

[32] *Rot. Curiae Regis*, i. 390 ; *Select Pleas*, i. no. 115. In the latter case the record is ordered by the bench.

[33] As in *Plac. Abbrev.* i. 58. In the eighth year of the reign a sheriff sends certification to Westminster that a certain accusation (*ibid.* 54) is *athia et per odium*.

[34] *Ibid.* 61.

[35] *Rot. Curiae Regis*, i. 261, 345, 372, 380.

[36] *Select Pleas*, i. nos. 60, 70.

[37] *Ibid.* no. 30 ; cf. *Bedford Eyre of 1202*, ed. G. Herbert Fowler, no. 234.

[38] Mitchell, 15 ; cf. *Rot. Claus.* i. 48 ; also p. 246 below.

[39] In *Pipe Roll* no. 51, 7 John, it is being entered as a part of the account for the year, but the amounts are not stated. These begin to appear three years later. In *Chancellor's Roll* no. 20, 10 John, the amount for Gloucestershire is 55 marks, for Worcestershire, 105s.

[40] The *visus compoti vicecomitum totius Angliae* upon a schedule attached at the conclusion of *P.R.* no. 54, 10 John, is for the Easter term. Most sheriffs make

Especially important in constitutional history was the part the sheriff played in regard to special impositions. He was still mentioned as the collector of tallage assessed by the justices.[41] The fortieth of moveables for the Holy Land was in 1201 collected by his appointed agents,[42] and he, like his successors of six or eight decades later, who were summarily directed to deliver the proceeds from special sources at a special time, was ordered with his collectors to have the money at the New Temple, London, fifteen days after Hilary.[43] The collection of the noted thirteenth of moveables levied in 1207, the first of a long series of national taxes of the kind, adhered to the general plan followed with reference to the carucage of 1195, the justices who made the assessment delivering a copy of their roll of particulars to the sheriff.[44] A special collector is also mentioned, as on the former occasion.[45] The scattered references appear to indicate that the sheriff still bore the primary responsibility in respect to various levies, although as early as 1217 special collectors of the familiar thirteenth-century sort had general charge of the assessment and collection of an aid.[46] Enforcement of collection from delinquents through the sheriff's office appears to be usual now just as it was later.

Distraint to collect the king's debts or debts which fell to him was indeed a duty frequently enjoined. The incipient Memoranda Roll of the first year of the reign records the familiar exchequer precept to the sheriff to distrain.[47] Special mandates to sheriffs exemplify the forceful language of royal writs in commanding these officials " as you love yourself and all yours have at the exchequer " on a specified day certain debts.[48] The sheriff still

---

a payment, and the account is marked *quietus*. The sheriff of Dorset and Somerset pays nothing and owes over £99. The sheriff of Warwick and Leicester pays £15 : 9 : 4 *blanch* toward his *ferm* and owes £30 : 19 : 0 *blanch*. The new order to the sheriff regarding the sufficiency of the lord's steward is recorded by Hoveden, iv. 152, and Walter of Coventry, ii. 179.

[41] Mitchell, 31-2, 61, 76. The *dona* of religious houses in 1203 were collected in part by the sheriff, some houses responding at the exchequer by their own representatives (*ibid.* 61).

[42] *Ibid.* 45-6.

[43] Hoveden, iv. 189.

[44] Stubbs, *Sel. Charters*, 284 ; Mitchell, 89-90.

[45] *Ibid.* 90, n. 27.

[46] *Ibid.* 122-3.

[47] Exchequer L.T.R. *Misc. Rolls*, 1, no. 3, m. 17 b (P.R.O.). Executive orders for distraint occur, as in *Lit. Pat.* i. 50, 51.

[48] *Ibid.* i. 70. This form in exchequer writs is of course much older (*Dialogus*, ii. i B).

took charge of lands in the king's name,[49] and at the great crisis of the reign was usually ordered to take seisin of the lands which were forfeited to the crown by the rebellion of the holder.[50]

His ancient duties as purchasing and disbursing agent of the crown found frequent and interesting illustration. In 1212 the king directed Henry de Braibroc to aid and counsel two special royal agents in buying at the market of Northampton.[51] The sheriff, moreover, was the usual fiscal agent upon whom the king drew all sorts of requisitions for money, supplies,[52] transport,[53] repairs upon his buildings or galleys,[54] the sustenance of his officials and men who were sent to the county,[55] and of his hostages,[56] as well as his grooms, horses and falcons.[57] Even table linen and delicacies for the king's Christmas feast [58] and wine, as well as liveries for his servants,[59] were sometimes provided in the same way.

The magisterial functions of the shrievalty were now prominent. In addition to the king's peace there was a sheriff's peace,[60] which was enforced in the county court. The sheriff sent the force necessary to maintain order and apprehend lawbreakers.[61] Forcible resistance to him is mentioned more than once [62] even in peaceful

[49] In Trinity term 1213 a sheriff is ordered to take *in manu domini regis* a half knight's fee held by a certain Avice who has been condemned to be burned for the death of her husband (*Placitorum Abbreviatio*, 85).

[50] As in the well-known writ of May 12, 1214 (*Lit. Claus.* i. 204), and another of April 6, 1216 (*ibid.* i. 259), directed to the sheriff of Lincoln.

[51] *Ibid.* i. 127.

[52] Thus three sacks of flour (*ibid.* i. 136), hay (*ibid.* i. 79), almonds, three frails of figs and thirty pounds of rice to go aboard the king's ship (*ibid.* i. 162).

[53] *E.g.* of the king's venison (*ibid.* i. 15), of fat hogs (i. 136), passage for the Chancellor to Flanders (i. 156), the carriage of prisoners (i. 74), of the utensils and equipment of the king's *balistarii* (i. 4, 102), of the equipment of the sumpter horse of the king's chapel (i. 65, 102). In 1207 a sheriff (*ibid.* i. 89) provides the ship and makes the arrangements for sending the king's treasure into Poitou. In 1215 (*ibid.* i. 190) another is directed to send on some eels that have been delivered to him.

[54] *Ibid.* i. 4, 61, 103.

[55] Savary de Mauleon and his men (*ibid.* i. 185) ; cf. i. 54, 79, 99, 152.

[56] *Ibid.* i. 51.

[57] *Ibid.* i. 66, 118, 138, 150, 156, 190, 240.

[58] *Ibid.* i. 127, 157. *Linea tela ad mappas faciendas* just before Easter (i. 66) and Christmas (i. 75).

[59] *Ibid.* i. 3, 51, 61, 71. Some dozens of dress ornaments are thus ordered (i. 167).

[60] *Select Pleas*, i. nos. 21, 31, 73 ; *Pleas of Crown for County of Gloucester*, ed. Maitland, 6, 25.

[61] *Select Pleas*, i. no. 45.

[62] A mêlée between certain persons and the sheriff of Rutland (*Lit. Pat.* i. 87) ; an accusation that a Neville has driven the sheriff of Lancaster from his county by force and arms (*P.R.* 5 John, under *nova oblata*).

times. One statement emanating from London already makes him the chief official responsible for keeping watch and ward at night.[63] It was the rule that no sheriff might hold pleas of the crown in his own county.[64] But he convened two different kinds of sessions, those of the county courts [65] and the great hundred courts or tourns,[66] the latter assembling twice a year, in which he received before the coroners presentments and appeals of such pleas for transmission to the justices. The more frequent meetings of the county attracted such business. At the Shropshire eyre of 1203 it was shown that the sheriff as well as the coroners kept a roll to record it.[67] This official in Staffordshire made use of a thief as an approver to accuse men of crime, and against one of these the approver fought a duel in the county court *coram vicecomite*.[68] The king's amnesty to prisoners, proclaimed in 1205 in memory of his mother, required the sheriff to bring them into full county court, where they were to be pledged to be loyal or else abjure the realm.[69] The outlawry of Robert fitz Walter in the county court of Essex in 1212, in response to the king's order to the sheriff after he had been demanded in but four sessions,[70] seems to have a bearing upon the history of one of the great guarantees of the thirty-ninth section of Magna Charta. Royal letters in 1207 ordered the sheriffs to seize the chattels of Baldwin Wake and John de Humet and banish these persons from the realm within three weeks.[71] In more usual course were royal precepts in special cases commanding the arrest of the king's enemies, the seizure of offenders against the peace,[72] or the abjuring of the realm by the latter.

The custody of the criminals of the county, so it has been shown earlier, rested with the sheriff a full generation before John's reign.[73] Now as well as in the earlier period he conducted prisoners before the justices,[74] and even to Westminster.[75] He admitted prisoners to mainprise except in cases of homicide and probably some other

---

[63] Liebermann, *Gesetze*, i. 490, sect. 6.
[64] Stubbs, *Sel. Charters*, 260.
[65] *Select Pleas*, i. nos. 3, 6, 35.
[66] *Ibid.* i. pp. 68-9 (A.D. 1214) ; *Rot. Chart.* i. 132 (1201).
[67] *Select Pleas*, i. no. 62.
[68] *Salt Arch. Soc.* iii. 97.
[69] *Lit. Pat.* i. 54. Similar procedure in a special case, *Lit. Pat.* i. 145.
[70] *Lit. Claus.* i. 165-6. Coggeshall (R.S. p. 165) says he was demanded in only three sessions.
[71] *Lit. Pat.* i. 69.   [72] *Lit. Claus.* i. 125.
[73] Assize of Clarendon, sects. 6, 7.   [74] *Lit. Claus.* i. 83.
[75] *Select Pleas*, i. no. 115 : from Cornwall by order of *Coram rege*.

grave offences.[76] His general duty was sometimes reinforced by special writs ordering him to receive from another sheriff offenders against the peace and guard them,[77] or to send prisoners to another county.[78] In the same manner he was sometimes ordered to release the king's prisoners or hostages.[79] Castellans also guarded these in their castles and received similar instructions,[80] and mayors and reeves of boroughs had charge of prisoners.[81] The sheriff's actual control of a jail was presumably determined by his possession of a royal castle. Already usage seems to determine whether the sheriff receives his county with the castle. In the reign of Richard the holding of a castle with a shrievalty was regarded as important and worth recording. Apparently the line between the counties wherein this did and those wherein it did not occur was already becoming clear.[82]

The sheriff in the capacity of custodian of the castle became an official of military importance. This could hardly have been regarded as his chief work, although in emergencies of this reign it sometimes became so.[83] His predecessor delivered the stronghold to him with implements, stores and arms.[84] In times of stress the king often sent a special agent to supervise his work in strengthening it.[85] But quite aside from this, his services were indispensable to the military system of the day. On numerous occasions John's writs ordered him to summon the military tenants of the crown to appear at a certain time and place equipped for the king's service— in 1205 ten knights from each shire,[86] in 1213 the earls, barons, knights, free men and sergeants owing military service,[87] in 1212 the

[76] McKechnie, *Magna Charta*, 2nd ed. 363. The *Bedford Eyre of 1202*, ed. G. H. Fowler, no. 234, shows the sheriff relying on a special warrant of the justiciar for releasing persons who regularly should be in prison.

[77] *Lit. Claus.* i. 89.   [78] *Lit. Pat.* i. 71.

[79] *Ibid.* i. 106, 125; *Lit. Claus.* i. 178.

[80] *Lit. Pat.* i. 96, 143. Sheriffs and castellans by order transfer prisoners back and forth (*Lit. Claus.* i. 91).

[81] *Lit. Pat.* i. 129.

[82] It is clear, for instance, that the castles of Canterbury (*Lit. Pat.* i. 144), Cambridge (*Rot. Chart.* i. 100; *Lit. Pat.* i. 92), Gloucester (*ibid.* i. 78), Launceston (*Rot. Chart.* i. 100; *Rot. Claus.* i. 155) and Carlisle (*ibid.* i. 150, 163) are often conferred with the county.

[83] Turner, *Trans. R.H.S.* N.S. 272, suggests that the chancery form conferring a castle " together with " a county implies this. But such a form is used in a minority of cases.

[84] *Rot. Chart.* i. 100; *Liberate, Misae. Praestita Rolls* (Record Com.), p. 7.

[85] *Lit. Pat.* i. 142. A good example in 1215.

[86] *Ibid.* i. 55; Stubbs, *Sel. Charters*, 281-2.

[87] Matthew Paris, ii. 538.

tenants by sergeanty who were to come with horses and arms.[88]
The sheriff might be directed also to come to the place of assembly
to certify who had [89] or who had not come.[90] In 1206 the royal
letters bade such a dignitary send to Nottingham all the king's
balistarios in the county.[91] Again, in 1214, sheriffs were called upon
to send the men they had promised for the king's galley service,[92]
and in 1212 men to handle shovels as well as choppers knowing
carpentry,[93] whose sustenance they were also to provide.[94]

Military levies and supplies were often dependent upon the
sheriff's action. The well-known inquest of the reign concerning
fees and tenements held by military service and sergeanty, and
those held by the king's progenitors but alienated, was conducted
by the sheriffs as directed by the king's writ.[95] Information regarding service due, essential in imposing scutage, was derived through
his office.[96] His further activity in the matter is well known,
whether in collecting directly or in aiding a lord to collect from his
tenants. The castellan was in charge of castle guard, but the
sheriff as such might be the agent of the law in enforcing this [97]
as well as other services. In the war of 1215–16 both sheriffs [98] and
castellans [99] levied a military charge known as tenserie, apparently
upon the estates of those who opposed the king. The sheriff was
also commanded to provide ships that were needed. Some of his
ordinary duties were closely akin to those of quartermaster. The
transport of military supplies and equipment by land,[100] or even of
troops by sea, fell easily within his province. The purchase of

[88] *Lit. Claus.* i. 131. Later the king directed the sheriffs to return home from
Chester *cum genti vestra* (*Lit. Pat.* i. 94).
[89] In 1205 to make known at London the name of the ten knights.
[90] In 1213 to be at Dover with the roll to certify as to absentees.
[91] *Lit. Claus.* i. 131.   [92] *Lit. Pat.* i. 106.
[93] *Lit. Claus.* i. 131, 132. The sheriff of Lincoln was to send one of his clerks
and four sergeants to see who came, and report the names.
[94] It is probable that the swearing in of the local levies was a function of the
sheriff now, as it was a little later (*Close Rolls*, 1229–31, p. 395).
[95] *Red Book*, iii. p. cclxxxv.
[96] Mitchell, 28, 60.
[97] In 1207 writs were issued to five sheriffs to enforce castle guard due from
knights and free tenants at the castle of Richmond (*Lit. Pat.* i. 73).
[98] *Lit. Claus.* i. 245, 246, 271, 282.
[99] *Ibid.* i. 249, 251, 253, 277 ; *Lit. Pat.* i. 106, 143, 167. A tenserie mentioned
in *Lit. Claus.* i. 259 was assessed by William earl of Salisbury and Fawkes de
Breaute. In Lincolnshire in 1216 certain of these were reserved to the castellan
as against the sheriff (*Lit. Pat.* i. 177). Round (*Geoffrey de Mandeville*, 414-16)
deals with the nature of tenserie.
[100] As the harness and equipment of *balistarii* (*Lit. Claus.* i. 102).

provisions, forage or materials, or the manufacture of the latter, was frequently enjoined. Thus the clerk of the sheriff in Berkshire, in 1215, bought hogs and grain for the castle of Wallingford.[101] The sheriff of Dorset in 1213 was ordered to come to Portsmouth with all haste, bringing as much hay and oats as he could find, and again to send to the same place ropes for the military engines, the *petrariae* and the mangonels.[102] At another time he was commanded to cause to be made at Bridport day and night cordage, great and small, for the ships.[103] In 1212 and also in 1213 sheriffs were directed to suspend the usual markets for victuals in their counties,[104] and to cause victuallers as they loved their chattels to follow the various sections of the army while they assembled.

The usefulness of the sheriff, as well as the exercise of the king's arbitrary power, is in this reign illustrated in what seem to be new directions. Two borough charters of 1200 established a process for electing the reeve responsible for town accounts, the burghers nominating two of the more discreet men, of whom the sheriff then selected one and presented him to the chief justiciar.[105] The prior and convent of Grimsby were in 1214 directed to assemble in their church to elect an abbot on a day designated by the sheriff of Lincoln.[106] A usage quite familiar later in the century appears when the sheriff causes proclamations to be made. In November 1204, as a preliminary to the stoppage of the circulation of metal, the sheriffs were notified to have it cried throughout their counties at fairs, weekly market days, and the mother churches on festal days that until the feast of St. Hilary following no one was to strike or have in his possession coined money.[107] The mandate of the same year that nine knights everywhere find a tenth for the king's service, the sheriff was ordered to proclaim with haste *foris per totam balliam tuam et in mercatis et nundinis et alibi*.[108] A similar process was repeated in 1213 [109] and in 1216, in the latter year to offer clemency, upon condition of submission, to persons in arms against the king.[110]

[101] *Lit. Claus.* i. 195.   [102] *Ibid.* i. 159.   [103] *Ibid.* i. 134.
[104] *Ibid.* i. 131 ; Matthew Paris, *Chronica Majora*, R.S. ii. 538.
[105] *Rot. Chart.* i. 45-6.   [106] *Lit. Claus.* i. 181.
[107] *Lit. Pat.* i. 47-8.   In 1205 the sheriff of Herefordshire was ordered to have it proclaimed that a new weekly market had been granted (*Rot. Chart.* i. 51).
[108] *Lit. Pat.* i. 55 ; Stubbs, *Sel. Charters*, 281-2.
[109] *Lit. Claus.* i. 132.
[110] *Ibid.* 1. 170.   The offer was to be cried in the vills, markets and hundreds of the county.

The king, again, employed the sheriff in the regulation of commerce. The notification in 1200 that merchants were to be safe to come and go with merchandise in England, was addressed to the heads of the principal seaboard counties and the mayors of several important cities and towns.[111]   The wardens of the ports and the sheriffs were at one time ordered to permit only two designated persons to buy or sell hawks or falcons [112] coming into the realm. In 1212 the king gave the bishop of Winchester an eight-days' extension of his fair, again bidding the sheriff have this cried throughout his bailiwick, so that merchants making the due and ancient customs might come and return safely and transfer their merchandise without impediment.[113]

Of great interest because of their bearing upon the idea of assemblies to deal with taxation are three precepts of the king issued in three successive years. By the first, in May 1212, sheriffs were to notify all the clerks and men of religion in their respective counties [114] not to come before the king at Northampton on the quindene of Trinity as had been ordered, but to come instead on the morrow of St. Peter ad Vincula. The document of 1213, to which reference was made above as one ordering sheriffs to make proclamation, commanded the debtors of the king's Jewish debtors, except earls and barons who had made fine, to come before the king with all haste.[115]   A mandate of 1214 directed the sheriff of Somerset to send twelve of the more discreet men of the county to discuss with the king the oblate which the knights of the county made at the pleas of the forest to have their liberties.[116]   Under these circumstances it was not strange that the two famous writs of 1213, calling up respectively four discreet knights of the county to discuss the affairs of the realm,[117] and the reeve and four men of each vill of the royal demesne to certify losses of bishops and abbots,[118] should go to the sheriffs for execution.   A logical sequel is seen in the provision of Magna Charta requiring them or the king's bailiffs to summon the lesser tenants-in-chief to the great council.[119]

The misrule of John's sheriffs, which is almost proverbial,[120] was recognized even in 1215 as in part the result of a vicious fiscal

[111] *Rot. Chart.* i. 60.   [112] *Lit. Claus.* i. 20.
[113] *Ibid.* i. 123.   [114] *Ibid.* i. 129-30.
[115] *Ibid.* i. 132.   [116] *Ibid.* i. 181.
[117] *Rept. on Dignity of Peer*, app. i. p. 2.   [118] Roger of Wendover, ii. 82.
[119] Magna Charta of 1215, sect. 14.
[120] McKechnie, *Magna Charta*, 2nd ed. 16, 311 ; Stubbs, *Const. Hist.* i. 575.

system. Both the farming of shires at an increment over and above the usual *ferm* and the exaction in addition of a sum for the privilege, were devices well known a century before John's time. His predecessor King Richard so extended them that in some cases he virtually sold the shires [121] to raise funds. John continued the practice. In his as well as his brother's time the most extreme cases are found in the north. William de Stuteville in 1203 owed a thousand pounds to have custody of Yorkshire so long as he served faithfully and rendered the *ferm* and other dues from the county ; [122] also five hundred pounds for the custody of the county of Northumberland in the same form as Yorkshire.[123] Such payments, of course, indicate large opportunities for speculative profit and constant temptation to extortion. This repetition of instances already notorious,[124] in a region wherein the most determined baronial opposition to John later originated, affords some explanation of the requirement in Magna Charta that with the exception of the king's demesne estates all counties, hundreds, wapentakes and trithings be let at the ancient ferms without increment.[125] The exaction of the *proficuum* had also contributed to these evils by increasing fiscal pressure on the sheriff.

Though driven by the demands of the exchequer, and of necessity often enforcing odious policies not his own, the sheriff was by no means the unwilling instrument of a bad regime. The holding of the office by ministers of state who acted by deputy conclusively shows that it was profitable. The shifting to the county as a whole of the amercement incurred through improper procedure in its court was a ground for much complaint.[126] It involved the sheriff's

---

[121] Stubbs, *Const. Hist.* i. 534.

[122] Madox, *Exchequer*, i. 461, n. (a). The king, however, reserved the right to commit to other custodians his castles and castellanies, his forests and his demesne, and to increase the render from these as much as he chose. William was to be supreme *custos* and respond as sheriff, but the king was to choose two persons to act under his directions. He was to take upon himself the making of emendations *de forisfectis* as pertained to the sheriff.

[123] *Ibid.* i. 461, n. (b).   [124] Stubbs, *Const. Hist.* i. 541.

[125] Magna Charta of 1215, chap. 25.

[126] By the king's agreement with the men of Cornwall in 1208 or 1209 (Madox, *Exchequer*, i. 410, n. l), the sheriff who incurs an amercement shall pay of his own means so the county be not involved. In 1201 this county owed a hundred marks for an amercement and for quittance of carucage (*Select Pleas*, i. no. 17). The charter to London in the first year of the reign (*Liber Albus*, ii. 250-51) concedes that amercements of sheriffs be limited to £20. In case they do not suffice for that amount, the citizens are not to lose. See also note 122. Exemption from *misericordia comitatus* (*Rot. Chart.* i. 164) occurs in a charter of 1206.

avoidance of the responsibility of his own carelessness and possibly implied his desire to increase the king's revenue at the expense of the community.   Just treatment at the hands of the sheriff was the end for which whole counties bargained with the king.   The latter conceded to the men of Devon in 1204 that if their sheriff oppressed them he should be at the king's mercy.[127]   There were also complaints of this official's conduct in admitting prisoners to plevin,[128] and plain intimation that he acted from improper motives.   The Devonshire charter grant just mentioned permits him henceforth to replevy prisoners only by the counsel of the county, so that no one may any longer be detained *per odium vel occasionem*.   Within forty years of the Assize of Clarendon the sheriff's custodianship of prisoners appears to be regarded as a perquisite.[129]

Furthermore, multiple exactions of a minor sort were clearly chargeable to his official policy.   The extinction of the bad customs of sheriffs and their servants was prescribed in Magna Charta.[130] The royal act looking to the enforcement of the section refers to these as the petty customs of sheriffs and their bailiffs.[131]   Some of them were undoubtedly related to the sheriff's conduct of the local courts and the holding of too frequent sessions.   Exemption from pleas and suit at shire and hundred courts, which was made in dozens of John's grants,[132] involved more than the time lost in attendance.   The sheriff had a reputation for extorting money at his tourn, probably for non-attendance.   The county of Devon obtained a grant that this session be held but once a year unless the sheriff must visit them oftener to preserve the king's peace, in which case he was to take nothing for his own use.[133]   The Magna Charta of 1217 declared that the sheriff should make his tourn but twice a year, that at his annual view of frankpledge he demand no exactions, and be content with what he used to have for making the view in the time of Henry II.[134]   The same document prescribed also that sessions of the county court be held only once a

---

[127] *Rot. Chart.* i. 132.

[128] *Select Pleas*, i. no. 40 ; *Placitorum Abbreviatio*, 63.

[129] In 1207 Richard de Mucegros and Gerard de Athée receive (*Lit. Pat.* i. 71, 78) the castle of Gloucester *cum prisonibus*.

[130] Chap. 48.   [131] *Rot. Lit. Pat.* i. 180 ; cf. i. 145.

[132] *Rot. Chart.* i. 6, 8, 9, 10 *et passim*.

[133] *Rot. Chart.* i. 132.   In the *Pipe Roll* of 5 John, under Bucks and Beds, the earl of Leicester pays for this year 45s. 8d. *ne vicecomes videat franciplegium in ballia sua*.

[134] Stubbs, *Sel. Charters*, 346-7.

month, and in any event not at briefer intervals than those which had been customary. There were also impositions upon communities in relation to the *scotale*, which was apparently held sometimes for the sheriff's benefit.[135] Exemption from this form of contribution was procured in some of the municipal charters of King Richard,[136] and in nearly a dozen of this reign [137] prior to 1213, when sheriffs were forbidden to make any *scotale* in the kingdom.[138] In the charter exemptions of the time, acquitting individuals of the necessity of rendering sheriff's aid, there is hint of the abuse of this older exaction, levied not only by sheriffs but also by their bailiffs.[139] It is shown also that sheriffs were acting illegally for bribes, demanding contributions and hospitality [140] and commandeering horses and carts.

The acts of the sheriffs of the time were perforce viewed from different angles. There can be little doubt that, when the early financial pinch of the reign was well over, John or his ministers took some steps to regulate their conduct. In 1208 appeared a new article of the eyre which called for inquest concerning prises made by sheriffs *contra voluntatem*.[141] The Annals of Dunstable relate that in 1211 the king instituted a very strict inquiry against the sheriffs as a result of which some were imprisoned and some fled.[142] The list of sheriffs, however, disproves the seriousness of

---

[135] But a charter of Richard I. grants quittance of the king's scotale (*Charter Roll*, i. 382). Reeves and bailiffs also held scotales; cf. *C.C.R.* ii. 167. For suggestions concerning the possible nature of the contribution, see Stubbs, *Const. Hist.* 6th ed. i. 672 and n. 1. In *Select Pleas* (no. 167) is mention of an actual case different from any suggested by Stubbs. Certain beadles set up taverns and distrain people to come to their ales. They are at mercy and are enjoined not to make such ales in the future.

[136] Stubbs, *Sel. Charters*, 266, 267.

[137] *E.H.R.* xiv. 102-3.

[138] Roger of Wendover, ii. 82-3.

[139] *Chron. Abingdon*, ii. 246 (*dono vicecomitum*); *Rot. Chart.* i. 50 (*de auxiliis vicecomitum et omnium ballivorum suorum*); *ibid.* i. 82 (*ab omnibus auxiliis et omni exactione vicecomitum et omnium ministerialium eorum*); *ibid.* i. 35 (*de auxiliis vicecomitis et praepositi hundredi*). So in the thirteenth century there was an aid of the hundred in addition to the sheriff's aid (*Unpublished Hundred Rolls*, Staffordshire, Chapter House Series, no. 2, m. 2); cf. hundred-scot (*Rot. Hund.* i. 470). Conditions seem to imply that both sheriffs' and bailiffs'· aid arose primarily because of demands made in holding the courts of shire and hundred. Cf. Davis, *England under Normans*, 311.

[140] The charter disafforesting Cornwall in April 1215 (*Rot. Chart.* i. 206) confers perpetual exemption *de donis, auxiliis et hospiciis vicecomitum*. Quittance of aids and gifts of sheriffs is granted in a charter of Richard I. (*Charter Roll*, i. 382). Cf. below, note 187.

[141] Cam, *Studies*, vi. 20, 192.   [142] *Ann. Monast.* iii. 35.

the outcome. Only two sheriffs were displaced out of regular course this year, of whom one was restored to office in about a month, the other in a year and a half.[143] Whatever the cause of complaint, it probably concerned merely the king's interests. A somewhat ineffectual attempt at general regulation followed. At the council held in St. Albans, in August 1213, the justiciar and the bishop of Winchester together with the magnates commanded sheriffs, foresters and other *ministri regis*, as they loved life and limb, to refrain from extortion, to do no injury and to make no scotale.[144]

The attitude of ecclesiastics in the beginning was apparently individualistic. The dispossession of the archbishop of York in 1200 and in 1207,[145] and the seizure of the property of the bishops who in 1208 fled from the wrath of the king,[146] all evoked excommunication of one or more sheriffs. These measures, together with one of 1212, forbidding religious houses to receive any papal mandate against the king, and to arrest any one who did,[147] associated sheriffs with what was regarded as illegal or even sacrilegious. Reflection upon the wrongs suffered by their order was likely to lead churchmen to entertain larger concepts of liberty. In September 1213 the papal legate and the bishops denounced to the king the captious exactions of sheriffs, declaring that they were intent only upon making the annual dues a means of obtaining money from the poor people of the counties.[148]

The baronial charges against the sheriff are deducible from Magna Charta. Nearly all of these and the demands resulting therefrom seem also to cover grievances of the church. The bad customs of sheriffs were to be abolished. They had made waste of lands committed to their custody and were to be displaced by other custodians if they did so in the future.[149] They and their bailiffs were no longer to impress for transport service the horses

---

[143] Robert fitz Roger superseded William of Huntingfield in the shrievalty of Norfolk and Suffolk at Easter 1211. The latter returned to the same position at Michaelmas 1212. Robert fitz Roger was himself superseded in Northumberland by three special agents of the king on Aug. 20, 1211, but was once more in office here at the ensuing Michaelmas.

[144] Wendover, *Flores Historiarum*, R.S. ii. 82-3.

[145] Hoveden, iv. 139 ; Walter of Coventry, *Memoriale*, R.S. ii. 169.

[146] *Ann. Monast.* ii. 261, 267.

[147] *Ibid.*

[148] Walter of Coventry, ii. 214-15.

[149] Magna Charta, chap. 4.

and carts of free men without the owner's consent.[150]   It may be added that the compulsory finding of such means of transport in 1210 was a special grievance of the gray abbots.[151]   Castellans, among whom were certainly some sheriffs, were not to make distraint to collect money for castle guard from knights who wished to render service instead.[152]   No vill and no person was to be compelled to minister to the king's sport, as sheriffs had clearly compelled them to do by constructing bridges over streams, except those which of old were under this obligation.[153]   The sheriff had been dealing with pleas of the crown in the absence of the coroners,[154] and was absolutely excluded from holding such pleas.[155]   So little confidence was placed in him where life, limb, property and the levy of monetary penalties are involved.   Finally, and this was in agreement with the promise of 1213 that just judgements were to be observed, the king was to appoint only justices, wardens, sheriffs and bailiffs who knew the law of the land and who meant to observe it.[156]   Here is recorded an objection to the foreign-born in office and also an imputation of wilful infraction of law, which in some cases was obviously justified.

This section of Magna Charta, in view of John's well-known favour to foreign adventurers,[157] tends to create the erroneous impression that sheriffs from overseas were numerous and consequently the source of much misgovernment.[158]   A chronicler names among the prominent grievances leading to Magna Charta the cruelty of perverse men whom the king loved and made justices, sheriffs and keepers of castles,[159] but does not state that these were foreigners.   John, of course, did freely employ foreigners of questionable character as castellans.   The section of Magna Charta, giving effect to the demand of the barons [160] that certain objectionable

[150] Chap. 30; Articles of the Barons, art. 20. In the Great Charter of 1217 (chap. 23) this power of impressment is forbidden to sheriffs and bailiffs unless they make the ancient payments, namely, for a two-horse cart ten pence a day and for a three-horse cart fourteen pence.

[151] *Ann. Monast.* iii. 33.

[152] Articles of the Barons, art. 19 ; Magna Charta, chap. 29.

[153] *Ibid.* chap. 23.   Cf. below, p. 219.

[154] Articles of the Barons, art. 14.   [155] *M.C.* chap. 24.

[156] Articles of the Barons, art. 42 ; *M.C.* chap. 45.   For exactions by occasion of pleas of the crown, *Pleas for Gloucester*, ed. Maitland, no. 108.

[157] *Rot. Lit. Pat.* i. p. xiv ; Powicke, *Loss of Normandy*, 337-41.

[158] McKechnie, *Magna Charta*, 312 : "The agents of his [John's] evil will, foreigners and desperados ".

[159] *Chronicle of Stephen*, ii. 518.

[160] Articles of the Barons, chap. 40 ; *M.C.* chap. 50.

M

persons be removed from their bailiwicks, names seven natives of Touraine. Three had served as sheriffs, namely Gerard de Athée, former seneschal of Touraine and valiant warrior there,[161] his kinsman Engelard de Cigogné, and Philip Mark.[162] All of these had been castellans,[163] and owed their standing to their military ability. Both Engelard and Philip were in 1215 active partisans of the king's cause. Gerard probably entered the king's employ as a mercenary, and was reputed to be of servile origin.[164] Engelard had been a justice in violation of the rule which forbade a sheriff to hold that position in his own county.[165] The prominence and the energy in opposition of these two men, no less than their dubious antecedents, marked them for denunciation. It is well known that as sheriffs both [166] were guilty of some extortion and bribe-taking,[167] but the hostility of the barons is hardly thus to be explained. Philip Mark, who was active enough [168] on the king's behalf to fall into the objectionable category, presumably also began his career as a mercenary. A fourth member of this group, André de Chanceaux, had been under-sheriff and acting sheriff of Hereford-shire.[169] The only other sheriffs prior to 1215 whose names indicate foreign origin [170] were Richard the Fleming, Hugh de Gornac,[171] a

[161] Concerning Gerard and this whole group, see *Trans. R.H.S.* xviii. 249-54. As to Gerard, Powicke, *Loss of Normandy*, 236 ; Ralph of Coggeshall, *Chronicon*, 146, 152 ; *Pleas for Gloucester*, ed. Maitland, p. xiv.

[162] In 1207, when he was granted lands in England (*Rot. Lit. Claus.* i. 79), called *socius* of Gerard. Roger of Wendover (ii. 60) calls him *nepos*.

[163] Gerard was in charge of Bristol castle in 1208 (*Rot. Lit. Claus.* i. 105) and held Hereford castle also with his shrievalty (*Rot. Pat.* i. 83). Engelard held Windsor castle at the crisis of the baronial struggle (McKechnie, *Magna Charta*, 445-6), and Philip Mark Nottingham castle, in 1211 (Wendover, ii. 60). Turner (*Trans. R.H.S.* N.S. xviii. 254) shows that despite Wendover (ii. 60) they are to be regarded as soldiers, not as courtiers or politicians.

[164] Powicke, *Loss of Normandy*, 338 ; Turner, *loc. cit.*, p. 249.

[165] McKechnie, *Magna Charta*, 319 ; Madox, *Exchequer*, ii. 146, n. (c).

[166] Gerard de Athée held the two counties of Gloucester and Hereford, 1207–1209, and that of Nottingham 1208-9 ; Engelard succeeded him in the former two shires and was sheriff until 1215. Philip Mark, Engelard's successor in Nottinghamshire, was in office until 1214.

[167] *Pleas County of Gloucester*, ed. Maitland, nos. 92, 100, 108.

[168] He was custodian of large amounts of the king's treasure both in 1213 and in 1216 (*Rot. Lit. Pat.* i. 99, 165-7).

[169] *Docs. from the Records of the Queen's Remembrancer*, Record Com., 1844, p. 236 ; *Trans. R.H.S.* N.S. xviii. 251.

[170] Hubert de Burgh, through his service as the king's chamberlain, and especially through his marriage to the daughter of William de Vernon, earl of Devon (*Rot. Chart.* i. 52), and the settlement upon him of the Isle of Wight, was fairly well identified with England when he held his first shrievalty in 1200–1201.

[171] *Rot. Selecti*, Record Com. p. 8 ; subsequently he had incurred John's displeasure by leaving his service (*Lit. Claus.* i. 65). He appears as a curial

Norman who in 1205 entered the king's service in England, and Philip de Ulcote, in 1204 constable of Chinon, a warrior who, after being taken prisoner in the struggle with France, was ransomed and employed in England.[172] Six or seven sheriffs out of a hundred in office between 1199 and 1215 form a very inconsiderable proportion, even though their activities did extend to half a dozen counties.

To appreciate the type of sheriff appointed by John it is necessary to distinguish between the sixteen years of the reign preceding Magna Charta and the sixteen months which followed. In the latter period the king in his war with the barons was clearly in straits. The sheriffs who had joined the Magna Charta movement were dismissed, and only persons active on the king's behalf held office. Since their accounts were seldom rendered at the exchequer after Easter 1215, the Pipe Rolls do not always give their names. William Brewer, sheriff of Hampshire, indeed accounted for that county the first half of the fiscal year 1215–16,[173] but he stood alone. The official record of appointments on the Patent Roll cannot tell the whole story. There is related an interesting tale of struggle between rival sheriffs in Norfolk and Suffolk which was probably enacted elsewhere. After Easter and until St. Peter ad Vincula, 1215, Roger de Cressy was sheriff *per barones*. Then the king committed these counties with the castle to the justiciar Hubert de Burg, whose representative, Walter of Elingham, had held only one county court in Norfolk and two in Suffolk when the barons violently intervened and reinstated de Cressy, who acted until Christmas. Then after the arrival of Prince Louis, the marshal of France held the office, Fulk Bagnard acting for him.[174]

To hold the counties at such a time required strong men. High officials of state as in the past were employed. After Magna Charta

---

attendant to the king abroad in 1199. Appointed sheriff of the counties of Bedford and Buckingham, Sept. 3, 1214 (*Lit. Pat.* i. 121), he could have been in office but a very short time. Cf. List of Sheriffs, *P.R.O. Lists and Indexes,* no. 9.

[172] One of the joint custodians of Northumberland, 1211–12, 1213–14, and sole sheriff from Michaelmas 1214 to 1216. His early service is mentioned in *Lit. Pat.* i. 40. He may have had lands in England, for he was entitled to scutage in 1205 (*Lit. Claus.* i. 55). In 1207 (*ibid.* 82) the king gave 200 marks for his ransom. He was a *custos* of the diocese of Durham in 1209 (*Lit. Pat.* 91), and by 1211 a very prominent figure. An inquisition of much later date accuses him of interfering with the chief sergeant in Northumberland, who was entitled to appoint and remove the holders of lesser sergeanties (*Cal. Inq. Misc.* Chancery, i. 129).

[173] *P.R.* no. 61.

[174] *K.R. Mem. Roll,* no. 3, m. 27.

the justiciar was commissioned for five counties,[175] Geoffrey de Neville, the king's chamberlain, for Yorkshire,[176] and Henry fitz Count, the earl of Cornwall, John's cousin, for that county. So were also a number of the king's former servants at court,[177] men who in the main had also counselled and aided the king in his struggle against the church.[178] A few of the sheriffs of June 1215 were continued in office.[179] John could hardly overlook military considerations in appointing any of these.[180] Nor could he afford to overlook energetic foreigners in his service. Peter de Maulay, who appeared in his employ in Normandy in 1203 and subsequently rose in station and estate in England, was in 1216 given two counties.[181] Among the sheriffs of this year were Fawkes de Breaute,[182] Walerand le Tyes [183] and Stephen Haringod,[184] employed

[175] Kent with the castle of Canterbury and Surrey (*Lit. Pat.* i. 144), Herefordshire (p. 149) and Norfolk and Suffolk (p. 150).

[176] He held this shire in Feb. 1216 (*Lit. Claus.* i. 247).

[177] John Marshal, Hubert de Burgh, William Brewer, William de Cantilupe (in Leicestershire, 1216—*Lit. Pat.* i. 177), Matthew fitz Herbert and Robert de Vipont (cf. *Rot. Chart.* i. 113) are of this type.

[178] See note 192.

[179] John de Wigenholt was sheriff and receiver of tenseries and fines belonging to Corfe castle in July 1216 (*Lit. Claus.* i. 277). He is designated as *clericus* (*Rot. Lit. Claus.* i. 303, 365). Philip de Ulcote and Robert de Cardinan also held office past the Magna Charta period.

[180] Ralph Musard, the successor of Engelard de Cigogné in Gloucestershire, was a baron though an alien (*Trans. R.H.S.* N.S. xviii. 254), who in 1214 owed the service of fifteen knights and furnished ten (Mitchell, 110, n. 71). John Le Strange (*Extraneus*), designated sheriff of Shropshire, Sept. 5, 1216, had been made a castellan in 1213 (*Lit. Pat.* i. 100); Robert de Cardinan, a former sheriff, appointed sheriff of Cornwall and castellan of Launceston castle, May 1215 (*Lit. Pat.* i. 142, 144), making way for Henry fitz Count the following Sept. (*ibid.* 155) but reappointed in Nov. (p. 159), in 1214 owed the service of seventy-four knights (p. 204). The appearance of the earl of Chester among the sheriffs suggests a similar consideration.

[181] Dorset and Somerset, June 26, 1216. The early mention of his Norman activity is in *Lit. Pat.* i. 25. He held land in Oxfordshire in 1205 (*Lit. Claus.* i. 23) and was soon the recipient of two grants (*ibid.* 59, 123). Mentioned as a counsellor of John in 1211 (Wendover, ii. 60), he was extremely active in administration and was custodian of important prisoners in Dec. 1215 (*Lit. Claus.* i. 215).

[182] A mercenary captain (Powicke, *Loss of Normandy*, 338), especially employed by John as an expert on military matters both in 1212 and 1215 (*Lit. Claus.* i. 119, 156). He was commissioned sheriff of Cambridge and Huntingdon, Mar. 9, 1216, and of Northampton on May 2 following.

[183] One of the king's leaders in June 1215 (*ibid.* i. 184 b), made sheriff of Herefordshire in December of that year (*Lit. Pat.* i. 161), and in Mar. 1216 castellan of Berkhampstead (*ibid.* 169).

[184] Also one of John's commanders in 1215, he had been authorized to take tenseries (*Lit. Claus.* i. 143, 184 b). He was in charge of Colchester castle (*ibid.* i. 151, 192) as well as the borough (*ibid.* 127, 184 b) before his appointment in

to hold important castles and apparently all notorious soldiers of fortune. Such were the leaders who prosecuted the struggle with inhuman cruelty, the foreigners whose banishment the barons had vainly demanded.[185] Furthermore, in defiance of Magna Charta John reappointed Engelard de Cigogné,[186] and at the very end Philip Mark.[187] To maintain himself the last year of his life, the king openly violated his promise, employed a number of foreign adventurers as sheriffs and broke with good administrative tradition.

But for the reign proper, the preceding sixteen years, in spite of several disgraceful figures, the shrievalty in general bore no such stamp. The administrative staff of John's predecessor had adhered to the traditions of Henry II., and John's despite some factional changes preserved continuity with that of his brother. This was particularly true of a half dozen great officials of state [188] who acted as sheriffs. These were in general the men who had controlled affairs from 1191 to 1193, some of them important feudatories, three of them now earls. In fully a dozen shires the sheriffs of Richard's time were not displaced at his brother's accession. A goodly proportion of John's earlier sheriffs were men who were fairly well known.[189] Indeed the list for five years, like that of Richard's just before 1193, was in general a list of notables. Some of these were long in office.[190] The period of

---

Mar. 1216 to the shrievalty of Essex (*Lit. Pat.* i. 172). In April 1215 the king ordered him to take no timber from the abbot of Colchester except by consent of the latter (*Lit. Claus.* i. 198).

[185] Articles of the Barons, no. 41 ; *Chronicle of Stephen*, 520-1.

[186] For Surrey, Apr. 22, 1216.

[187] For Nottinghamshire, Oct. 1216 (*Lit. Pat.* i. 199). He held Newark castle prior to June 1216 (p. 187). The men of Nottingham paid him a hundred shillings for his goodwill that he should not enter their liberty and should maintain it (*Inq. Misc.* Exchequer, i. 90).

[188] Geoffrey fitz Peter, William Marshall, Hugh de Neville, Hugh Bardolf, William Brewer, William earl of Salisbury.

[189] In addition to those named in the preceding note, William de Braiose, Robert de Harcourt, Robert fitz Roger, William de Stuteville, Gilbert Basset, Reginald Basset, Robert Belet, Ralph de Grafton, Reginald of Cornhill, and William fitz Alan.

[190] De Braiose was sheriff of Hereford, except for a half year, from 1191 to 1200. Fitz Alan held Shropshire continuously from 1190 to 1201, being succeeded by the Justiciar until 1204, and he by Thomas de Erdington, who held both Shropshire and Staffordshire until 1215. Reginald of Cornhill held Kent 1192–1215 ; Robert de Vipont, Westmoreland, 1203–28 ; Geoffrey fitz Peter, Staffordshire, 1198–1204 ; Robert de Turnham held Surrey with but one year's interlude from 1194 to 1207. William de Cantilupe was sheriff in Worcestershire 1200–1215, and in Warwickshire and Leicestershire 1200–04 and 1209–15. Thomas Basset held Oxfordshire 1202–13. John de Wichenholt, after two years of service as under-sheriff in

crisis [191] in Normandy absorbed the energies of some of the greater figures. After 1204 there were more obscure sheriffs, but up to 1215 less variation in the personnel of the office than in Richard's time. The sheriffs of the year 1214 – 15, when John had fully developed his policies and the barons were known to be near revolt, were often men of lesser standing and supposedly of the subservient official type.[192] But they were far from being nobodies, though the greater number were at one time or another to be found in attendance at a *curia* where nobodies sometimes throve. Their names represent the orderly element in the government and not its negation. No less than five of them appear in Magna Charta as members of the king's council and his advisers in this crisis.[193] Yet three of these sheriffs actually joined the baronial movement.[194] Of the committee of twenty-five barons chosen to secure the enforcement of Magna Charta four were former sheriffs.[195] Among the baronial leaders

Berkshire, was made sheriff in 1204 and served until 1215. Henry de Braibroc was sheriff of Bedfordshire and Buckinghamshire 1204–14, and of Rutlandshire 1211–15, William Marshalf of Sussex 1193–1204.

[191] Robert de Turnham, sheriff both before and after the fiscal year 1204–5, was in that year actively engaged in the war in Poitou where he was taken prisoner (Coggeshall, R.S. 146, 152; cf. *Chron. and Memorials of Reign of Richard I.*, ed. Stubbs, p. 551; Benedictus, ii. 172; Hoveden, iii. 206). Hugh de Neville, William Brewer, and William Marshall gave up shrievalties at this time, Hubert de Burg two, and Geoffrey fitz Peter several.

[192] They include, besides the four foreigners named on pages 160-61, Reginald of Cornhill, Henry de Braibroc, John de Wichenholt, Robert de Cardinan, Matthew Mantel, Gilbert fitz Renfrid, Robert fitz Roger, Robert de Percy, Robert de Ros, William earl of Salisbury, John fitz Robert, William Brewer, Hubert de Burgh, John Marshal, Thomas Basset, Thomas de Erdington, Ralph de Bray, Matthew fitz Herbert, and William de Cantilupe. All but the first eight are found on the Charter Rolls at some time during the reign as witnesses of the king's grants. Ralph de Bray was constable of Normandy (*Lit. Claus.* i. 165). Hugh de Gornac also appears in the curial circle. Roger of Wendover (ii. 60) names among the *consilarii iniquissimi* of John who in 1211 aided him against the church, Hugh de Neville, Robert de Vipont, William de Cantilupe, Henry de Cornhill, the brother of Reginald (Madox, *Exchequer*, i. 458-9), Robert de Braibroc, Philip de Ulcote, Philip Mark, Peter de Maulay, Gerald de Athée, Engelard de Cigogné, William Brewer, Peter fitz Herbert and Thomas Basset. Matthew Paris (ii. 531) adds the name of Reginald of Cornhill.

[193] Earl William of Salisbury, Hubert de Burgh, John Marshal, Matthew fitz Herbert and Thomas Basset.

[194] John fitz Robert, Robert de Ros (Stubbs, *Const. Hist.* i. 582) and Reginald of Cornhill. John was with the king's enemies in Sept. 1216 (*Lit. Claus.* i. 289); Reginald turned Rochester castle over to Robert fitz Walter (Coggeshall, 174, 175-6) and was taken prisoner there in Dec. 1215.

[195] Matthew Paris, i. 262: Robert de Ros, John fitz Robert, William of Huntingfield and Richard de Muntfichet.

of the North who began the struggle two had held the same position.[196]

Both the great and the petty shortcomings of the sheriffs weighed heavily with the leaders who shaped Magna Charta. From their tyranny as well as the king's was relief demanded. But so far as evidence exists for the period prior to 1215, the cruelty and oppression of those who were native-born yield nothing to the few foreigners in the office. Moreover, misconduct on the part of administrators in the king's confidence and long in office was especially vicious. The worst known offender, William Brewer, stood high in the counsels of both Richard and John and was trained in the school of Henry II.[197] John had his circle of shrewdly chosen officials, most of whom followed [198] blindly to the end. But with few possible exceptions his sheriffs before 1215 were not adventurers like some of his other servants. After 1207 they were a better class of men than the average of the king's immediate *entourage*. Indeed the more respectable members of this group were often sheriffs. Nor does the oppressive rule of the sheriffs denote executive weakness,[199] but rather a design on the part of the exchequer to share the extortionate gains which provide their chief reward. The deliberate selection of cruel, dishonest and lawbreaking sheriffs [200] was inconsistent with the policy and achievement of a reign now known to have been marked by organizing genius and substantial administrative progress.[201] The continuation in office after John's death of some of the foreign group, among them men detested by the barons of 1215, shows that they were not devoid of ability and even merit. In the ordering of the shrievalty is disclosed the

[196] Robert de Ros and Thomas of Multon. The latter was the son-in-law of Hugh de Morvill (*Book of Fees*, part i. p. 266). He was taken prisoner at Rochester with Reginald of Cornhill.

[197] Sheriff of Devon in 1204 and of Dorset and Somerset in 1210. See above, and also the *D.N.B.* According to the *Gesta Regis Henrici Secundi*, i. 5, the sheriffs who were replaced after the inquest of 1170 were much more cruel than those before that time.

[198] The story told long afterward (*Inq. Misc.*, Chancery, i. 461), near the end of the century, of a sheriff of Northampton, who was persuaded not to take into the king's hand property of one of his enemies, and who is thus classed as a traitor, refers to the period of John's earlier war on the Continent.

[199] Maitland (*Pleas for Gloucester*, p. xvii), seems to regard the misgovernment of John's sheriffs as indicating a regaining of their former power; McKechnie, *Magna Charta*, 2nd ed. 318, presents the matter clearly.

[200] Turner, *Trans. R.H.S.* N.S. xviii. 250, shows that there is no evidence for placing Gerard de Athée in this category.

[201] Hilary Jenkinson in *Magna Charta Commemoration Essays*, 244-300.

work of men like Hubert Walter and Geoffrey fitz Peter, and it may be of John himself. The written instruments by which the sheriff of the time was appointed and installed, his functions in the issue of special summons and proclamation, his duties in respect to purveyance and carriage, his relations to the peace, to the king's courts, to the exchequer and to taxation, and some of his vices as well, were common with those of the reign of Edward I. Standing only at the threshold of the century, the sheriff of John in many respects belonged to its later decades. To the administrative organization of his office even the work of the great Edward owed much.

VII

THE APPOINTMENT, STAFF, AND JUDICIAL DUTIES
OF THE SHERIFF, 1216–1307

In the interval between the accession of Henry III. in 1216 and the death of his son Edward I. in 1307 the sheriff lost much of the semblance of immediate subordination to an absolute king. By the end of this period he appears, not so much the personal servant of the monarch, as the holder of an office the activities of which, in so far as they are controlled or directed by the chancery, the exchequer and the justices, are subjected in general to fixed rules and forms. The military functions in any proper sense of the word have vanished, the discretionary powers largely so, and the office, once conferred upon great officers of state, is usually held by members of local families who are in theory knights, and who sometimes bear the honorary title master.[1] The legal requirement that they have land enough in the county wherewith to answer the king and the people,[2] dates only from the time of Edward II. Until that time the appointment of men from other shires was not unusual. The occasional employment of a bishop and the life tenure of a few great feudatories appear only as isolated personal arrangements. Thus the persons usually in the office are chosen from a class within the county experienced in the conduct of the king's business. At the death of Edward what has been said of the constitution in general may be said of the shrievalty. It has not attained its ultimate form, but this is already indicated in the rough. Nor will novel development of a radical nature take place in the future. The office as it stands to-day differs chiefly from that of six hundred years ago, not because it has assumed added functions, but because

---

[1] In the case of John Gerberd, recently sheriff of Somerset (*L.T.R. Mem. Roll* 77, m. 27, P.R.O.). In other cases the incoming sheriff is *magister*.
[2] Statute of Lincoln, 1315–16, *Stat. of Realm*, i. 174.

it has lost many of the older powers, which with the growing complexity of government have been transferred to other agencies. Its functions are old but few, decidedly fewer than in some Anglo-Saxon lands across the seas where the office was introduced long ago. For the purposes of this study the history of the shrievalty may be discontinued at the beginning of the fourteenth century when its future character is fairly obvious, but before its great decline has begun.

The transition from the sheriff of John to the sheriff of Edward I. was necessarily gradual. It has more than once been pointed out that prominent and even notorious sheriffs of John[3] continued in office for some years during his son's minority. To the group of loyal barons, headed by William Marshal, who had the child Henry crowned king, fell both the problems and the governmental machinery of John's reign. In prosecuting the struggle against the rebels, now aided by the French prince, they found the military strength and skill of those employed by the late king most useful. Two of the worst sheriffs of the mercenary type, Stephen Harengod and Walerand le Tyes,[4] were at once displaced. But Fawkes de Breauté retained half a dozen counties. Hubert de Burgh held three other important ones. The names of Peter de Maulay, Philip de Ulcote and Philip Mark[5] upon the list further prove the retention of the foreign group. Of John's other sheriffs Matthew fitz Herbert, William de Cantilupe and Robert de Vipont retained the shrievalties which they had held for some time.[6] The earls of Chester and Salisbury, with Ralph Musard, Walter de Lascy and Geoffrey Neville, controlled eight counties and further perpetuated the former system. A dispute in 1218 between William of Warenne, Earl of Surrey, and Engelard de Cigogné over the shrievalty of Surrey was judicially considered by the council, and the commissions of John's sheriffs were held to be valid during the minority of his son.[7] There were, however, other new sheriffs. For a short time the Earl Marshal himself was of the number, and William de

[3] McKechnie, *Magna Charta*, 2nd ed., 312 ; Turner, *R.H.S.* N.S. xviii. 273-7.

[4] Walerand remained constable of Berkhampstead (*Bracton's Note Book*, ed. Maitland, no. 1406).

[5] Peter held Dorset and Somerset ; Philip de Ulcote, Northumberland ; and Philip Mark, Nottingham and Derby. See *P.R.O. Lists and Indexes*, no. 9.

[6] In Sussex, in Warwickshire and Leicestershire, and in Westmoreland respectively.

[7] Turner, 275-6 ; Baldwin, *King's Council*, 17. But strange claims were sometimes made in this reign concerning sheriffs of John's time. See below, p. 181.

Beauchamp now re-established the rule of his family in Worcestershire.

The period between 1219 and 1226 substituted for some of the military figures required in the office by William Marshal agents of the curial group now in control. The insolence and revolt of Fawkes de Breauté led to the loss in 1224 of his numerous charges. A very general change of sheriffs in this year and the next displaced some of the more powerful feudal figures.[8] Hubert de Burgh, the justiciar, continued to act as sheriff of Kent and at the same time to exercise his judicial functions,[9] although Magna Charta had forbidden sheriffs to hold pleas of the crown. Hubert's rival, Peter des Roches, bishop of Winchester, made way in his shrievalty for the bishop of Salisbury, and Stephen Segrave, Peter's adherent, disappeared from the list at practically the same time. Other well-known curials first held the office in this period.[10] A trend toward the revival of hereditary shrievalties, to be noticed later, was lessened through the surrender of the office by the earls of Ferrers, Surrey and Salisbury [11] and by the downfall in 1232 of Hubert de Burgh, who subsequent to 1226 was earl as well as sheriff of Kent. The final onslaught made on Earl Hubert by Peter des Roches and his associates was accompanied by a very general change of sheriffs. At this time twenty-one counties were conferred for life on Peter of Rievaux,[12] who fortunately held them but a very brief term.

So well was the political and administrative importance of the office recognized, that the beginning of the baronial reforms in 1258 was accompanied by measures designed to control the conduct of sheriffs and to assure their loyalty to the reformers. The sheriff was the king's agent in enforcing a system which the representatives of national interests were seeking to break down. By the Provisions of Oxford four knights were appointed in each shire to watch the sheriffs [13] and to cause complaints against them and against hundredors and bailiffs to come before the justices. The reputation and conduct of sheriffs were still much the same as in John's

[8] Notably the earl of Chester, William de Cantilupe, and a little later, Ralph Musard.

[9] *P.R.* 62 (2 Henry III.), Norfolk and Suffolk. Ralph Musard was a justice in eyre in 1221, though not in his own county (*Pleas for Gloucester*, Maitland, p. x).

[10] Walter de Pattishull (Buckingham and Bedford, 1224–8), William Brewer, Junr. (Devon, 1224–55), Hugh, bishop of Lincoln (Lincoln, 1223–6), and Walter, bishop of Carlisle (Cumberland, 1223–33).   [12] *C.P.R.*, 1225–32, pp. 488-9.

[11] In the period 1223–26.

[13] Stubbs, *Sel. Charters*, 387 ; *Const. Hist.* ii. 84.

time, if exception be made of the mercenary captains. Indeed
some of John's sheriffs had again been in office in comparatively
recent years.[14] The sheriff's hope of remuneration still lay in more
or less questionable impositions. High payments, occasioned by
his subletting bailiwicks at heavy *ferms*, and oppressive insistence
upon attendance at local courts are mentioned as the baronial
grievances of the day against the sheriff.[15] The usual tenure
extending over several continuous years gave opportunity for
peculation, and the old-time desire for sheriffs who were men of
the county, and not ignorant of its customs and careless of its
interests, was still alive.

A general plan of reforming the office was included by the baronial
committee in the Provisions of Oxford. This required that sheriffs
should be loyal, substantial men and land tenants, vavassors of the
county in which they served. They were to treat the people of the
county well, loyally and justly, to hold office only for a year, and
within that time to give up their accounts to the exchequer and
answer for them. They and their bailiffs were forbidden to take
any fee and were made punishable for so doing. The discontinuance
of the farming of shires was possibly intended, it being specified that
the king should grant to sheriffs of his own according to their contributions so they might rightfully keep the county.[16]

A parliament held by the new council at Michaelmas adopted
further regulations concerning the sheriffs as anticipated in the
scheme of reform.[17] These provided that sheriffs be designated
per electionem, that is to say under direction of the baronial council.
Those of 1258 were actually so chosen.[18] The letters patent in the
king's name putting these measures in effect ordered none of the
sheriffs or bailiffs to continue in his bailiwick more than one year, so
that wrongs done might then be redressed.[19] According to the
Annals of Dunstable [20] it was specified that sheriffs were to receive
a commission (*chartam*) written in English and sealed with the great

---

[14] Engelard de Cigogné (Berkshire, 1234–6); Walerand le Tyes (Herefordshire, 1246–49. Walerand's predecessor, Emery de Chancaux, seems to belong
with Engelard to the family proscribed in Magna Charta.

[15] *Ann. Monast.* i. 441 ; Stubbs, *Sel. Charters*, 384.

[16] *Ann. Monast.* i. 451, 455, 504, 506-7 ; Stubbs, *Sel. Charters*, 391.

[17] *Ann. Monast.* i. 396.

[18] *K.R. Mem. Roll*, no. 32, m. 5 a. The sheriff of Lancaster in Feb. 1259
was chosen by four *electi* (*P.R.O. Lists and Indexes*, no. 9, p. vi).

[19] Madox, *Exchequer*, ii. 148.

[20] *Ann. Monast.* iii. 210.

seal, which was to be read several times in the year before the county court so its provisions might be brought to the attention of all. As to what these were to be there is no hint except in the general scheme outlined above and in the new oath of office, now prescribed by the council for sheriffs when they were admitted to office.

This form of oath, designed to impose restraints upon the conduct of sheriffs, and forming part of the movement to bind officers of state generally by similar restriction, has been preserved in more than one record.[21]   It was being administered to the new sheriffs prior to October 20, and is eloquent concerning the abuses for which their predecessors were responsible. By its terms the sheriff undertook to serve the king loyally ; to do right to rich and poor alike and not to omit this for love, hatred, fear or covetousness ; to take nothing by occasion of this jurisdiction either by himself or others, except only the meat and drink by custom brought to his table and that for one day at most ; to have not more than five horses [22] in the place where he lodged by reason of his office ; to lodge with no one who had less than forty pounds value of land, nor in any religious house which had less than the value of a hundred marks yearly in land or rents ; not to lodge with any of these more than once a year, or twice at most, and then only with their prayer and consent, not drawing this into a precedent ; not to take presents of anything else worth more than twelve pence in case it be expedient to lodge there longer ; not to have more bailiffs than were reasonably needed to maintain his office and these persons whose trustiness could be vouched ; and finally to farm to no one the counties, hundreds, wapentakes and other bailiwicks of the realm.

A general change of sheriffs [23] in the autumn of 1258 marks the attempt to introduce the reforms. In practically every case the sheriff was appointed *ut custos* [24] to carry out the new fiscal

[21] The exchequer copy is on a strip annexed at the beginning of *K.R. Mem. Roll*, no. 32, P.R.O. The copy preserved in the patent roll of the year is printed in Madox, *Exchequer*, ii. 148, note (k), and also in *Royal Letters of Henry III.*, R.S. ii. 130-2. A third copy, appearing in the *Annals of Burton* (*Ann. Monast.* i. 454, 506) is reprinted in Stubbs, *Sel. Charters*, 398-400.

[22] The official Memoranda Roll MS. has *cink*; the copy in the *Royal Letters* has *cints*, apparently a copyist's error for *cink*, which the translator has mistaken for *cent*; but the original in Madox is *cinc*. The version in the *Annals of Burton* (p. 454) would permit the sheriff to have only his own horse in the place where he lodges. Cf. p. 282.

[23] *C.P.R.*, 1247-58, p. 655.

[24] Cf. list of sheriffs in *Deputy Keeper's 31st Report*.

arrangement by which he was to avoid the evils incident to his farming the shire and letting its subdivisions at farm. This plan was obviously unsuccessful, no doubt on account of financial results, for the next year witnessed a return to the older one.[25] A royal act of November 18, 1258, directed the barons of the exchequer to write certain knights [26] to come and receive custody of their counties in the form recently provided by the magnates of the King's council, and to commit such custody to them when they had received their oath.[27] The original plan involved only an annual term for sheriffs and the selection of new ones every autumn. Seventeen of the sheriffs were changed about Michaelmas 1259, and five more before the year was half gone.[28] Sheriffs were again required to take the prescribed oath.[29] They were presumably appointed under an agreement of 1259 by which for the current year the chief justice, treasurer and barons of the exchequer were to choose. For the future four good men were to be designated in the county court, one of whom should be appointed by the barons of the exchequer.[30]

The king soon succeeded in shaking off his tutelage under the council of fifteen and in regaining control of the sheriffs. But baronial control continued for the next two years, nearly all the sheriffs of 1259 remaining in office. Before Michaelmas 1261 [31] Henry III. was unwilling longer to concur in the rule of the committees that had been thrust upon him, and a struggle had all but begun. On September 11, the king ordered the sheriffs to send to him at Windsor the three knights of each county who had been summoned by certain magnates to meet at St. Albans.[32] On October 28, he ordered the sheriff of Kent to keep the sea coast between the Cinque Ports so that no foreigner might enter without licence.[33] At Michaelmas, possibly even in the preceding summer,[34] he appointed

[25] The list for 1259–60 shows that in a considerable number of counties the sheriff is again *firmarius*.

[26] Fourteen in number.

[27] *C.P.R.*, 1247–58, p. 655.

[28] *P.R.O. Lists and Indexes*, no. 9.

[29] *K.R. Mem. Roll*, no. 33, m. 4 d, margin : *de vicecomitibus juratis* and the accompanying list of sheriffs of 27 counties who were sworn.

[30] *Annals of Burton*, R.S. 478 ; Stubbs, *Const. Hist.* ii. 217.

[31] According to *Flores Historiarum*, R.S. ii. 463, the king took the initiative in February.

[32] Stubbs, *Sel. Charters*, 405 ; Close Roll 77, m. 6 d.

[33] Close Roll 79, m. 19 (P.R.O.).

[34] July appointees in Cumberland, Cambridge and Huntingdon, Buckingham and Bedford, and apparently in Devon, Dorset, Somerset, Gloucester and Kent (*P.R.O. Lists and Indexes*, no. 9) are royalists.

sheriffs by letters patent. After the translation of St. Edward the barons removed these and once more designated *custodes comitatuum*, who would not let the itinerant justices perform their office.[35] In his letters patent of October 18, addressed to the men of two counties, the king complained of this keeping of counties and interference with justice on the part of his adversaries, and commanded obedience only to those to whom he had committed the counties.[36] The adjustment of the general issue without bloodshed in the course of a few months more is a familiar story.[37] Early in 1262 it was agreed that this year Earl Richard, the king's brother, was at Michaelmas to appoint a sheriff of each shire from among four knights selected as before from the counties, but that henceforth the king should appoint whomsoever he wished.[38] The king's appointees retained their places until the battle of Lewes more than two years later.[39] His right to appoint these, as well as ministers of state and household officials, was reaffirmed by the decision of the king of France in arbitrating at Amiens, in January 1264, the issues between the two parties.[40]

The result of this contest over the mode of appointment, which seems to have gone in the king's favour almost from its beginning, was soon reversed. After Simon de Montfort defeated King Henry at Lewes in June, 1264, and reimposed upon him the control of a special baronial council, the royal nominees were removed from the control of the counties. This change of sheriffs was very general except in three counties held as of fee, and in London and Middlesex, where they were nominated by municipal authority. The new sheriffs, most of whom were appointed at or near midsummer, in many cases bore the title *custos* as at the beginning of baronial

[35] Stubbs, *Const. Hist.* ii. 87 ; *Liber de Antiquis Legibus,* Camden Soc. 49 ; *Flores Historiarum,* R.S. ii. 473 ; *Ann. Monast.* iii. 217.

[36] *C.P.R.,* 1258–66, p. 178. In 1268 allowance was made by the exchequer to the heir of Alexander de Hampden, sheriff of Bedford and Buckingham in this period, for the time he was deterred from holding shires and hundreds, and freely performing his office (*L.T.R. Mem. Roll* 42, m. 11).

[37] Stubbs, *Const. Hist.* ii. 88, 217.

[38] *Annales Osney et Chron. Wykes,* R.S. 130-1 ; *Royal Letters of Henry III.* ii. 198.

[39] Eustace de Balliol, appointed by the king, Oct. 20, 1261, to keep the county of Cumberland and castle of Carlisle (*C.P.R.,* 1258–66, p. 179), actually accounted for several years. The king also committed other counties with castles (*ibid.* 179, 204), and sheriffs in office generally remained until 1264.

[40] Rymer, *Foedera,* i. 433-4 ; *Liber de Antiquis Legibus,* Camden Soc. 60 : Stubbs, *Sel. Charters,* 408.

control.[41] They evidently were chosen by the barons of the king's council to hold office during pleasure.[42] In the following October Earl Simon's government adopted the policy of restoring to sheriffs the custodianship of certain castles [43] for the conservation of the peace and the keeping of prisoners. The reformers of 1258 had assigned these to special guardians, even in counties the sheriffs of which usually appear as castellans.[44] The new government continued to make wide use of special keepers of the peace concurrently with sheriffs,[45] in June 1265, calling on both officials in various counties to send horses and arms to Gloucester to aid De Montfort in his struggle with Gilbert de Clare.[46] As early as June 30, 1264, the king's letters to the sheriff of Lincoln were made to announce that he was at peace with the barons and to order the sheriff to proceed against disturbers of the peace.[47] In the summer and autumn of this year, various sheriffs and keepers of the peace were directed to call out the knights and free tenants of their counties who owed military service, to ward off the danger of invasion from abroad.[48]

At the renewal of the struggle between the two parties, in the spring of 1265, King Henry seems to have appointed a few sheriffs in regions favourable to his cause.[49] The importance of such officials was a lesson too obvious to be lost. When the battle of Evesham in the following August reversed the tables and ended baronial rule, Henry at once began turning out the sheriffs and installing those of his own choosing.[50] But one or two of the group appointed by

[41] Fifteen are so designated. Cf. list in *31st Report of Deputy Keeper*.
[42] *C.P.R.*, 1258–66, pp. 340.
[43] *Ibid.* 373.
[44] *Ann. Monast.* i. 453. Cf. *P.R.O. Lists and Indexes*, no. 9.
[45] *Foedera* (1816), i. 442; *C.P.R.*, 1258–66, pp. 361, 426. These were permitted to call upon the sheriffs for aid, but not to interfere with their fiscal duties (*ibid.* 405).
[46] *Ibid.* 487.
[47] *Ibid.* 362.
[48] *Ibid.* 360-61, 372; *Liber de Legibus*, 67-9.
[49] Roger de Acle, appointed for Gloucestershire, Feb. 1, 1265, was probably nominated by the baronial council through the influence of Gilbert de Clare. But the king regarded him as a friend, for he remained in office five years after the battle of Evesham. John de Grey, appointed sheriff of Nottingham and Derby, Apr. 14, 1264, and reappointed, Aug. 8, 1265, was clearly a royalist (*P.R.O. Lists and Indexes*, no. 9). In Herefordshire a change comes at midsummer, and there appear to have been rival sheriffs for a time. In Kent Roger de Leyburn serves from Aug. 8. In 1266 he is pardoned his arrears for the laudable service done the king for the whole period of his shrievalty (*K.R. Mem. Roll* 40, m. 4).
[50] New commissions are recorded in *C.P.R.*, 1258–66, pp. 455, 471, 490.

the barons could have been in office beyond Michaelmas.[51]   There was henceforth no restraint upon the king's power of appointing. He took the city of London into his own hands because of its adherence to De Montfort.   When the mayor and citizens according to custom proceeded to the exchequer at Michaelmas to present the sheriffs newly chosen for the city, they found no one, and the nominees were not admitted to office.[52]   It required a grant of twenty thousand marks to regain favour, and the right of choosing sheriffs was again conceded only on May 1, 1266.[53]   One change touching the shrievalty made by the reformers and now permanently retained, affects the sheriff's tourn.   From attendance here great prelates, earls, barons, persons in religion and women were exempt except when the presence of any of these was especially required. Those holding tenements in various hundreds were excused except in the hundred where they dwelt.   Furthermore the tourn was to be held according to the form of Magna Charta and the usage of the reigns of Richard and John.   The prohibition upon the levy of fines for beaupleader in the courts of shire and hundred was also reaffirmed.[54]   The sheriff was of course especially useful to the king from 1265 to 1267 in restoring peace and order, and for this the *custos pacis* also continued to be employed.[55]

The usefulness and importance of the sheriff in the thirteenth century are also shown by his appearance in the parts of Ireland and Wales which became subject to the king's authority.   A writ of Edward I., directed to the justiciary of Ireland in 1274 conferred upon the latter the power of appointing and removing sheriffs. The sheriffs of Tipperary, Limerick and Waterford are especially mentioned as included in this arrangement.[56]   King Henry I. and his son Earl Robert of Gloucester each had a sheriff in Wales,[57] the one in Pembroke the other in Cardiff.   A century later in 1223 a writ of Henry III. was addressed to his sheriff of Carmarthen and Cardigan.[58]   After the second Welsh war of Edward I. the Statute of Wales in 1284, in providing for the government of the subjugated regions, erected the Welsh counties as they were to remain for three

---

[51] New appointments for Southampton and Wiltshire are made the next spring.   [52] *Liber de Legibus*, 77.
   [53] *Ibid.* 80-2, 85.   [54] *Stat. of Marlborough*, x. xi.
   [55] *C.P.R.*, 1258–66, p. 671. See below, pp. 200, 221.
   [56] *Ibid.*, 1272–81, p. 57.   [57] Above, p. 108 and *P.R. 31 Henry I*. 136.
   [58] *Rot. Lit. Claus*. i. 576. Sheriffs regularly appear in Carmarthen from 1241. in Cardigan from 1279, in Merioneth from 1283.

N

centuries and a half.   By this act the office of sheriff was established for Anglesea, Caernarvon, Merioneth, Flint, Carmarthen and Cardigan.[59] Coroners and bailiffs of commotes were also designated, and the latter were to discharge their offices according to what was given them in charge by the sheriffs and justices.

The sheriffs of the English counties in the thirteenth century were still appointed and removed by the king, but as the century wore on less and less by his direct action.   The fiscal importance of the office, the sheriff's semi-annual appearance at the exchequer, and the fact that this body attended to the detail of farming the shire,[60] tended to give an increasing control of the office to the treasurer and barons of the exchequer.   Before them the sheriff took the oath of office, a usage mentioned in 1252 as customary,[61] and evidently continued during the rule of the barons beginning in 1258, for both the form of oath prescribed this year and the record of the swearing in of the sheriffs the next year are preserved in the Memoranda Rolls.[62]   The ordinary form of oath other than this one does not appear in the records so far as the writer has observed. It is clear that there was a regular form and that it was sometimes administered outside the exchequer.   Upon the death of the sheriff of Surrey and Sussex in the Easter term, 1303, the treasurer and barons sent an exchequer clerk to the Tower of London with commissions for two persons who were designated to administer the office, and who were sworn according to a form of oath which the clerk had with him.[63]   When King Edward in 1305 removed for negligence Milo Pichard, sheriff of Herefordshire, who seems to have been in service in Scotland, the oath of his successor was taken by the king in person before his appointment.[64]   Some variations of the form of oath, including the form administered to an hereditary sheriff in 1298,[65] have been preserved, but concerning the ordinary

[59] Stats. of Realm, i. 55-6.
[60] In the time of Edward I. it is mentioned as fixing the amount (Madox, Exchequer, ii. 143).
[61] Ibid. ii. 68, note (w) ; cf. Mapes, De Nugis Curialium (Camden Soc.), Distinc, i. x.
[62] Above, p. 171.   At Michaelmas 1259 a sheriff fecit sacramentum sic provisum est per magnates de consilio regis (K.R. Mem. Roll, no. 32, m. 5).
[63] L.T.R. Mem. Roll 73, m. 7.
[64] P.R.O. Lists and Indexes, no. 9, Herefordshire.
[65] For the oath taken by the sheriffs of London about the fourteenth century see Liber Albus, i. 306-7, for that administered in 1298 to Guy de Beauchamp, earl of Warwick, as hereditary sheriff, Madox, Exchequer, ii. 149, note (kk).   The latter form has to do chiefly with the preservation of the rights of the crown,

form nothing is certain beyond the fact that the sheriff is said to swear to do well and faithfully what pertains to his office.[66]

The naming and dismissing of sheriffs was in every respect a royal prerogative. It seems probable, as has been shown, that sheriffs of Henry III.'s childhood even questioned the right of his ministers to depose them. The government of the young king's minority, presumably the council, in 1220 removed Henry fitz Count, earl of Cornwall from the shrievalty of that county because he had withdrawn from the *curia* without the licence of the council, and had declined to carry out the royal precepts.[67]   Henry III. more than once prevented the acceptance at the exchequer of the sheriffs chosen by the citizens of London in accordance with their chartered privilege.[68]   The baronial council of the De Montfort supremacy affected to act in the king's name when they appointed sheriffs.[69] A writ of Edward I. in November 1303, removed a sheriff because of his delay in following the king with horses and arms into Scotland, whither he had been ordered.[70]   Yet Henry III. upon occasion directed the barons of the exchequer to fill vacancies in the office,[71] and Edward I. seems usually to have employed them for this purpose. Philip de Bath and Philip Lovel in 1257 or 1258 came before the barons and acknowledged that on the king's behalf they had committed two counties to custody ; and at the beginning of Edward's reign Reginald de Aclea was commanded by writ of exchequer to take the office of sheriff of Gloucestershire into the king's hand and to commit it to a trusty person until the king gave further orders through the barons.[72]   By writ of great seal Edward some-

---

lawful treatment and justice for the people of the shire, and the appointment of bailiffs who will loyally serve the king and the people. See also *Liber de Legibus*, 85, and *C.P.R.*, 1247-58, p. 250.

[66] *Gilbertus eodem die praestitit sacramentum de fideliter faciendo ea que ad officium vicecomitis pertinent in comitatu Westmeriland* (*L.T.R. Mem. Roll* 62, m. 5 d, *anno* 1290). So also *K.R. Mem. Roll* 76, m. 48, *anno* 1303.   The oath of the sheriff in the seventeenth century preserved a few features which appear in the one framed by the barons in 1258 (Greenwood, *Bouletherion, or a Practical Demonstration of County Judicatures*, London, 1663, 230-1).   Cf. the oath in Madox, *Exchequer*, ii. 149, note (kk) and the earlier portion of that in Dalton, *Office and Authority of Sheriffs* (London, 1700), pp. 9-10.

[67] *Pat. Rolls*, 1216-35, p. 231.

[68] *Liber de Legibus*, 8, 12.

[69] Above, p. 172.   *Rex commisit Petro Foliot Comitatus Oxon et Berks in forma qua superius* (*K.R. Mem. Roll* for 1259, no. 32, m. 5).

[70] Milo Pichard, sheriff of Herefordshire (*ibid.* no. 77, m. 45 d).

[71] As in Madox, *Exchequer*, ii. 68, note (w).

[72] *Ibid.* ii. 68, and notes (u), (y).

times directed the barons to commit a county to a certain person during the king's pleasure.[73]

In the last year of King Edward's reign the treasurer and barons issued an order of very general purport that sheriffs found unfit be removed and others put in their places.[74] Here as elsewhere it is seen that the treasurer is regarded as the official who directs those acts,[75] for in this instance it was ordered in the exchequer that the barons might do what was necessary without staying for the treasurer's presence, even though he did not happen to be in the exchequer at the time. There were instances in which a sheriff was replaced by the treasurer's *locum tenens* attended by several barons.[76] Progress is being made toward later usage in appointing sheriffs, but the matter as yet seems exclusively an exchequer function. The rule long in force was established in 1315 or 1316, when it was required by statute that sheriffs be assigned by the chancellor, treasurer, barons of the exchequer and justices, or in the absence of the chancellor, by the others.[77]

The assumption of the office by the person designated at the exchequer was regarded as compulsory.[78] The knights chosen by the baronial council in 1258 to act as sheriffs were enjoined to come and receive office as they loved all they had.[79] The fiscal duties of the office might involve the incumbent in difficulties for the rest of his life, and were likely to bring his estate into the custodianship of the exchequer if he died in office.[80] Some persons sought and obtained the king's grant of the privilege that they should not be made sheriffs against their will. This was a very common form of exemption in the time of Henry III.[81] The person installed in the office might put an under-sheriff in his place, but appearances at the exchequer in person would still be required unless its officials consented

[73] Madox, *Exchequer*, ii. 142.

[74] *Ibid.* ii. 109.

[75] Cf. *R.P.O. Lists and Indexes*, no. 9, p. iii.

[76] *K.R. Mem. Roll*, no. 76, m. 48 ; for negligent and improvident conduct in office. Commissions were issued to two persons to take up the duties of the position.

[77] *Stats. of Realm*, i. 174 ; cf. 14 Edward I. cap. 7.

[78] Thus, an exchequer writ of 1295 commanded Robert de Woddeton as he loved himself and all his goods to come before the barons with all haste, and when he came they committed a county to him (Madox, *Exchequer*, ii. 143, note (d). Cf. *P.R.O. Lists and Indexes*, no. 9, under Kent, *anno* 1304.

[79] *C.P.R.*, 1247–58, p. 655.

[80] Below, p. 256.

[81] Usually joined with exemption from service on juries, assizes and recognitions, and from the obligation to be made escheator or coroner (as in *C.C.R.* i. 413).

to receive a deputy. Moreover, when a sheriff committed his county to another person, the king's consent was required.[82] In the earlier years of the minority of Henry III. a county was sometimes committed to one person for another.[83]

To these arrangements the sheriffs in the palatinates of Chester and of Durham, the officials respectively of the earl in the one case and the bishop in the other,[84] were of course exceptions. But there were two other highly interesting classes of exceptions, one relating to the term of service, the other to the mode of choice. The first of these arose from appointments for life sometimes found in John's time, but much more frequently after Henry III. attained his majority. Some of these came to nothing, like the many life grants of shires made in 1232 to Peter of Rievaux. In one case the grant was made to a sheriff of two counties on petition of the men of those counties, the grantee living but a short time and being succeeded for a brief period by a son.[85] In another instance a life grant of the office meant only a term of two years.[86] But not everywhere was a life grant a passing arrangement of such brief duration. In the year 1300 there were five shrievalties held as of fee, namely those for the counties of Worcester, Cornwall, Rutland, Westmoreland and Lancaster.

The oldest of these hereditary shrievalties, Westmoreland, had since 1203 been held by Robert de Vipont, one of John's followers, and by his descendants. The king gave this county to Robert and his heirs.[87] In August 1219 the county of Cumberland was also granted him in lieu of the castle and honor of Tickhill,[88] to be held without paying *ferm* or *proficua* until a month after the feast of St. Hilary 1222. Since the county of Westmoreland rendered no *ferm* at the exchequer, Robert no doubt held it under some similar

[82] Madox, *Exchequer*, i. 463, n. (q), records the justiciar's assent. Henry III. consented (*ibid.* ii. 152-3) that Walter, archbishop of York, depute the keeping of two counties to another.

[83] *Pat. Rolls*, 1216–25, p. 418.

[84] Some complaints concerning the usages of the sheriff of Chester in 1249 while that county was in royal wardship are recorded in *Cl. Rolls*, 1247–51, p. 185. Concerning the sheriff of Durham, see Lapsley, *Durham*, 80-86.

[85] Ralph fitz Nicholas, sheriff of Nottingham and Derby in 1232 (*C.P.R.*, 1225–32, p. 472). His total tenure extended from 1224 to 1236. Hugh fitz Ralph held the office three and a half years more.

[86] Godfrey de Craucumbe, who received custody of the county and castle of Oxford in 1230 or 1231 (Madox, *Exchequer*, ii. 141, n. (u)), had been sheriff since 1225, and could have held the position only until Oct. 1232.

[87] *Assize Roll* 982, m. 34 d, P.R.O.

[88] Madox, *Exchequer*, ii. 68, n. (t).

arrangement. After his death in 1228 it was placed in charge of Hubert de Burgh, the justiciar, who had the custody of Robert's land and his heir, John, at that time a minor.[89] This John appears as sheriff from 1234. His lands were seized and in 1242 given into the custody of the bishop of Carlisle,[90] but in 1247 he was again sheriff. On the eve of the barons' war a second Robert de Vipont appears on the list of sheriffs. The latter joined the barons against the king,[91] and was superseded after the battle of Lewes by John fitz John, presumably his brother. There is another sheriff on the list in 1275. But in 1278 or 1279, this official is already said to hold the office for the two daughters of Robert de Vipont and their husbands.[92] In 1290, Isabella, wife of Roger Clifford, the elder daughter, disagreed with her sister Idonea, wife of Reginald de Leyburn, over the appointment of an under-sheriff to perform the duties of the office. The king then took the office temporarily into his own hands, and nominated an incumbent until the two agreed on the person to be appointed.[93]

The restoration in the thirteenth century of hereditary succession in Worcestershire was undoubtedly effected through the influence of the older rights and claims of the line of Urse d'Abetot. In the reign of Henry II., and again in that of Richard I., William de Beauchamp, the head of this house, for some years held this shrievalty. The government of the Earl Marshal in 1217 conferred on Walter de Beauchamp, son of William, the county and castle of Worcester with the forest of this county to be held during good pleasure.[94] The extent of the grant was thus practically the same as that held by the first Walter de Beauchamp in the reign of Henry I. The county and castle were subsequently conferred until the king should attain the age of fourteen.[95] Walter was then, if the king so pleased, to resign them in as good condition as he received

[89] Pat. Rolls, 1225–32, pp. 177, 187.

[90] Ibid., 1232–47, p. 284.

[91] Cl. Roll 87, m. 13. In May 1256 he was in the service of Edward, the king's son (ibid. 71, m. 12 d). In 1260 he was held to the king in several debts of his predecessors, and sheriffs were ordered not to distrain him to take arms until they received special mandate (ibid. 75, m. 11 d).

[92] Assize Roll 981, m. 19. It is clear from this roll that Roger Clifford was sheriff still earlier.

[93] L.T.R. Mem. Roll 62, m. 5 d. From 1295, Robert de Clifford, son and heir of Isabella, and Idonea, her sister, jointly presented the sheriff (as in Chancellor's Roll 94, Westmoreland; K.R. Mem. Roll 74, m. 51; L.T.R. Mem. Roll 67, m. 7 d).

[94] Pat. Rolls, 1216–35, p. 37.

[95] Madox, Exchequer, ii. 149-50, n. (n).

them.   He remained sheriff, but was disseised in or just before 1230, when he made fine to have the county as before.[96]   In 1238 he made the claim that he held the county as of fee and the still more surprising assertion that sheriffs of Worcestershire in the reign of John, none of whom were of his lineage, had so held it, rendering thirteen pounds and not more as *proficuum*.   The question was left to the barons of the exchequer for determination [97] and apparently was decided in Walter's favour, for at his death in 1235 the office passed to the fourth William de Beauchamp, who late in the reign of Henry III. became earl of Warwick.[98]   His son, the noted Earl Guy, succeeded in turn in 1298.   The oath of the latter, taken upon assuming the office of sheriff at Michaelmas of that year, is carefully preserved on the Memoranda Roll of the Exchequer,[99] and shows the importance attached to safeguards against a power which was likely to be maintained for a generation.

The three remaining life shrievalties by the reign of Edward I. had all become attached to appanages held by the king's near relatives.   The oldest of these was Cornwall, which had been controlled practically as a palatinate in turn by Count William of Mortain, Earl Reginald and Earl John, none of whom apparently made accounting at the exchequer for its revenues.   Richard, the brother of King Henry, when still but a child, in 1217, was appointed sheriff of Berkshire, obviously as a device for providing his support.[100]   But in 1225 he was created earl of Cornwall, and the county was made over to him together with the king's homages and *ferms*.[101]   From this time the sheriffs of Cornwall were not regularly named in the public records until after the death of Earl Richard in 1272.   His son Earl Edmund regained the sheriffdom in 1278, until his death in 1300 appointing a sheriff [102] who still accounted to the king for debts in the county, but not its *ferm*.[103]   The earl of Cornwall

<hr/>

[96] Madox, *Exchequer*, i. 463, n. (s).   [97] *Cl. Rolls*, 1237–42, p. 70.

[98] Doyle, *Baronage*, iii. 578.   He was accustomed to appoint and remove the under-sheriff who responded for him at the exchequer (*K.R. Mem. Roll* 43, m. 6).

[99] Printed, Madox, *Exchequer*, ii. 149, n. (kk).   An under-sheriff, whom he nominated at the exchequer (*L.T.R. Mem. Roll* 70, m. 14, and 77, m. 25), acted for him.

[100] Henry de Scaccario acted for him : *Deputy Keeper's 31st Report*.

[101] *Pat. Rolls*, 1216–25, p. 507.   The sheriffs are those of Earl Richard : *Pipe Roll* 26 Henry III., ed. Cannon, p. 108.

[102] Described in 1305 as sheriff and seneschal of Cornwall (*K.R. Mem. Roll* 78, m. 67).

[103] The detailed account of the income from hundred and county courts of the sheriff and seneschal of Cornwall for part of the year 5-6 Edward I., however,

also gained the shrievalty [104] of Rutland before the death of Henry III., appearing himself as sheriff in 1288. In 1300 it was transferred with the county to his wife, the countess Margaret, *nomine dotis*. She continued to hold it into the reign of Edward II., constantly appearing at the exchequer by an attorney.[105]

The county of Lancaster originally became hereditary in the family of William of Lancaster, who was sheriff from 1232 to 1246, holding the county and castle at the king's pleasure in the usual form.[106] In 1255 Patrick de Ulvesby was commissioned for five years to keep not only these but also the honor of Lancaster, rendering as *proficuum* sixty marks and keeping the castle at his own expense.[107] He was ousted in February 1259 by a *custos* appointed by the barons. After 1265, Roger of Lancaster held the county as *custos*. A little later the king committed the county to him for life on condition that he render a hundred marks a year at the exchequer. But in June 1267 the sheriffdom was given to Edmund, the king's son, who at this time was created earl of Lancaster.[108] For the first twelve years of the reign of Edward I. he accounted at the exchequer for debts, but not for *ferms*. Possibly the two sheriffs who served for the succeeding fourteen years were his agents. When the famous Earl Thomas succeeded his father in 1298, he once more assumed the office of sheriff for the county.

The other exceptional mode of filling the office was through election. Of this London had long afforded an example. The privileges of election continued to be exercised in the south-west after King John's death. In 1225 the men of Devon made fine in the sum of two hundred marks for the privilege of choosing a sheriff *de se ipsis* to serve for three years from Michaelmas of that year. In 1230 this was repeated.[109] The Pipe Roll of 1222 [110] mentions

---

shows £48 : 11 : 11 *ad perficiendam firmam corpus sic comitatus post terras datas*. In 12 Edward I. the earl made the claim that he held the whole county of the king in chief with all pleas of county and hundred courts, and also all income arising from the view of frankpledge and the semi-annual tourn. The sheriff and none other had custody of parks and escheats (*Assize Roll* 111, m. 30 d, P.R.O.).

[104] *Rot. Hund.* ii. 1, 6, 53. The county had for a time pertained to that of Northampton.

[105] *Deputy Keeper's 31st Report*, 332.

[106] *C.P.R.*, 1232–47, p. 239.   [107] *Ibid.*, 1247–58, p. 448.

[108] *Deputy Keeper's 31st Report*, 301 ; Doyle, *Baronage*, ii. 309.

[109] *C.P.R.*, 1216–25, p. 554. The men of the county were to name three knights, one of whom the king was then to appoint. For the later instance, Madox, *Exchequer*, i. 417, n. (g).

[110] Madox, *Exchequer*, i. 508, n. (n).

the fact that the men of Dorset and Somerset owe the same amount for having Roger de la Ford as sheriff. The king's letters patent of 1226 inform these two counties of the royal assent to the election of William fitz Henry *ad vicecomitem nostrum faciendum*, and directed them according to the usual formula to be intendant to him as sheriff.[111] In 1235 or 1236 it was again agreed by the same counties that a payment should be made [112] *pro habenda electione vicecomitis sui*.[113] Election had presumably come to be regarded and claimed as one of the liberties of those counties to be exercised permanently. It is interesting to note in the Hundred Rolls the claim of the men of Cornwall that they used to elect their sheriff by virtue of a charter of King John, but that Earl Richard took away this liberty from the county and Edmund his son still deprives them of it. They certainly made fine for the privilege with King Henry III. at some time prior to May 1221, engaging to pay five hundred marks for this and other liberties.[114] The payment in Devonshire to have as sheriff a man of the county forecasted an attempt at reform which recurred from time to time until its object was attained by the statute almost a century later, requiring sheriffs to be land-holders of the county. The baronial reformers of 1258 sought to attain this very end.

In the autumn of 1274, after King Edward began his actual rule in England, he dismissed about three-fourths of the sheriffs, probably on account of complaints of misrule similar to those levied against sheriffs during the inquest of 1274 and 1275 and recorded in the Hundred Rolls. The results of this searching inquiry have been so well formulated in the work of Miss Cam that it is unnecessary to notice them here in detail. But no general change of sheriffs occurred again until October 1278, when practically every sheriff in England except in the counties held as of fee was superseded.[115] The Annals of Dunstable represents this as a removal of sheriffs who were clerks and foreigners and a substitution of knights of their own counties.[116] The names of the removed show that they could have been foreigners only in the sense that they were men from other

[111] *Pat. Rolls*, 1225–32, p. 45. Another had first been elected.
[112] Madox, *Exchequer*, i. 420, n. (t).
[113] The person chosen seems to have been the sheriff of the time, Thomas of Cirencester, in office 1228–32, and again 1234–37.
[114] *Rot. Hund.* i. 56 ; *Lit. Claus.* i. 457.
[115] *P.R.O. Lists and Indexes*, no. 9. Wiltshire is an exception.
[116] *Ann. Monast.* iii. 279.

counties. This demand for a sheriff who dwelt in the county must have had due weight in securing the remarkable concession, made some twenty years later, by which counties were permitted to elect their sheriffs.

In the *Articula supra Cartas* of 1300 the king conceded to the men of each shire wherein the shrievalty was not of fee the right to elect the sheriff if they so desired.[117] A subsequent section of the same document appears to disclose a purpose of the grant in directing that at this election the commons of the shire choose sheriffs who will not charge them for rewards, nor lodge too often in one place, nor with poor persons nor men of religion.[118] The letters close [119] putting these provisions in force were addressed in 1300 to the coroners and the whole community of the counties, and specified that they elect, if they wished, one who could best know and execute the office of sheriff. They were then to present him, on the morrow of Michaelmas next, before the treasurer and barons of exchequer, to take the usual oath and do what pertained to the office. This was in accord with the usage long employed in respect to London, except that the mayor and citizens presented the sheriffs chosen by the city.[120] The sheriffs elected by the counties were to be presented by some lawful and circumspect man with letters patent under the seals of six of the more discreet and upright knights of the county.

This experiment under the terms of the *Articula supra Cartas* marks a crucial point in deciding how the English sheriff of the future is to be chosen. The privilege of electing sheriffs was an aspiration traceable for nearly two centuries before this document was issued. The grant upon the face of it aimed to prevent current abuses on the part of the sheriffs and to secure suitable men for the position. It ought virtually to have settled the question of resident sheriffs. The privilege theoretically at least remained open until 1311,[121] and yet one must agree with Stubbs that it was treated with indifference. Not even the men of the south-western counties, who had once been willing to pay well for the opportunity, seem now to have embraced it. The exchequer records mention one shrievalty in which election was adopted.[122] There is but the one mention of

[117] *Stats. of Realm*, i. 139, cap. 8.   [118] *Ibid.* cap. 13.
[119] *C. Cl. R.*, 1296–1302, p. 362.
[120] *Liber de Legibus*, 69, 77, 85, 86.
[121] Stubbs, *Const. Hist.* ii. 217-18.
[122] But *C. Cl. R.*, 1296–1302, p. 362, possibly shows interest in the matter elsewhere.

presentation of a sheriff under the seals of the six knights certifying to election. That the sheriff is to-day appointed and not elected may be regarded as in no small measure due to this result.

The evidence which does exist is for Shropshire and Staffordshire, and it throws some light on local conditions and on thirteenth century electoral ideas. Here in 1301 or 1302 Richard de Harley, who elsewhere appears as steward in this region for Amadeus, count of Savoy,[123] was elected and installed as sheriff in accordance with the new arrangements. But in April 1303 letters close,[124] directed to the coroners and community of the two counties, informed them that Richard was insufficiently qualified, and ordered that another be chosen and presented at the exchequer on the morrow of Michaelmas. The result of the election showed a division of opinion in the counties. When the election of the person formally designated was certified at the exchequer under the seals of six knights as required, a protest was lodged before the treasurer's *locum tenens* and the barons, and directed also to the chancellor and certain judges. In this two coroners of Salop and certain knights of both counties declared that they could not assent, and asked that the king appoint his sheriffs *pro voluntate sua*. The exchequer next ordered a second election to designate a suitable and capable person, with the result that the same individual as before was reported chosen. Against this the opposition now certified that several magnates of the county, including abbots and priors as well as barons and other knights of the counties, had not consented. The argument that quality rather than numbers should have weight prevailed. The treasurer and barons, so the record states,[125] agreed with others of the king's council that they could not uphold this election, whereupon a certain Walter de Baysi, a knight present in court, was deputed to the office. Thus concludes the only thirteenth century report of a county election to choose a sheriff.

The sheriff's usual relation to the county is expressed as custodianship. In relation to its issues he was either *firmarius* or *custos*.[126] The appointment of a sheriff for the East Riding of Yorkshire was protested by the sheriff as an act diminishing his *ferm*.[127] A special form of custody was often entrusted to the

[123] *C.P.R.*, 1302–7, p. 187. Cf. p. 207.   [124] *C. Cl. R.*, 1302–7, p. 84.
[125] *L.T.R. Mem. Roll* 73, m. 40 d; *K.R. Mem. Roll* 76, m. 31.
[126] As in *Chancellor's Roll* 53 (1259–60), P.R.O., especially under Lancaster, Shropshire and Staffordshire.
[127] *Royal Letters of Henry III.* ii. 325.

sheriffs of Devon, namely the keeping of the king's stannary and answering therefor at the exchequer.[128] The sheriff in the time of Edward I. was the usual custodian of the Jewry of Oxford.[129] In 1228 the king committed to the sheriff of Norfolk and Suffolk the coast of these two counties.[130] The sheriff of Herefordshire, in receiving the county and castle in 1231, was designated *custodem ad partes marchiarum tuendas et conservandas ab incursionibus Walensium,* and the sheriffs of Gloucestershire and Worcestershire were ordered to lend him the aid he needed.[131] A still more curious obligation attached to the holding of shrievalties is mentioned in 1253, when a grant of the king to Simon de Montfort involved the sheriffs of six counties.[132] Simon was to receive six hundred marks a year from the revenues of these counties. If a sheriff failed to pay, the earl or his heirs might distrain him by his lands or goods. No one was to be made sheriff in these counties without the assent of the earl and his heirs until provision for six hundred marks of land was made them.

In the period under consideration a royal castle was usually committed to the sheriff in about half of the counties. It is easy to distinguish the counties in which this does occur from those in which it does not, for the formulae employed in designating the sheriffs and in giving notice of their appointment were slightly different in the two cases.[133] In both of the reigns under consideration a castle was occasionally transferred by special order from a sheriff to another guardian.[134] By the time of Edward I. failure to grant the castle to a sheriff in a county where this had become the rule was regarded by him as a distinct grievance. It meant both inconvenience and loss. In 1306 the sheriff of Oxford induced the exchequer to grant him recompense for a three years' deprivation of Oxford castle, wherein, as he asserted, sheriffs had dwelt and had had the custody and profit of the jail.[135] In 1275 the *corpus* of Colchester castle was

---

[128] *C.P.R.,* 1247–58, pp. 173, 364.
[129] *C.P.R.,* 1272–81, p. 186.
[130] *Pat. Rolls,* 1225–32, p. 183 (Orewell, Oxford, Dunwich and Yarmouth to Len).
[131] *Cl. Rolls,* 1227–31, p. 601.
[132] *C.P.R.,* 1247–58, p. 249. This sum was granted to the earl in the name of a fee out of the issues of these counties.
[133] Cf. *P.R.O. Lists and Indexes,* no. 9, p. vi. The record of appointments occurs in the Memoranda Rolls ; for the time of Edward I. occasionally in the Fine Rolls.
[134] *C.P.R.,* 1247–58, p. 447 ; *L.T.R. Mem. Roll* 74, m. 8.
[135] *Ibid.* 76, m. 39 ; 75, m. 40.

committed to the sheriff, so he might keep prisoners in the jail there and do other things pertaining to his office with greater security.[136] The monks of Barnwell kept a list of minor payments and suits due from their men, so they might find protection against the sheriffs *ministri*, and so it would no longer be necessary to go to the castle and consult the sheriff's roll there to ascertain the justice of such claims.[137]   The sheriff of Wiltshire is mentioned as holding an inquest at the castle of Sarum.[138]   By Bracton's time the castle was proverbially the meeting-place of the county court.[138a]   Of course it might still be a fortress, but when in custody of the sheriff it was a prison and a public building for county use rather than a stronghold, as in the time of John and the earlier years of Henry III.   Its upkeep requires the consideration due a royal building, not usually that of an important stronghold.   Henry III., soon after attaining his majority, often made a fixed allowance to sheriffs to meet this expense.[139]   Later it was sometimes specified that the sheriff was to keep the *corpus* of the castle at his own cost, but was to receive a fixed annual sum for so doing.[140]

Since the sheriff as a rule held office during good pleasure, to effect a change it was necessary only to notify him to deliver the county or the county and the castle to his successor.   The incoming sheriff was sometimes required to find pledges for good service in office and the payment of his *ferm* and other debts due the king.[141] The old sheriff took a receipt from the new by an indenture made between them for the rolls, summonses, writs and other records which he gave over.[142]   The chirograph between the sheriffs of Herefordshire on the Wednesday before the feast of All Saints

[136] *C.P.R.*, 1272–81, p. 127.

[137] *Memoranda de Bernewell*, ed. J. W. Clark, 238.

[138] *Wilts Inquisitions Post Mortem*, Index Library, 123-4.

[138a] Bracton, *De Legibus*, R.S., v. 360.

[139] For the custody of the castles of Hereford and Paincastle, the hundred of Wormelowe and certain royal manors the sheriff of Herefordshire in 1232 was allowed the *proficuum* of the county, the things pertaining to the castellanies, and in addition a hundred marks in time of peace and a hundred pounds in time of truce or war (*C.P.R.*, 1232–47, p. 4).

[140] *Ibid.*, 1247–58, pp. 416-17.   Cf. the twenty marks awarded the guardian of Launceston castle for *custodia* (*L.T.R. Mem. Roll* 74, m. 8 (1303–4)).

[141] Madox, *Exchequer*, ii. 149-51.

[142] As in *K.R. Mem. Roll* 70, mm. 59, 60 (25 Edward I.); Madox, *Exchequer*, ii. 143, n. (e).   In 1303 the outgoing sheriff of Cornwall was directed to take an indenture (*L.T.R. Mem. Roll* 74, m. 4) for the county and castles and what appertained to them, whether rolls, men, summonses, arms, victuals, cattle, plows or other things.   In the same year the exchequer directed the executors of the testament of a deceased sheriff to deliver his rolls, etc. (*ibid.* 73, m. 7).

1278, for example, specifies the delivery of the county and castle with their appurtenances, among which are statutes under the king's great seal and letters patent confirming these; also certain writs returnable; a writ concerning the tallage of the Jews; writs of novel disseisin and of mort d'ancestor; likewise a virgate of land at that time in the sheriff's hands; and rolls of the county and of a certain hundred together with writs touching these.[143]

The sheriff's control over a numerous body of administrative subordinates beyond all question established his hegemony in local government. The selection of these, as Miss Cam has well declared, " not only affects the well-being of the county but also brings credit or discredit upon the king's service." [144]   From 1258 on it was the concern of king and council to induce the sheriff so far as possible to appoint only bailiffs who would accord to the people fair treatment and to the king loyal service. This word bailiff is a general term which includes both the heads of the lesser territorial divisions of the county and the staff attached directly to the sheriff's office. The Latin word *serviens*, most literally translated as sergeant, is often used as an equivalent title.

Except in the franchises, the territorial bailiffs were the sheriff's officials. The bailiff of the hundred or the wapentake, who held the sessions of its court and played an active part in suppressing crime,[145] farmed his bailiwick of the sheriff and at his direction executed writs, collected the king's debts and summoned juries.[146] From the time of Magna Charta to that of the Hundred Rolls there was complaint that the sheriff exacted from him a high *ferm* and thus whetted his propensities for extortion. In Kent the sheriff had

[143] *Sheriff's Accounts*, Exchequer, K.R., 18/2, P.R.O. The indenture between the sheriffs of the same county in 19 Edward I. accounts for fifteen writs returnable in specific cases, in Trinity term, also writs *ad primam assisam* (*ibid.*).

[144] *Studies*, 180.

[145] These duties of the *serviens* of a hundred are illustrated in *Select Pleas*, nos. 20, 119, 198. In *Rot. Hund.* i. 202, 204, 206, one Hamo de Forestall is designated both as *ballivus* and as *serviens*. In a sheriff's account the annual *ferm* received for a wapentake is headed *De Firma Servientis* (*L.T.R. Misc. Rolls* 5/8, P.R.O.).

[146] The *capitalis serviens* of each hundred (as, for instance, in *Assize Roll* 700, m. 13, Oxfordshire) aids in choosing the presentment jurors at the eyre. Sometimes the person who does this is called *ballivus* (as in *Assize Roll* 701, m. 31; in Britton, Nichols ed. i. 22, the bailiff of the sheriff). In *Inq. Misc.* i. 343, is mention of a letter of the sheriff of Essex to the bailiff of a hundred ordering him to cause a jury to come. Cf. Cam, *Studies*, 160-61. The sending of estreats to bailiffs of hundreds is shown in *Feudal Aids*, i. 23, 25, where bailiffs of liberties are immediately concerned.

a *ballivus* over the lathe, and the sheriffs of Yorkshire and Lincoln-shire [147] were accustomed to farm the ridings to a similar agent.

Of the sheriff's personal staff the chief was the under-sheriff, who is frequently mentioned as performing one or another of the numerous duties of his superior.[148] It was, however, only in unusual cases, and then with special permission of the exchequer, that he might altogether take over the functions of the office. The office force in addition to the under-sheriff must have been fairly numerous, rather more so than in the seventeenth century. When the sheriff travelled in pursuit of his duties, it was conceded that he might take with him to receive the hospitality of the shire a retinue which called for five or six horses. No doubt some of these attendants were grooms, for it is clear that sheriffs employed grooms in various capacities.[149] But far more important were the clerks and the office bailiffs.

The clerks, indeed, in legal reckoning seem to have been included among the sheriff's bailiffs.[150] The most prominent among the sheriff's clerks was his *receptor* or treasurer [151] whose prototype, so it has been shown earlier, is mentioned in the reign of Richard I. This seems to be the same clerk who made payments on behalf of the sheriff.[152] Another served as his custodian of writs (*receptor brevium*).[153] Clerks were also sent on the sheriff's official errands,[154] and their services were utilized in empanelling jurors,[155] as well as

[147] Magna Charta of 1215, sect. 25 ; cf. *Rot. Hund.* i. 130, 337. For the bailiff of the Kentish lathe, *ibid.* i. 204, 225.

[148] Thus in the absence of the sheriff he holds pleas of the county (*Chancery Miscellanea*, 138, P.R.O., writs and returns, 19-20 Edward I.). With the sheriff's clerk and others he is sent to convene four neighbouring hundreds (Bracton's Note Book, ii. no. 530). He was sometimes the jailer (p. 231 below), and his financial irregularities (as in *Rot. Hund.* i. 2) show that he had other duties as well.

[149] At London in the fourteenth century the sheriff's *garciones* took an oath of office (*Liber Albus*, i. 319), and had to do with the exaction of carriage and with food brought into the city by vehicle. The *garcio* of the sheriff of Bedfordshire is mentioned in *Rot. Hund.* i. 2.

[150] *De clericis et aliis ballivis vicecomitum* (*Rot. Hund.* introd. i. 14 ; *Foedera*, 1816, i. pars ii. 517. Cf. Miss Cam, *Studies*, 148-9). For the oath of office taken in London by clerks of under-sheriffs, see *Liber Albus*, i. 317-18.

[151] A fine of £20 was paid *ad opus Rogeri* [*de Clifford*] *Willelmo de Cumbe receptori suo in castro de Appleby* (*Assize Roll* 981, m. 27 d).

[152] Thus, ten marks of silver *de finibus et transgressionibus monete et lanarum*, received from the sheriff of Cumberland *per manus Ricardi de Langewathby clerici sui* by precept of the treasurer (*Exchequer K.R. Accounts*, 544/10, P.R.O.).

[153] *Fleta*, liber ii. cap. 67, sect. 18. Cf. *Bracton's Note Book*, no. 1741 ; Bracton, R.S., iii. 208.

[154] In *Bracton's Note Book*, no. 1052, one speaks for the sheriff before the justices. Cf. note 131.

[155] According to *Rot. Hund.* ii. 306, the clerk of the sheriff of Nottingham removes jurors who will not levy high amercements on their neighbours.

in keeping rolls and accounts and in drawing up instruments. The sheriff's clerk is also mentioned as having the power of arrest and as being in charge of the jail in the sheriff's absence.[156]

Some of the sheriff's *subballivi* who are mentioned may be sub-ordinates of the bailiffs of the hundreds,[157] but it is clear that the sheriff had directly under his control officials, like the bailiffs errant of later days, who were employed in making distraints and summonses,[158] a function sometimes performed in the twelfth century by the holders of sergeanties. These appear not only in Essex, but in the northernmost counties wherein territorial hundreds were lacking. In 1231 the king granted for the relief of the men of Cumberland that the sheriff should reduce the number of these, so the county should perpetually be in the custodianship of four *servientes* only, each to have two *pedites*.[159] According to the *Liber Albus*[160] none of the sheriffs of London was to have more than eight sergeants. The king and his council in 1280, in order to prevent unjust attachments, fines and summonses, made an order which regulated the conduct of sergeants in Westmoreland,[161] specifying, however, that they were still to have power to take thieves. The sheriffs of London also appointed officials, one for each ward, who here and elsewhere in the time of Edward I. were known as beadles (*bedelli*).[162] But it appears that sheriffs had both beadles

---

[156] *C. Cl. R.*, 1272–79, p. 487. As to his being in charge of the jail, *Inq. Misc.*, Chancery, i. 538. The story here is one of rascality. In *Rot. Hund.* i. 209 is complaint concerning a sheriff's clerk who took 12*d.* to relax a distraint; in *Feudal Aids*, i. 90, the sheriff's clerks and *ministri* take false pledge of prisoners.

[157] *Rot. Hund.* ii. 215. From *Feudal Aids*, i. 87, it is clear that hundredors had *subballivi*. In Kent the bailiff of a lathe also has a *subballivum* (*Rot. Hund.* i. 226).

[158] Cam, *Studies*, 149 (bailiffs itinerant in Essex). Cf. Smith, *De Republica Anglorum* (ed. Alston), 77. In Northumberland in 1198 there was a sergeanty *per servicium breviandi placita corone versus vicecomitem et faciendi summoniciones*, another *per servicium breviandi et faciendi distracciones* (*Book of Fees*, part i. p. 5). In Herefordshire at the same date was a sergeanty for making summonses and bearing the king's treasure (*ibid.* 6).

[159] *Pat. Rolls*, 1225–32, p. 456.   [160] iii. 5.

[161] *C. Cl. R.*, 1279–88, p. 109. In the barony of Kendal the sheriff was supposed to have two *servientes equites* and two *pedites* who were to have charge of matters touching the office of sheriff *sine denariis capiendis pro hospitiis vel aliis malis toltis faciendis* (*Assize Roll* 982, m. 22).

[162] Sharpe, *Letter Books City of London*, Book C, p. 57. Their oath of office is in *Liber Albus*, i. 318-19. In Nottinghamshire the farmers of the wapentakes had *subbedelli* (*Rot. Hund.* ii. 305). There is also mentioned a *bedellus* of a wapentake (ii. 316). In Gloucestershire in 1221 a certain Thomas who is a *serviens* is also included among the *bedelli* (*Select Pleas*, i. no. 167). Cf. *Salt. Arch. Soc.* iv. 37 ; also Mapes, *De Nugis Curialium* (Camden Soc.), *Distinc.* i. x, and *Materials for Life of Becket*, R.S., i. 54.

and bailiffs.[163] The sheriff's bailiffs made arrests, attachments, distraints, and summonses, and served as keepers of the jail,[164] although a special jailer was sometimes appointed.

These minor peace officials were obviously unpaid, and they long subsisted by exaction. In 1221 the men of Shropshire complained that from the time of Henry II. twelve or more *servientes* were set by the sheriff to guard the county and that they in reality lived off its men.[165] The baronial reformers of 1258 required sheriffs to take pledge in their oath of office to appoint no more bailiffs than were reasonably needed to maintain the office, and these persons [166] of assured trustiness, so the land might not be too much burdened by their eating and drinking. These were to be required to swear when put in office not to ask nor take of any man or religious house or town any lamb, sheaf, corn, wool nor any manner of moveable, nor money, nor anything of value, as they were wont to do in times past. The taking of unauthorized sums under colour of fines and amercements was also at a later time charged against them.[167] The Hundred Rolls reflect something of these same grievances. For instance in the hundred of Bassetlaw, Nottinghamshire, where there were formerly only six foot sergeants, there were now two riding beadles with two grooms in the North Clay division and two foot beadles in South Clay and also in Hatfield, one mounted and one on foot. The country was much aggrieved by their collecting hay, sheaves and grain and making other exactions through which it was said that they levied twenty-three marks a year. It is clear that the king's justices had power to deal with such cases, for in 1302, while on eyre in Cornwall, they forbade the prior of Bodmin in the capacity of the king's *hundredarius* of his hundred to keep one mounted bailiff and two *garciones*, directing that he retain only the one riding bailiff and one foot beadle, who in the past had been the

---

[163] *Feudal Aids,* i. 87. The beadles seem quite distinct from the bailiffs of the hundred. The well-known cacherell or catchpole appears in one place (*Rot. Hund.* ii. 207) as the subordinate of the *serviens* of the sheriff of Sussex. In 9 Edward I. the king granted the beadlery of one of his hundreds for two marks annually (*Fine Roll* 79, m. 11).

[164] *Select Pleas,* nos. 170, 194 ; *Rot. Hund.* i. 41, 269. Concerning the pursuit of thieves, cf. *Select Pleas,* no. 173 ; *Northumberland A.R.* 70.

[165] *Select Pleas,* no. 168. The sheriff had been directed to reduce the number of these to twelve, but to retain that number (*Pat. Rolls,* 1216-25, p. 240-41).

[166] P. 171, n. 21. Cf. the king's orders to the sheriffs in 1250 (Madox, *Exchequer,* ii. 102.)

[167] *Liber Albus,* i. 318 ; *C. Cl. R.,* 1279-88, p. 109.

o

only subordinates of the hundredor.[168]   These were subordinates of the chief bailiff of the hundred, but the sheriff's responsibility for his good conduct still remains.

The sheriff's position in local government is obscured by a dependence upon the exchequer which gave to many of his activities a fiscal significance out of keeping with their actual importance, and thus lent undue emphasis to derived aims in government.   The sheriff's *proficuum* which he rendered to the exchequer capitalized incidents in the ordinary course of local government.   Indeed, little as the fact is usually appreciated, the great bulk of the records relating to the sheriff's activities discloses him, not as a factor in local government, but as an agent of the king who carries out his orders and those of various central organs of his government. Since the Norman Conquest the correlation of local and central elements of government had meant a constant emphasis of the latter to the exclusion of the former.   There is traceable little definite interest in local government for its own sake.   Toward the end of the thirteenth century the almost inevitable decline of the old system of local government in the hands of the sheriff and his agents was at hand.   One of the great tasks of the fourteenth century was to fashion new machinery to do the work of the old.   Always the dark side of the sheriff's activity, his older place in local government is hardly illustrated in any detail until it is lost.

The sheriff's position as presiding official of local courts probably affords the easiest avenue of approach to this popular aspect of affairs.   In the county court he supervised a judicial system resting upon custom and also carried into effect useful administrative regulations ; in his semi-annual tourn of the hundreds of his county he maintained order and enforced a local police system still of great importance at the beginning of the thirteenth century, but almost broken down at its end.

At intervals of four weeks, in some northern counties six, the sheriff convened the county court.[169]   The session proper according to the usage of the time was concluded in one day.[170]   In the time of Edward I. there is mention of a *retrocomitatus* [171] on the day follow-

---

[168] *Rot. Hund.* ii. 307.   Cf. Cam, *Studies,* 147.

[169] See *Sel. Coroners' Rolls,* Selden Soc., 20, 23, 63, 65.   Cf. Magna Charta of 1217, cap. 42.   The well-known forty-day period between sessions in Lincolnshire (Pollock and Maitland, 1899, i. 538) was actually six weeks.   So in Yorkshire, Lancashire and Northumberland (*E.H.R.* xl. p. 5).

[170] Magna Charta of 1215, cap. 19.   [171] *Fleta,* bk. ii. cap. 67, sect. 18.

ing, when there was indeed no assembly of the county, although on this occasion writs were presented to the sheriff for execution and he collected sums due, probably attending also to some of the formalities of judicial business.[172]

The interests of public business, not to mention those of the sheriff himself, were best promoted by a large attendance. Even villains, who probably could not act as judgement finders,[173] were present,[174] nor is it conceivable that any one was turned away. The duty of attendance, suit at the county court as it was called, rested upon certain men of the shire and was enforceable by the sheriff through distraint.[175] The principles upon which the obligation rested were so diverse [176] that in its enforcement the sheriff could hardly have followed anything but the rule of usage. Sometimes a legal decision was necessary to settle disputes concerning the matter.[177] Despite the fact that a lord might be and often was represented by his steward,[178] holders of baronies in some counties did suit at county court after county court.[179] Knights figured prominently in the proceedings.[180] A more popular constituent element attended by virtue of the fact that whole manors or vills were represented by a given number of suitors.[181] Even small parcels of land sometimes owed suit.[182] Along with freeholders, certainly were present some who were only free men, while the old principle of representation of vills by the reeve and four men clearly introduced villains, at least for some purposes. Mr. Maitland has shown that in some counties were held each year two more largely attended sessions corresponding to the semi-annual county courts of King Edgar's day.[183] Casual reference to this *magnus comitatus* or *comitatus generalis* occurs in a fair number of counties. Sheriffs could some-

---

[172] *Bracton's Note Book*, no. 212. A party to a suit who appears on this day is not in default.

[173] *Leges Henrici*, 29. 1 a (Liebermann, *Gesetze*, i. 563).

[174] *Ibid.* 7. 7 (Liebermann, *Gesetze*, i. 553).

[175] *Rot. Lit. Claus.* i. 430; *Cl. Rolls*, 1234-37, p. 457; *Memoranda de Bernewell*, 238.

[176] Pollock and Maitland, i. 537-47.

[177] *Rot. Parl.* i. 12. The issue here is whether the sheriff may exact suit from each of the heirs to an estate or whether only one suit is due.

[178] *Leges Henrici*, 7. 7 a ; Pollock and Maitland, i. 543.

[179] *Northumberland A.R.* 327-8.

[180] *Bracton's Note Book*, no. 1730.

[181] *Leges Henrici*, 7. 7 b. A bailiff sometimes did suit even for a hundred or half hundred (Pollock and Maitland, i. 558 ; *Rot. Hund.* ii. 207).

[182] *Ibid.* i. 25, 34, 37, 331.

[183] Pollock and Maitland, i. 540, n. 1.

times be induced to accept a composition to acquit a suitor of attendance at the courts of the shire.[184] Just before the barons' war it was believed that for years many persons had been excused without the king's licence by arrangement with sheriffs and their bailiffs.[185] It is clear from the Hundred Rolls that withdrawal of suit did occur in numerous instances. On the other hand, one of the grievances of the barons in 1258 was the exaction of new suits of shires.[186]

In addition to the customary demand special attendance was required of some persons in matters affecting crown pleas. Presentments of such pleas were made in the hundred court, but they might also be made before the coroners at the sessions of the county.[187] Thither the sheriff might summon the men of a community for an inquest.[188] Thither also came men, at least four, from each of the four vills nearest the place where the crime was committed to make presentment.[189] Prior to 1259 sheriffs were claiming that all above twelve years of age in the vill concerned should be present in such cases. The articles of that year forbade the itinerant justices to amerce vills for not appearing before coroners and sheriffs upon the death of a man or other matters pertaining to the crown if enough came to make full inquest.[190] The power to impose amercement in such cases rightfully belonged only to the king's justices. In the time of Edward I. it was easy to show that the sheriff who amerced vills or individuals in such cases was conducting himself improperly, for he was inflicting a second penalty.[191] It seems clear that many of these questionable demands for appearance had to do with inquests which were not held in the county court.

The relation between the sheriff who presided and the doomsmen of the county who found the judgements was sometimes fraught with difficulty. The rare cases which afford any detail show that the suitors in the foreground were the knights and stewards.[192] The active group who found judgements professed to know the law and sometimes severely criticized the sheriff's conduct. The latter

---

[184] William de Camvill gave half a mark to successive sheriffs for the suit of a vill (*Rot. Hund.* ii. 226).   [185] *Ann. Monast.* i. 331.

[186] Stubbs, *Sel. Charters*, 385 (sect. 24).

[187] *Assize Roll* 786, m. 1, P.R.O.

[188] An inquest concerning a death (*C.P.R.*, 1247–58, p. 607–8).

[189] As in *Rot. Hund.* ii. 29 ; *Sel. Coroners' Rolls*, pp. xxx-xxxi.

[190] *Stats. of Realm*, i. 11 ; Stubbs, *Sel. Charters*, 404–5.

[191] As in *Rot. Hund.* ii. 295, where this is said to be *contra formam articuli*.

[192] *Bracton's Note Book*, nos. 212, 1730.

upon occasion attempted to hold them a second day to complete the business at hand.[193]   It is well known that the men of Lincolnshire in 1226 refused suit under such conditions and plainly told the sheriff that he alone could not hear nor judge suits.   When he declared the unfinished business adjourned to the respective wapentakes they cited Magna Charta against him, declaring not only that in this county he might not hold a county court more frequently than once in forty days, but also that he might not hold more than two tourns a year.[194]   In the county court of Somerset in 1225 the sheriff was believed to be favouring one of the parties in a land suit, and all withdrew except two or three.   With these nevertheless he proceeded at the hour of vespers to declare judgement and to award seisin.[195]   In disputed judgements, as in elections at a later time, it is clear that sheriffs might take liberties.   An article of the eyre from 1254 on required justices to inquire whether sheriffs maintained causes which suppressed justice.   In 1285 sheriffs were charged with the enforcement of a statute of maintenance, by the terms of which it became their duty under threat of grievous penalty to suffer no barretor or maintainer of quarrels in the shire to take part in the finding of judgements except as attorney for his lord.[196]

Such a popular body was, generally speaking, no court of record. The coroners, to be sure, had to keep a roll of inquests held here concerning pleas of the crown, and the sheriff, probably for his own convenience, came to keep such a record.[197]   This was a useful check on the coroner.   The first Statute of Westminster required the keeping of such counter rolls by the sheriff.[198]   Sheriffs and coroners were also required by writ read in county court to send up under their seals record and process of matters in inquests before them.[199]   But generally speaking the sheriff as yet seems to keep no roll of ordinary proceedings in the county court and apparently no written memorandum except that of fines and amercements for fiscal purposes.   When the king's courts wished an account of proceedings in a case which has passed to them, even though the

[193] In 1212 adjourned pleas of the county were heard by the sheriff's precept at York castle (*Select Pleas*, 106).   Cf. the story related in Pollock and Maitland, i. 549-50; also Magna Charta, cap. 18-19.

[194] Pollock and Maitland, i. 550; *Bracton's Note Book*, no. 1730.

[195] *Somerset Pleas*, Somerset Rec. Soc. 61-3.

[196] *Stats. of Realm*, i. 35, cap. 33.

[197] Above, p. 150 ; Bracton, Rolls Ser. ii. 430.

[198] *Stats. of Realm*, i. 29, cap. 10.

[199] As in *Sel. Coroners' Rolls*, 70.

matter is of some years' standing, they had to order the sheriff to have made in the county court what they called a record. This was a recital of what was agreed to from the memory of the suitors present on the former occasion, and was transmitted by four or six knights of the county.[200] As the century advances it is sometimes stated that this is attested by the sheriff's seal.

The jurisdiction of the county court extended to transgressions included in the sheriff's peace. Cases of mêlée, beating and wounding in default of action in the courts of lords, were disposed of by the sheriff,[201] unless the accuser specified that the king's peace had been broken. To the sheriff's peace belonged also certain minor thefts.[202] The Assize of Clarendon conferred a responsibility in much graver criminal matters when it required the sheriff to have presentments made before him in pleas of the crown. This was done in most cases in his tourn.[203] But in the thirteenth century the men of four neighbouring vills might appear in the county court before the sheriff and coroners to make inquest concerning crime.[204] Moreover, crime might be prosecuted in the county court by the old-time process of private accusation or appeal.[205] In the county court the accused might wage duel against his accuser.[206] There also persons who threatened breach of the peace might be placed under pledges for its observance.[207] The king's writ might even authorize the sheriff to dispose of a criminal case by jury in full county court.[208] Despite Magna Charta he sometimes proceeded without such authority to hold crown pleas which belonged to the justices.[209] Sheriffs and sergeants in Westmoreland continued thus as late as 1280 to deliver prisoners and jails.[210] On the other hand, proceedings leading to the ancient process of outlawry in the county court were initiated by order of the itinerant justices or by the

---

[200] *Northumberland A.R.* 195-6; *Bracton's Note Book*, nos. 40, 212. In 1221 a sheriff was ordered to have record made *in suo comitatu* of a cause heard there in King John's time (*ibid.* no. 1513).

[201] Bracton, ii. 540.

[202] *Pleas for Gloucester*, ed. Maitland, p. 6.

[203] Cf. *Royal Letters*, i. 450.   [204] As in *Rot. Hund.* ii. 29.

[205] *Ann. Monast.* iii. 308; *Northumberland A.R.* 92, 93, 350.

[206] *Select Pleas*, nos. 126, 172.

[207] As in *Coroners' Rolls*, no. 3, m. 1, P.R.O. In *Select Pleas*, no. 73, the sheriff gives peace in the county court for the protection of a complainant.

[208] *Select Pleas*, no. 186.   [209] *Ann. Monast.* i. 330.

[210] *C. Cl. R.* 1279-80, p. 109. In the palatinate of Durham also (Lapsley, *Durham*, 83) the sheriff delivered the jails in the county court. An inquest concerning jail delivery by sheriffs without warrant is recorded in *Feudal Aids*, i. 90.

king's writ to the sheriff.[211]   After each eyre the names of fugitives
were placed upon a roll, one copy of which was left in custody of
the coroners and sheriff so these persons might be demanded at the
first county court thereafter, and so on until they appeared or were
outlawed.[212]

The civil causes which came before the sheriff in this assembly
were still quite numerous, although in the time of Edward I. the
interpretation of the king's courts was limiting this jurisdiction
in personal actions to cases in which the amount involved was forty
shillings or less.[213]   The prospect of trial by ancient customary
process in one's own county without jury still had its attractions.
Moreover, in the county court one might place himself on the assize [214]
and thus have his case removed to the king's court.   The land
cases which still came up before the county were fairly important,[215]
and such cases might still be decided by duel.[216]   Pleas of trespass
were held in the county court,[217] and lords resorted to replevin here
to collect their rents, although by 1285 need was found for new
process by which such cases might be removed before the justices.[218]
Proceedings under the writ *justices*, ordering the sheriff to institute
process to collect debt or enforce service were normally held here.
Fugitive villains with their chattels and *sequela* were upon writ
claimed before the same tribunal, although the suing out of the writ
de libertate probanda by the defendant automatically transferred
jurisdiction to the justices.[219]   To the sheriff's jurisdiction also
belonged the case of the beast wrongfully taken and detained by
way of pledge, the ubiquitous plea *de namio vetito*.[220]   Properly
speaking this belonged to the king's courts, but in the days of the

[211] As in *Sel. Coroners' Rolls*, 62, 72.   Since the question whether one might be
put on the exigent to be outlawed except by the writ of the king or the mandate
of his justice was judicially raised in London in 14 Edward II. and submitted to
the council (*Munimenta Gildhallae Londoniensis*, iii. part i. 336, 337), there must
have been doubt about the matter.
[212] Britton, ed. Nichols, i. 134.
[213] Pollock and Maitland, i. 553-4 ; Britton, Nichols ed. i. 155.
[214] *Bracton's Note Book*, no. 535.
[215] In 1269 claims are made to 14 and 16 acres of land (*Northumberland A.R.*
186).
[216] *Bracton's Note Book*, nos. 15, 1436.
[217] Stat. of Gloucester, cap. 8 (*Stats. of Realm*, i. 48).
[218] *Ibid.* i. 72, cap. 2.
[219] *Yorkshire Assize Rolls*, Yorkshire Archæolog. Soc., Record Series, xliv. 18 ;
Northumberland A.R. 46, 159, 195-6, 242. Cf. Glanville, v. 1.
[220] Pollock and Maitland, ii. 577.   By the time of Fleta (ii. cap. 43, sect. 3) it
was being held that this belonged to the sheriff under the king's writ.

Hundred Rolls it was one of the best-known pleas in the sheriff's hands.[221] It was often a difficult matter to deliver the beast taken for debt and retained when other gage and pledge was offered.[222] The sheriff's authority to do this was strengthened by statutes of 1259, 1267 and 1275.[223]

In some of these matters it is seen that the sheriff assumed jurisdiction only after the king's writ had issued at the instance of the plaintiff. In Bracton's time it was being held that when the sheriff acted by virtue of such authority he acted as a justice.[224] Fleta declares that the court of the county was two-fold. There is the king's court baron wherein the sheriff was bailiff and judgements are found by suitors ; but the king also had a court wherein the sheriff is justice with jurisdiction delegated by writ through which he has record.[225] Even a suit between parties under the king's writ of right might go to the county court,[226] and this involved the procedure concerning essoins mentioned in Glanville. Thus when a party to a suit bore to the county court a writ for viewing his adversary who claimed illness, the sheriff then and there appointed knights for the purpose.[227] Judicial as well as original writs might order the sheriff to hold special pleas or inquests or make record of proceedings in the county court.

Amercements and fines levied before sheriffs were recorded on their rolls of *proficua* or income for the year. Such sums were occasioned by failure to attend, others for having an inquest or being admitted to mainprise,[228] others for default, unjust detention, trespass, licence for making concord,[229] contempt, false claim or concealment.[230] The petition of the barons in 1259 complained that sheriffs were imposing amercements out of keeping with the offences involved,[231] and the Provisions of Westminster forbade fines for beaupleader (*pro pulchre placitando*) whether at eyres,

---

[221] *Rot. Hund.* i. 33, 72, 125 ; cf. *Placita de Quo Warranto*, 177.

[222] *Pleas for Gloucester*, ed. Maitland, no. 25. Mr. Maitland shows (*Hist. Eng. Law*, ii. 577) that it even necessitated calling out the *posse comitatus*.

[223] *Stats. of Realm*, i. 10, cap. 21 ; 24, cap. 21 ; 31, cap. 17.

[224] Bracton, Twiss ed., ii. 542.

[225] *Fleta*, ii. cap. 43.

[226] As in *C. Cl. R.* 1247–51, p. 37. Glanville, lib. i. cap. 4 : *ubi curiae dominorum probantur de recto defecisse.*

[227] *Bracton's Note Book*, no. 1052. Essoins *de malo veniendi* and *de malo lecti* and *de servicio regis* were also made in the county (*ibid.* nos. 1019, 1513).

[228] *L.T.R. Misc. Rolls*, 5/29.

[229] *Ibid.* 5/77.

[230] *Ibid.* 5/74.

[231] Stubbs, *Sel. Charters*, 384, cap. 16.

counties or baronial courts.[232]   In 1280 King Edward and his council ordered that when the men of Westmoreland were amerced the amercement should be taxed in full county court by good and lawful men, as ought to be done by the statute,[233] and not at the will of the sheriff and his bailiffs.[234]

The sheriff was also responsible for making public announcement of executive acts or orders, new laws or even judicial summonses by having them read or proclaimed in the county court.   This was not the only method of proclamation.   But sheriffs were directed to have read *in pleno comitatu* various charters and grants of the king,[235] including the Magna Charta of 1217 [236] and some at least of its successors.[237]   In the time of Edward I. at least two statutes [238] were published in this way.   In 1300 sheriffs were ordered to have the Charter of Liberties and the Charter of the Forest read four times a year *in pleno comitatu*.[239]   The king's writ calling out the tenants-in-chief for military service,[240] was often proclaimed in this way ; so was an order citing certain persons to appear before the king's council.[241]   Before the writ of mesne became effective the sheriff was required, according to the Statute of Westminster, to cause solemn proclamation to be made in two full county courts that the mesne lord at the day set in the writ should come to answer his tenants.[242]

Useful publicity through the medium of the county assembly was also attained when public officials were installed.   Thus the coroners chosen in this body naturally took the oath of office at once before the sheriff,[243] though sometimes in the presence of the county as assembled before the king's justices.[244]   County escheators in the reign of Henry III. were also installed in office by taking the oath of office in the county court.[245]   Special wardens of the waters

---

[232] Stubbs, *Sel. Charters,* 402, cap. 5. Re-enacted by the Statute of Marlborough, cap. xi.

[233] Presumably the first Statute of Westminster.

[234] *C. Cl. R.* 1279–88, p. 109.

[235] Below, p. 218.   [236] *Foedera* (1816), i. 147.

[237] *Cl. Rolls,* 1234–37, pp. 421, 541 ; Close Roll, 69, m. 12 d (1256).

[238] *Cl. Rolls,* 1288–96, p. 380 : *Stats. of Realm,* i. 152.   The Statute concerning jury summons was proclaimed generally in cities, boroughs and market towns.

[239] *Foedera* (1816), i. pars. ii. 919.   [240] As in *Dignity of a Peer,* iii. 10.

[241] Dugdale, *Summons of the Nobility,* 3.

[242] *Stats. of Realm,* i. 78, cap. 9.

[243] *Lit. Claus.* i. 364, 402, 414 ; *Cl. Rolls,* 1231–4, p. 7 ; *Placita de Quo Warranto,* 159.   [244] *Sel. Coroners' Rolls,* p. xxxvii.

[245] Close Roll, 65, m. 17 ; *Cl. Rolls,* 1247–51, p. 476.

chosen in the county court of Northumberland to protect the salmon were likewise sworn in before the sheriff,[246] as were the special wardens appointed in 1299 to enforce the law against bad money.[247] The Statute of Lincoln mentions in the reign of Edward II. bailiffs of the hundreds as sworn and known in full county court, but does not pretend to originate this usage, which was clearly older.[248]

Another duty of the sheriff in the county court was the holding of various elections. Directions concerning the permissive election of sheriffs were in the time of Edward I. sent to the coroners of the county. But in other cases the regular procedure was by writ to the sheriff ordering him to cause election to be made. One of the earliest forms of election in the county court to be mentioned is that of four knights of the county who designated in each hundred two knights to fill up the list of jurors to make presentments before the itinerant justices. . Dating from 1194,[249] this procedure was continued throughout the next century,[250] but a century later it is seen to consist of nothing but nomination by the bailiff of the hundred,[251] a fact which may well lead to reflection upon the earlier meaning of the word *eligere*. The sheriff caused to be chosen reguarders, twelve in each reguard,[252] who performed the function of presentment jurors before the justices of the forest. The verderers, assigned to guard the king's venison in each forest, were regularly thus elected, at least after 1219.[253] Stubbs shows [254] also that conservators of the peace were sometimes elected. The election of coroners in the county court is far better known. When this is first mentioned, in 1194, four were to be chosen in each county.[255] After 1219 vacancies in the office as they occurred were filled by the king's writ to the sheriff directing that he cause election to be made in full county court by assent of the whole county.[256] The

[246] *Northumberland A.R.* 208-9.

[247] *Stats. of Realm*, i. 133. In exceptional cases before chief wardens.

[248] *Stats. of Realm*, i. 175. These alone were to have execution of the writs that came to the sheriffs. *Assize Roll*, 982, mm. 22, 23, shows the same usage in 1280. In Gloucestershire (*Feudal Aids*, i. 23) the earl had a bailiff *juratum vicecomiti* who responded for certain summonses.

[249] Stubbs, *Sel. Charters*, 260.

[250] As in *Assize Rolls*, 135, m. 1 ; 669, m. 16 d ; 915, m. 1.

[251] Britton, i. 22.

[252] *Royal Letters*, i. 345.

[253] *Lit. Claus.* i. 409, 493, 529 ; *Cl. Rolls*, 1231-4, p. 50 ; *ibid.* 1272-9, p. 145.

[254] *Const. Hist.* ii. 239.

[255] Stubbs, *Sel. Charters*, 260, sect. 20.

[256] *Lit. Claus.* i. 402, 409, 414 ; *Cl. Rolls*, 1231-4, p. 7. Sometimes this was done before the king's justices (*Sel. Coroners' Rolls*, p. xxxvii).

sheriff had power to compel the person chosen to assume the office.[257]
If poor judgement were used in choosing him and the king regarded
him as unqualified, he removed him.   King Edward I., who insisted
that coroners must have lands in the county,[258] in letters close of
the year 1304 to the sheriff of Northumberland complains of a report
that the latter has spared the rich men who have lands and elected
as coroner a certain William of Tynemouth who has none.   Because
the king regards William as insufficient for the office he orders the
sheriff, if William has no lands, to remove him from office without
delay and cause another to be elected in his place.[259]

Election of the knights of the shire appears later than the other
forms.   There is·no definite reference to election of any kind in the
county court in John's reign, if allowance be made for the obvious
exception of the selection of a sheriff of Devon.   But the principle
was already known.   The various groups of persons whom the
king desired to convene, however, including four discreet knights
from each county in 1213, he merely directed the sheriff to send.
In 1220 occurs an election *de voluntate et consilio omnium* in each of
the various county courts of two knights who are to assess and levy
a carucage.[260]   But the first clear case of the choice of men by
county courts to speak for the county at large in any matter occurs
in 1227.   To terminate certain differences with sheriffs, four
knights of each shire were chosen to appear before the king
at Westminster.[261]

The election for the first time in 1254 of knights in all the shires
to appear before the council at Westminster and make a grant to
the king may be regarded as another general application of the
procedure and principles of 1227.   Except in 1264 [262] it is possible
that knights were merely returned by the sheriff and not elected [263]
to the other parliaments prior to the second parliament of 1275.
The knights who came on behalf of their communities with full

[257] In the case of a man who is exempt by letters patent the sheriff of Bedford
is ordered *quod eundem Willelmum ad officium coronatoris assumendum in praedicto comitatu non compellat donec rex aliud inde percepit* (Close Roll, 77, m. 16,
43 Henry III.).
[258] *C. Cl. R.* 1279–88, p. 443.   [259] *Ibid.* 1302–7, 226.
[260] *Rot. Lit. Claus.* i. 437.
[261] *Dignity of a Peer*, app. i. p. 4 ; Stubbs, *Sel. Charters*, 357 ; *Rot. Lit. Claus.*
i. 212, 213.
[262] Stubbs, *Sel. Charters*, 412 ; Rymer, *Foedera*, i. 442.
[263] For the writs of 1261 and 1265, *Dignity of a Peer*, app. i. 23, 34 ; Stubbs,
Sel. Charters, 405, 415.   The words in the writ for the Easter parliament of 1275
are *venire facias* instead of *elegi facias* (*E.H.R.* xxv. 234, 236).

power [264] to the parliaments of January 1283, were probably so chosen. Election to successive parliaments is of course the rule. But not until 1295 does the writ to the sheriff direct him to cause to come also elected citizens of each city and elected burghers of each borough of the county.[265] From the same year it became customary to add a clause which called for the return of the writ at the specified time and place with the names of the knights, citizens and burghers designated.

The sheriff was regarded as head of the system of hundred courts remaining in the king's hands. To be sure the ordinary sessions were held by the bailiff, but he held his position from the sheriff at *ferm*. The reason assigned for issuing the well-known writ of 1234, declaring that hundreds and wapentakes are to meet every three weeks, is to inform the sheriff how these courts ought to be held.[266] The under-sheriff of Wiltshire in the fifty-fourth year of Henry III. summoned a hundred court to meet at a place in another hundred and took of various persons sums because they did not come,[267] but he was obviously committing an irregular act. The convening of a hundred and with it the suitors of four neighbouring vills, to make inquest concerning a death of a man or other plea of the crown was in 1234 quite the rule.[268] It was not, then, the convening of such sessions by sheriffs and bailiffs which was in question before the justices in 1254,[269] but rather the mulcting of persons who failed to come. Exactly this grievance against the sheriffs of Shropshire and Sussex is reported in the Hundred Rolls.[270] By this time there was a feeling that this business ought to be reserved to the two great hundreds held each year, and that the calling of these special sessions was unjust. The sheriff evidently had the power to commit cases to the ordinary sessions of the hundred court,[271] and he might be directed by the chancery to receive an attorney appointed to sue for a party to a case at one of the king's hundreds.[272] A sheriff was in 1235 ordered by the court Coram Rege to direct the bailiffs of a hundred to cause record to be made before them concerning a matter determined there.[273]

---

[264] *Dignity of a Peer*, app. i. 46 ; Stubbs, *Sel. Charters*, 465.
[265] Town officials in 1283 returned these (*ibid*. 468).
[266] *Royal Letters*, i. 450.   [267] *Rot. Hund*. ii. 256.
[268] Cf. the writ of 1234, *Royal Letters*, i. 450.
[269] *Ann. Monast*. i. 330.   [270] *Ibid*. ii. 101, 203.
[271] *Bracton's Note Book*, no. 1730 ; cf. *Leges Henrici Primi*, 7, 5.
[272] *C.P.R.* 1247–58, p. 25.   [273] *Bracton's Note Book*, no. 1159.

Twice a year the sheriff made his tourn, assembling the hundreds in especially full sessions in which he received presentments of offences against the king's peace, exercised an ancient jurisdiction in matters of local police and of civil causes, and at one of these sessions each year made view of frankpledge.[274]   Here he claimed to be *justitiarius quoad diem*.[275]   Prior to the Provisions of Westminster [276] the barons were complaining accordingly, that if earls and barons with tenements in various places did not come personally he amerced them *sine consideratione et judicio*.   One of the few permanent reforms of the period was a relaxation of this requirement in the case of barons, magnates and men of religion.   The exemption of a hundred or manor from the tourn was of course a prized, and indeed the most common, liberty of the period.

Some regions, such as Wales and the northern counties of England, were in the time of Edward I. receiving the tourn for the first time though without view of frankpledge.[277]   In Northumberland, one of the northern counties which had no hundreds, there had been no general tourn but only inquests with one coroner *per certas personas*, until William Heyroun, sheriff in the period 1246-58, distrained all *liberos tenentes et villatas* to come to two tourns a year.[278]   The king and his council in Westmoreland in 1280 to avoid " grievances occasioned by assemblies and congregations the sheriff calls his tourn ", at which he assembled in divers places the four men and reeve of every town without warrant of the king, ordered that henceforth one tourn and no more be made each year at a certain day after Easter to indict outlaws, thieves and felons.[279] Persons taken by bailiffs for less serious offences were to be under plevin of their neighbours until the sheriff and coroners knew by such general inquisition whether they had been rightfully grieved.   A

[274] Pollock and Maitland, i. 558-60.   For greater detail see the writer's *Frankpledge System*, chap. 4.
[275] Stubbs, *Sel. Charters*, 384 (sect. 17).
[276] *Ibid.* 384.
[277] Morris, *Frankpledge System*, 43, 48-51.   The tourn of the franchise held at Wakefield included no view of frankpledge (*Wakefield Court Records*, i. 33-6, 42-4).
[278] *Northumberland A.R.* 163-4 ; *Inq. Misc.* Chancery, i. 123.   He levied heavy amercements because they did not come.
[279] *C. Cl. R.*, 1279-88, p. 109.   But from *Chancery Misc.* bundle 133, file 7, Westmoreland, it appears that the sheriff protested, holding that Magna Charta authorized two tourns a year, and that indictments, now required upon the oath of twelve men, might be taken only in these.   It was accordingly ordered on May 12, 1290, that two tourns a year be held in due and accustomed places. *Assize Roll* 982, m. 23, reviews the early history of the tourn in Westmoreland.

tourn of much the same character was also introduced in Cumberland in this period.[280]

Except in privileged towns and private franchises it is clear that the sheriff's headship of local courts, like his control over the bailiffs of the county, was sufficient to give him a pre-eminent position in local affairs. Judgement by suitors was effective only through a capable administrator. The general sanctions of the community no less than the orders of the king found expression in the action of the sheriff or his agents. Moreover his prestige in enforcing judicial decree was reinforced by his fiscal and peace powers. Indeed one feels that at least from the time of William Rufus the fiscal motive is more and more warping the ancient communal constitution. Amercement for default and technicality was certainly not a device instituted in England as late as Norman times. But payments for beaupleader in local courts dissappeared only in the thirteenth century. There was a tendency for the sheriff in discharging some of his judicial duties to regard himself as *justiciarius*. It was to his pecuniary advantage to adopt the fiscal devices which marked much of the work of the king's courts, and to pattern after the justices in levying many amercements. The local system was of course preserved by measures to continue due attendance at its courts, but its spirit was often perverted. The burdensome summons of towns and hundreds to make inquests in criminal cases frequently seems to spring from a desire to have excuse for amercement as much or possibly more than from a desire to ferret out crime and see justice done. The function of the four neighbouring vills in presenting crime and making inquests, imported apparently from Norman usage into the local constitution, was sometimes turned in the same fashion. Upon local institutions that spirit, which in the king's courts so constantly makes for the king's profit, has a more or less subversive effect.

The union in one office of the powers of judge, summoner, constable, and executive official naturally became less and less effective as time advanced. All were likely to be more or less vitiated by the sheriff's necessity of finding remuneration for his own labours in additional profit which he might retain. The local constitution, weakened by two centuries of systematic centralization and expansion of central government, and deprived of the vitality which enabled it to meet expanding needs, before the death of Edward I.

[280] *Assize Roll* 137, m. 15 d ; *Chancery Misc.*, bundle 133, file 7, Westmoreland.

was becoming ineffective and decadent. Frankpledge police was all but useless. The courts of shire and hundred were losing importance as civil courts. The maintenance of order and the repression of crime were so ineffective that various devices, including that of special custodians of the peace, were being employed. It was only by retaining the sheriff's strong police powers and at the same time developing in the counties a new jurisdiction, dependent largely on central power and in the hands of the justice, that order was once more to be restored.

VIII

THE EXECUTIVE AND PEACE FUNCTIONS OF
THE THIRTEENTH CENTURY SHERIFF

THE duties of the sheriff with which this chapter deals have to do with legal process, general administration and enforcement of the peace. They were not constant nor even periodical. Except in matters of peace enforcement practically all were performed in response to writs or royal letters. Even in matters of police such orders to sheriffs were not infrequent. The king's administration, exclusive of certain fixed usages in finance to be noted later, so far as it involved the sheriff, seems to have rested upon specific order. Furthermore, legal process and the operation of the common law were summed up in the sheriff's execution of chancery writs and his enforcement of judicial orders and decrees.

The administration of justice thus imposed upon the sheriff an exacting, complicated and, in the main, an unremunerative array of duties. The person who obtained a writ at the chancery for a half mark, as the Fine Rolls show in hundreds of cases, usually took it to the sheriff to institute proceedings. Because of complaints concerning failure to execute writs the second Statute of Westminster provided that these be presented publicly to the sheriff or under-sheriff,[1] who was then required to issue a bill in evidence of the fact. The next step was the acceptance on the part of the sheriff of pledges for the prosecution of the case.[2] Even in cases which came before the county court, most of them without writ, the finding of pledges to prosecute was the rule. This usage appears in criminal[3] as well as civil cases. Original writs also directed

---

[1] In the county court or the *retrocomitatus* (*Stats. of Realm*, i. 90, sect. 39).

[2] Britton, Nichols ed. i. 288-9. Specific cases, *ibid.* i. 384-5, and Bracton, *De Legibus*, R.S., iii. 138, 498.

[3] Thus strangers, the victims of assault, are carried by the sheriff to the castle until they can find the necessary sureties (*Northumberland Pleas*, Newcastle-upon-Tyne Records Soc., Records Ser. ii. p. 5).

P

the sheriff to summon defendants by good summoners,[4] and, when the situation so demanded, he summoned a jury. At the institution of proceedings he was sometimes called upon to make attachments. The man accused of crime he probably as a rule apprehended without order, but a writ to this effect might be forthcoming. Some civil processes also called for attachment.[5]

When the justices held their session, whether in the county or at Westminster, they ordered the sheriff to summon jurors, to hold inquests, to ascertain specific facts in question. But before this stage was reached the most important object was to get the defendant into court. He had certain recognized essoins or excuses, some of which the sheriff might be called to investigate, such as the essoin *de malo lecti* or bed sickness. In this case the sheriff might be directed to send men to view him and report whether his indisposition amounted to a real *languor*.[6] If without essoin he failed to make his appearance the sheriff would be ordered to distrain him by his land and chattels to be present on a day set,[7] and for default the great *cape* might order the sheriff to take the land of the defaulter by view of legal men.[8] In making lesser distraints, as for debt, according to Britton[9] a rule of law enforceable by amercement forbade the sheriff to take more than the amount involved, or to seize beasts unless necessary, or riding horses, household utensils or other goods within doors, when sufficient distress might be found out of doors.

The visit of the justices to the shire multiplied these duties. Prior to their appearance for the great eyre the sheriff received the well-known writ bidding him convene the county court before them at a given time and place.[10] This ordered him to cause to come before them all the unpleaded pleas of the crown arising since the last eyre, with the attachments pertaining to them, and also all assizes and pleas placed *ad primam assisam* before the justices,

---

[4] As in Bracton, iv. 108 ; cf. Glanville, i. 30.
[5] Pollock and Maitland, ii. 592 ; Bracton, iii. 160 ; Britton, Nichols ed. i. 58. If a cleric had no lay fee, a writ was sent to the bishop to cause him to appear, and only in case the latter failed was writ sent to the sheriff (Bracton, vi. 498).
[6] Glanville, i. 19.
[7] Newcastle-on-Tyne Records Ser. ii. p. 70.
[8] Bracton, R.S. v. 374. As to the little *cape*, *ibid*. 424.
[9] Nichols ed. i. 89.
[10] *Lit. Claus.* i. 403, 1218. For other writs of summons, *ibid.* i. 473 (1220). *C.P.R.*, 1232–47, p. 77, and Bracton, *De Legibus*, ii. 188. Cf. Britton, Nichols ed. i. 19.

together with the writs. This particular mandate furthermore ordered the sheriff to proclaim throughout his bailiwick that all pleas and assizes not terminated before the justices at Westminster (sometimes also before justices of assize or jail delivery) should come before the justices in eyre.[11] As early as 1218 it was the rule to summon all who had been sheriffs [12] since the last eyre to be before the justices, with the writs of pleas and assizes which they had received in their time, to answer as they ought. Later in the century the attendance of those who had been coroners since the last eyre was similarly required.[13] Britton describes a ceremony by which, on the appearance of the justices, they took the wand of the sheriff until he swore to execute duly the lawful commands of the justices and to keep secret the counsels of the eyre ; then, after the wand was delivered back to him, he presented his bailiffs, clerks and others by whom the precepts of the justices were to be executed, that they might take the same oath.[14] This usage was even leading lawyers in the reign of Edward II. to put forth the claim that the justices had power to remove the sheriff from office. In some instances a sheriff unable to attend the eyre gained authorization that his clerk act for him.[15]

In relation to the lesser sessions held by itinerant justices the sheriff's responsibility was much the same. To the sheriffs of the counties containing royal forests were dispatched writs, ordering summons to the eyre of the pleas of the forest.[16] Prior to the arrival of the justices of the forest, the sheriff was also directed to summon before him those who were to make the reguard, to see that the knights were chosen to fill vacancies in the number of the twelve reguarders, to receive the oath of the foresters, to lead these to view offences, according to a list of *capitula* forwarded the sheriff, and to cause these knights to swear to perform their duties.[17] The sheriff was also instructed to cause pleas of assizes to come before justices

[11] The eyre was also prorogued by the king's order to the sheriff, as in Gloucestershire in 1256 (Close Roll 69, m. 21 d). In 1262 the sheriff of Surrey was ordered to proclaim that the eyre for *common pleas* would sit not at Bermondsey as previously commanded, but at Guildford (Close Roll 78, m. 5 d). Similarly the sheriff of Essex is ordered to make known the postponement of the visit of the justices of the forest (Close Roll 88, m. 7).

[12] *Rot. Lit. Claus.* i. 403-4.   Under-sheriffs are also included.

[13] *Sel. Coroners' Rolls*, p. xxviii.

[14] Britton, Nichols ed. i. 21.   [15] *Cl. Rolls*, 1234-7, p. 146.

[16] *Select Pleas of the Forest*, Selden Soc., p. li. ; *C. Cl. R.*, 1279-88, p. 363.

[17] *Pat. Rolls*, 1216-25, p. 482 ; *ibid.*, 1225-32, p. 286 ; *Cl. Rolls*, 1231-4, p. 31 ; *C. Cl. R.*, 1272-9, p. 494 ; *Royal Letters of Henry III.* i. 345 ; Close Roll 66, m. 12 d.

of assize who were sent to the county,[18] and to cause all prisoners in jail and their attachments to come before justices assigned for jail delivery, together with knights and men of the townships by whom, if necessary, inquisition might be made concerning prisoners.[19] In the earlier years of Henry III. it was not unusual for the sheriff himself to be appointed one of the justices of jail delivery for the county [20] or one of the justices of oyer and terminer, designated to deal with a specific cause.[21]

Among his many other functions in relation to all this work of the justices, that of supplying men [22] to serve on assizes, juries and recognitions was not least. This was a form of duty for which men sought and obtained release by special grant of the king.[23] A statute of 1293, in order to cease burdening persons of little means, enacted that persons empanelled by sheriffs, under-sheriffs and bailiffs for such service outside the county should have lands of the value of a hundred shillings, and those within the county forty shillings. This rule, however, was subsequently qualified by the provision that former usage should hold respecting common pleas in the eyres of the justices and assizes, and also for juries and recognitions in cities and towns.[24] The obligation of the community to be represented in court was also abused. The barons complained in 1258 that, when any *justitiaria* was ordered before an assigned justice, sheriffs by proclamation in the markets bade all knights and free tenants come to hear and do the king's precept and, if they failed to come, amerced them at will.[25]

In the execution of writs and judicial orders lay the sheriff's

---

[18] *Pat. Rolls*, 1216–25, pp. 343, 390; *ibid.*, 1225–32, pp. 83, 206, 444; *C.P.R.*, 1232–47, p. 127; *C. Cl. R.*, 1247–51, pp. 536-7. In 1285 a sheriff was instructed to cause all assizes, juries and inquisitions that remained since the preceding Michaelmas to come before the justices with writs and all other aids (*C. Cl. R.*, 1279–88, p. 365).

[19] *Pat. Rolls*, 1225–32, p. 183; *C.P.R.*, 1232–47, pp. 79, 127; *Lit. Claus.* i. 437 (*vocatis coronatoribus*).

[20] *Pat. Rolls*, 1225–32, pp. 159, 161, 365.   [21] *Ibid.* pp. 71, 73, 81.

[22] See *C.P.R.*, 1247–58, p. 51. A grand assize might be chosen before the justices at Westminster (*Northumberland A.R.* 24). In *Bracton's Note Book*, no. 278, a sheriff causes a jury to be chosen *pluries*, removing some. See p. 279 below. For the grand assize the sheriff had summoned four men who chose the twelve knights (as in Close Roll 89, m. 2 d).

[23] As in *C.P.R.*, 1232–47, p. 89; *Lit. Claus.* i. 598; *Cl. Rolls*, 1247–51, pp. 6-7; *ibid.*, 1272–9, p. 178. In *C.P.R.*, 1247–58, p. 16, all those associated with the work of the escheators in the north are excused. In Close Roll 65 m. 19 (36 Henry III.), the sheriff of Cambridge is to excuse a man who is coroner and escheator.

[24] *Stats. of Realm*, i. 113; *C. Cl. R.*, 1288–96, pp. 380, 381.

[25] Stubbs, *Sel. Charters*, 385, sect. 19.

ministerial function. Besides executing original writs, he was bound also to carry out the precepts conveyed in all kinds of writs issued to him by the court of king's bench, the court of common pleas and the itinerant justices; and he received in addition a vast number of exchequer writs. He was subject to amercement if returnable writs were not before the judges at the proper time,[26] and by the second Statute of Westminster was made liable to action for damages, if he neglected to have served properly a writ presented to him by a party to a suit. When he left office he was required to take of his successor receipt by indenture for unexecuted writs on hand.[27] In many cases the mandates which found expression in writs were made operative only ineffectively or after considerable delay. When the bailiff of a franchise possessing the right to return writs failed in this duty, the sheriff was especially enjoined to enter the liberty and attend to the matter.[28] By trickery on the part of sheriffs, in collusion with parties to suits at the eyre, some writs were held back for an unduly long period to make causes go by default. This evil the Statute of Westminster sought to abolish, specifying that the court should appoint a time when the sheriff must certify to the chief justice in eyre how many and what writs he had, so thereafter no more might be received.[29] The state trials of 1289, however, still show complaints of the same nature. All three of the central courts are found issuing a writ *alias*, or even a writ *pluries*, to secure desired action. A sheriff might show reason why he could not execute a writ.[30] But sometimes he excused inattention to duty by the fictitious return that the writ came too late to be served; or, when ordered to distrain a person by chattels, he fraudulently caused to be written on the back of the writ that there were no such chattels.[31] Moreover, a sheriff might fail completely

[26] *Placitorum Abbreviatio,* 261. A sheriff might also be amerced for making return without impression of his seal. In *Northumberland A.R.* 158, the sheriff is ordered to levy the amount of a debt and costs. According to Sir Thomas Smith he is " the generall minister and highest for execution of such commandments according to law as the judges do ordaine " (*De Republica Anglorum,* ed. Alston, p. 78).

[27] Above, p. 188.

[28] Britton, Nichols ed., i. 126-7; Bracton, vi. 488, 490. The Memoranda Rolls are replete with such orders to sheriffs. Cf. *Rot. Hund.* i. 130.

[29] *Stats. of Realm,* i. 80, cap. 10. Cf. *State Trials of Edward I.,* R.H.S., app. ii. nos. 10, 13, 81, 343, 365; app. iii. no. 1.

[30] *Northumberland A.R.* 276.

[31] Bracton, R.S., vi. 464; *Fleta,* part ii. 67, sect. 18; cf. *Northumberland Pleas,* Newcastle-upon-Tyne Records Ser., ii. 258. See also *British Record Soc.* vol. 19, p. xviii.

through negligence in making attachment, and sometimes did not send back the writ as ordered after he had made summons or carried out some other precept contained in it.[32] For upon the back of each returnable writ he was supposed to make return unless the lord of the franchise had this peculiar privilege, which the agents of the crown after about 1250 were very carefully safeguarding.[33] Another source of difficulty seems to have been the unsatisfactory character of the agents whom the sheriff often employed for this work. The Statute of Lincoln in the reign of Edward II. required that writs be executed by the hundredors, who were generally known.[34]

It would be easy, however, in dwelling upon the law's delay through the sheriff's dereliction, to depreciate the large amount of useful service which he rendered. All the king's courts had power to call sheriffs to account by amercement for neglect or for false or insufficient return of writs.[35] All depended upon them to get much of their business done. In cases involving pasturage, for instance, sheriffs had to execute writs of admeasurement[36] and in some other cases choose twelve knights to make partition of lands.[37] They had to make a multitude of necessary inquisitions to get facts required in civil as also in criminal matters.[38] Large bodies of law, such as those affecting the disposition of land of deceased tenants-in-chief and of the property of offenders against the crown, could be made operative in no other way.[39] Sheriffs also

[32] Bracton, vi. 480.
[33] See Madox, *Exchequer*, ii. 103, for one of the earliest instances. At Appleby bills of summons and writs were delivered to the burgesses to serve (*Inq. Misc.*, Chancery, i. 347). Cf. Pollock and Maitland, i. 583, 644.
[34] *Stats. of Realm*, i. 175.
[35] As in *Assize Roll*, 982, m. 34 d, where a sheriff and former sheriff are at mercy for contemning execution of the king's writs. A sheriff who reports that a man has property which may be levied upon and then reports otherwise is subject to amercement (Rigg, *Cal. of Pleas, Exchequer of the Jews*, ii. 178).
[36] For the form of the writ, Britton, i. 384-5. Cf. *Liber Memorandorum Ecclesiae de Bernewell*, ed. J. W. Clark, 118.
[37] Bracton, R.S., i. 593.
[38] Bracton (ii. 292-3) represents it as the sheriff's business to determine by writ if one imprisoned were accused *de odio et atia*, and if he were not culpable to dismiss him under sureties; also (ii. 206) the calling of a jury of knights to decide if one accused of tort were guilty. So also the sheriff is ordered to make special inquest concerning a murder or a burning (as in Close Roll 66, m. 9 d). For inquests held by the sheriff in cases of redisseisin, Statute of Merton, cap. 3.
[39] Thus the writ of *diem clausit extremum*, the writ *de aetate*, the writ to inquire whether an heir is an idiot and his lands are consequently in wardship of the king (as in *Cal. Inquis.* ii. 498). See Bracton, ii. 394, 400-2, for matters affecting criminals' property; for the form of the writ *de aetate*, C. Cl. R. 1272-9, pp. 548-549.

had to execute writs of *certiorari* and to dispatch before the justices at Westminster men to bear record of causes removed from the county court ; to make attachments, distraints and summonses ; to send in long procession knights and law-worthy men who knew facts and could recite them upon oath, as well as prisoners to answer accusation and debtors of the king to make satisfaction. Sheriffs were also ordered in some instances to send complainants,[40] defendants or inquests [41] before the king's council ; or even to send before the king in chancery men by whom the truth concerning a suit might best be learned.[42]

The sheriff was of course an important agent of the executive power quite independently of judicial action. He had a power to inhibit certain acts [43] as well as a power to enjoin others. In the time of Henry III. he was still regarded by the king and council as their agent in the maintenance of popular liberties and private rights ; [44] and they instructed him from time to time to keep safely the rights and prerogatives of the king.[45] A vast field of government might still be regarded as the king's personal affair concerning which he had but to direct. Thus the sheriff received royal mandates to take and give seisin of lands or chattels [46] upon which the king had a claim, to make inquisition to determine his rights, to seize his

[40] As in *Lit. Claus.* i. 358, 405. In the earlier instance prisoners who complain that they are being held to ransom contrary to the treaty of 1217 are sent to show their grievances. In 1289 sheriffs were to send to Westminster those who had suffered wrongs at the hands of the king's ministers during his absence (*Cl. Rolls,* 1288–96, 653).

[41] *C. Cl. R.* 1302–7, p. 483 ; *Curia Regis Rolls,* 151, m. 25, P.R.O. ; *Memoranda de Parliamento,* ed. Maitland, p. lxxii ; Close Roll 67, m. 14 (a man accused of falsifying the king's letters) ; *ibid.* 75, m. 11 d (a bishop to appear before the king for citing an abbot before the court of Rome) ; *Placitor. Abbreviatio,* 133.

[42] *C. Cl. R.* 1302–7, p. 344: So he is directed to summon a person to appear before the exchequer of the Jews at Westminster (*Docs. from Records of Queen's Remembrancer,* 285).

[43] A minor example occurs in *C. Cl. R.* 1247–51, p. 115.

[44] Thus the sheriffs in 1250 are admonished to preserve the liberties of the church (Madox, *Exchequer,* ii. 102) ; in 1241 to place in safe custody money derived through papal taxation (*Cl. Rolls,* 1237–46, p. 361) ; the sheriff of York is to maintain and defend the prior and convent of Fountains in seisin of the manor of *Reminton* (Close Roll 66, m. 23 d, 1252) ; the sheriff of Dorset and Somerset to see to the keeping of the great charter of liberties (Close Roll 66, m. 9 d).

[45] See, for example, Madox, *Exchequer,* ii. 102-3.

[46] *E.g.* a writ directing a sheriff (*Lit. Claus.* i. 424) to take into the king's hands the lands of a deceased tenant-in-chief. The lands of the rebels of 1215 were seized and restored as they returned to their allegiance, by order conveyed through writs to sheriffs (*Lit. Claus.* i. 310–11, and *passim* to 1218). As to chattels, *Rot. Lit. Claus.* ii. 59.

enemies, to admit to bail his prisoners, to enforce his ancient dues, to maintain his highways, to make extents of lands,[47] or perambulations,[48] to establish boundaries by fixing metes.[49] This personal interest of the king is possibly to be taken literally when the sheriff of Nottingham is instructed to carry to some house a man wounded in Hazlewood ; [50] or the sheriff of York to find by counsel of the justices a suitable clerk who after ordination shall be presented to the archbishop to fill a certain ecclesiastical post ; [51] or the same sheriff to permit draft animals of the king's demesne to be brought within the castle of Pickering.[52] It is quite clear when in 1303 sheriffs send representatives of cities and boroughs to take part in negotiations with the merchants of the realm for new duties and imposts ; [53] and also when Edward I. directs the sheriff of Dorset and Somerset to place in defense all the preserved rivers within his bailiwick, to proclaim that no one hawk in them while the king is out of the realm, and to punish persons thus transgressing.[54]

Two subjects especially within the king's power to which he often called the sheriff's attention were the Jews, prior to their expulsion in 1290, and the coinage. Jewish affairs were often regulated by letters to the sheriff.[55] Thus Henry III. commanded the sheriff of Northampton to send before him certain influential Jews from each vill of the county to treat with him.[56] The disposition of Jewish property and the reception into a Jewish community of a family of their co-religionists from another town were similarly authorized.[57] In 1268, as a result of a well-known episode, the sheriff of Oxford was ordered to arrest the Jews who threw down and broke the cross carried in solemn procession on Ascension day, and not to permit them to have administration of their goods till they gave

[47] As *C. Cl. R.* 1279–88, p. 36. For an example of the writ *ad quod damnum, Placitor. Abbreviatio*, 151.
[48] In 1225 a perambulation of the forest (*Pat. Rolls*, 1216–25, p. 568) by the knights of the shire. Other orders for perambulations, Close Roll 65, m. 11 d, and *ibid.* m. 13, in the latter instance by order of the itinerant justice.
[49] *Bracton's Note Book*, i. no. 249 ; *C. Cl. R.* 1231–34, p. 143 ; *ibid.* 1237–42, p. 428 ; *ibid.* 1272–79, p. 572 ; *ibid.* 1302–7, p. 219.
[50] Close Roll 65, m. 26 d.
[51] *C. Cl. R.* 1279–88, p. 39.
[52] *Cl. Rolls*, 1234–7, p. 290.   [53] *Ibid.* 1302–7, p. 89.
[54] *C. Cl. R.* 1237–42, pp. 346–7.
[55] *Cl. Rolls*, 1234–7, pp. 13, 425.   [56] *Ibid.* 1237–42, pp. 346–7.
[57] As at Winchester in 1253 (Close Roll 65, m. 22) and at York in 1279 (*C. Cl. R.* 1272–9, p. 577). Cf. *ibid.* 432.

security that two new crosses should be made at their expense, one of marble and one of silver.[58]

The royal control of the coinage occasioned sheriffs more difficulty. Thus, in 1222, the king's council ordered that henceforth only round half-pennies and farthings should circulate, and bade the sheriffs make proclamation that after Easter no one upon the king's *forisfactura* might buy or sell by any other.[59] In 1247 a precept sent to many sheriffs called to their attention a former order against clipped money. They were to cause this to be cried and, if such coins were found thereafter, to have them pierced and all those who uttered this money arrested.[60] The behest was not everywhere carried out, for in 1248 Walter of Bath, sheriff of Devon, was ordered under arrest until he should satisfy the king concerning his transgression in permitting bad money to run in his county against the king's prohibition.[61] The prevention of the circulation of both false and clipped money was especially committed to the sheriffs in 1278.[62] An enactment of 1299 was directed against the importation of bad coins, known as cockards and pollards, and it was ordered that after Christmas they should be received at the rate only of a half-penny for the penny.[63] The task of enforcement seems now to have been too great for the sheriffs. The next year they all received mandate, along with the justice of Chester, to place the matter in the hands of two men whom they were to have selected in each market town and in other towns as they saw fit. These men they were to cause to swear openly in the presence of the people that they would have the statute observed.[64] Later in the same year the attention of sheriffs was called [65] once more to the ordinance against counterfeit money.

The king's orders to sheriffs as in the past also exemplify methods by which thirteenth-century commerce was regulated. The estab-

[58] Close Roll 86, m. 12, 53 Henry III. The same attitude is sometimes shown in dealing with Jewish property. In 1266 the sheriff of Kent was ordered to have pulled down the house of a Jew of Canterbury next to the chapel of Omer of Canterbury, and the site to be given to Omer (*ibid.* 83, m. 10). Some years later the mayor and sheriffs of London were to see that a Jewish school was located elsewhere than next to the oratory of the Friars Penitent, so it would be less of a *nocumentum* to the Friars (*ibid.* 89, m. 3).

[59] *Lit. Claus.* i. 516.

[60] *Cl. Rolls*, 1247–51, pp. 8-9.   [61] *Ibid.* p. 41.

[62] *Ann. Monast.* iii. 280 ; *C. Cl. R.* 1272–9, p. 516.

[63] *Stats. of Realm*, i. 131 ff. ; *C. Cl. R.* 1296–1302, p. 331.

[64] *C. Cl. R.* 1296–1302, p. 331.

[65] *Ibid.* p. 412.

lishment and holding of fairs and markets [66] called for the intervention of the sheriffs. Special mandates to prevent merchants from being impeded in coming to fairs [67] were sent to them. Letters close were also conveyed to them when a fair was no longer to be held, and goods and merchandise no longer to be taken thither to be bought and sold.[68] Change in the market day,[69] as well as its interruption [70] or resumption,[71] was authorized in the same way. Sheriffs were ordered to proclaim the price at which wines should be sold in London,[72] and also to proclaim the ancient assize regulating the width of cloth, that it should be observed at fairs.[73] They were also ordered to break up putrid wines found in their counties.[74] In 1307 they were generally charged to hold inquest into this matter in towns and boroughs, to destroy corrupt wines of Gascony, and to arrest those who exposed them for sale.[75] In one locality the ardour of one form of commercial enterprise was discouraged by directing a sheriff to proclaim that no baker should impress on his bread the sign of the cross or the lamb of God or the name of Jesus Christ.[76]

Communication with enemies or others might be cut off or kept open by the same process, as when the sheriffs of several western counties were ordered to permit for the duration of a truce, until Pentecost, 1260, free communication for merchandising between the king's lands and Griffin, son of Llewellyn.[77] Edward I., more than once, commanded sheriffs to prevent the export of wool,[78] in 1276 directing that none be taken to Flanders except by the king's license until the king's merchants had satisfaction for goods seized there.[79] In 1279 the sheriff of York had the king's letters [80] ordering him to proclaim in the county court and in the city of York that Jews might there traffic in lawful goods and merchandise both with Christians and Jews, and might buy victuals and other necessities.

[66] As in *C. Cl. R.* 1302–7, p. 230 ; 1227–31, p. 475.
[67] *Lit. Claus.* i. 486-7.
[68] *Cl. Rolls*, 1232–4, p. 241.   [69] *Lit. Claus.* i. 366.
[70] *Cl. Rolls*, 1231–4, p. 321.   [71] *Cl. Rolls*, 1227–31, p. 598.
[72] 42 Henry III., Close Roll 73, m. 12.
[73] So at Boston, 55 Henry III. (Close Roll 88, m. 3). In 1217 the sheriff had a special assistant to aid in enforcing this at the fair of Worcester. Those who after the proclamation sold dyed cloth of width other than that prescribed were to be taken into custody (*Pat. Rolls*, 1216–25, p. 153).
[74] Close Roll 67, m. 12, 38 Henry III. ; *ibid.* no. 75, m. 18.
[75] *C. Cl. R.* 1302–7, p. 526.
[76] Close Roll 65, m. 6 d, 36 Henry III.   [77] *Ibid.* 75, m. 16.
[78] *C. Cl. R.* 1272–9, pp. 119, 423.   [79] *Ibid.*   [80] *Ibid.* p. 577.

The sheriff was expected to prevent unauthorized commerce. Thus in 1234 the king gave permission to a certain Fulbrith to take a ship-load of grain from Yorkshire to London, the sheriff to see that security was accepted not to take it elsewhere.[81] In the preceding year the sheriff of Salop was told to permit no merchandise to be taken from the county to Blancminster to be sold there other than as accustomed in past reigns.[82] The sheriff of Kent was once instructed to permit the men of the bishop of Coventry and Lichfield to bring stone by the Medway for his chapel in London.[83] In 1220 the mayor and sheriffs of London were directed not to distrain, nor permit to be distrained, merchants who brought salt fish to the city by ship, so that they might sell the fish as in the time of Henry II. and King Richard.[84] In 1307 the sheriff of Kent received a precept to proclaim that all persons of his bailiwick who wished to sell grain and other kinds of victuals to strangers and aliens, or to take them to parts beyond the seas, might lawfully sell them. At the same time the sheriff was informed that the ordinance was to be firmly observed which was lately made by the king and his council against the taking out of the realm of money, silver vessels or silver in mass without his license.[85] Special promises of good sale, prompt satisfaction of price and harmless status in all things, were made in 1301 at the king's instance by proclamation of various sheriffs, to induce merchants to send north by land and by water to Berwick-on-Tweed provisions and other necessities for the king's army.[86] A curious instance of royal control in an industrial process occurs in 1307 in an order to the sheriff of Surrey and the mayor and sheriffs of London, commanding that those who wish to use kilns shall burn brushwood or charcoal and not sea coal, from the use of which an intolerable smell diffuses itself, greatly infecting the air to the annoyance and injury to bodily health of resident magnates and citizens.[87]

Many of these executive precepts would have been ineffective without the sheriff's proclamation. The student of arbitrary government in the Tudor and Stuart periods may well give more attention to the extensive use of this device in the three reigns of

[81] *Cl. Rolls*, 1231–4, p. 414.
[82] *Ibid.* p. 226.
[83] Close Roll 77, m. 14.
[84] *Lit. Claus.* i. 425.
[85] *C. Cl. R.* 1302–7, 530. Cf. a similar order to the barons and bailiffs of Winchelsea in 7 Edward I. (Fine Roll 77, m. 19).
[86] *C. Cl. R.* 1296–1302, p. 490.
[87] *Ibid.* 1302–7, p. 537.

the thirteenth century and the usage in this respect which unites the earlier with the later era of strong monarchy. There were two methods by which the sheriff was ordered to give publicity to a multitude of mandates. He caused them to be read in the county court, or he caused them to be proclaimed generally in public places throughout his bailiwick, as was being done extensively in King John's time. Some acts required both forms of publication.[88] Quittances and privileges granted monasteries or individuals, as well as military writs of summons[89] and other matters already noticed, were likely to be proclaimed in the county court.[90] Some examples of matters cried *per totam ballivam* are the writ of 1234, requiring those holding a knight's fee or more *in capite* to receive knighthood,[91] a decree of 1226, requiring French subjects to leave the realm with their chattels and merchandise,[92] regulations for the local control of toll,[93] the manufacture of cloth,[94] restriction upon the taking of salmon,[95] prohibition of fishing rights in the king's reserved streams,[96] and announcement of the grant of a new fair or market.[97] An order of 1297, permitting Flemish merchants to buy war supplies in England, was proclaimed in all ports and market towns,[98] also another for the release of the men of the count of Flanders taken by reason of the war. Notice to Jews as to how they were to make claim for certain debts was sometimes proclaimed in the synagogues.[99] Sheriffs' proclamations thus convey announcement, including postponement of administrative acts,[100] but also warning or positive command. The enforcement of law, the preservation of the peace, the conduct of administration, the observance of popular liberties and private rights,[101] the maintenance of the king's interests and prerogatives, as well as the effectiveness of his most arbitrary and

---

[88] *E.g.*, prohibition in 1251 in Northumberland of the circulation of Scottish money or of any except the king's new money (*Cl. Rolls*, 1247–51, p. 549) ; also orders in 1233 for a special watch in each Kentish vill (*ibid.* 1231–4, p. 310).
[89] *Cl. Rolls*, 1237–42, p. 239 ; *C. Cl. R.* 1302–7, p. 86 ; *Royal Letters*, ii. 300.
[90] As in *Cl. Rolls*, 1231–4, p. 21 ; *C. Cl. R.* 1288–96, p. 69.
[91] Madox, *Exchequer*, i. 510, note (d).
[92] *Lit. Claus.* ii. 155.
[93] *Pat. Rolls*, 1216–25, p. 499.
[94] *Rot. Lit. Claus.* i. 378.
[95] *Ibid.* i. 385.
[96] As in Essex in 1254 (Close Roll 66, m. 2 d).
[97] *Rot. Lit. Claus.* i. 391 ; *C. Ch. R.* i. 219 ; *Cl. Rolls*, 1247–51, p. 40.
[98] *C. Cl. R.* 1296–1302, 13, 15.
[99] Rigg, *Cal. Plea Rolls*, Exchequer of Jews, i. 89, 199.
[100] *E.g.* eyres (*Lit. Claus.* i. 405 ; *C. Cl. R.* 1279–88, p. 424).
[101] As in *Cl. Rolls*, 1234–7, p. 426.

unwarranted demands, often depended upon the publication by the sheriff of the appropriate acts of government.

In relation to public works the sheriff enforced obligations which rested in the main upon local usage or common law, by sanction of judicial decree, executive order, or, in some instances, mere sanction of custom. Service with plows was being rendered under a sheriff's direction in some communities in the Norman period.[102] Presumably he had to do even earlier with the building or repair of bridges and fortifications. In the thirteenth century it was the duty of the itinerant justices to inquire what broken bridges and causeways there were, what communities they traversed, and who was responsible for repairing and sustaining them.[103] Upon the basis of the information thus derived the justice amerced the delinquent community and ordered the sheriff to distrain it to make the necessary repairs.[104] This, for instance, was the process once employed to compel the county to repair the stone bridge at Nottingham.[105] Sometimes the king's writ authorized the sheriff to make inquest by twelve men to ascertain by their oath whose duty it was to repair a given bridge.[106] An old-time duty of keeping the king's park enclosed was, in some places, also enforced by writ to the sheriff, ordering distraint.[107] The royal mandates of 1234, ordering sheriffs to build and repair bridges so that the king with his birds could freely pass the rivers,[108] in the light of Magna Charta [109] must be understood to apply to vills in which this was the ancient rule. In 1241 the sheriff of Berkshire was instructed to take five marks of each vill which had failed in this duty.[110] In some communities customary levies of murage and pontage are mentioned. The Hundred Rolls complain of some abuse on the part of sheriffs in this matter. Thus the sheriff took for the pontage of Cambridge two shillings from each hide in the geldable parts of the county, whereas he formerly took but six pence, and now used but a third of this sum to mend the broken bridge.[111]

[102] Above, p. 67.   [103] *Fleta*, i. cap. 20, sect. 41.

[104] Cf. *Public Works in Medieval Law*, Selden Soc., i. pp. xxiii, 53-54.

[105] *Assize Roll* 664, m. 51 (8 Edward I.). At Huntingdon users of the bridge who carried certain commodities across it were held responsible for keeping it up (*Placitor. Abbreviatio*, 148).

[106] *Ann. Monast.* i. 373 ; *Plac. de Quo War.* 284.

[107] *Cl. Rolls*, 1231–4, p. 39 ; *ibid.* 1247–51, p. 58.

[108] *Public Works in Medieval Law*, Selden Soc., pp. xxiv-xxv ; *C. Cl. R.* 1234–1237, pp. 9, 33 ; but cf. 34 : *sicut reparari consueverunt tempore domini Johannis regis.*

[109] Cap. 23.   [110] *C.P.R.* 1234–42. p. 375.   [111] *Rot. Hund.* i. 50, 52.

New public works, for which custom made no provision, introduced a different problem. In the earlier years of the thirteenth century local communities, especially counties,[112] sometimes financed these. But later writs tacitly imply that this is to be done by the king's authority.[113] Thus the sheriff who in 1291 held in custody the town of Hereford, lately taken into the king's hands, was authorized by patent to appoint two persons to collect the pavage conceded by royal authority to the bailiffs and good men of the town.[114] Ten years later the sheriff at Oxford was directed by letters close to cause the pavements of the suburbs to be repaired, inducing landholders to cause pavements to be made before their tenements, and, if need be, to compel them to do so.[115] A royal mandate to the sheriff of Norfolk, in 1258, ordered him to compel tenants in Marshland and elsewhere in the county to keep up walls against the inundation of the sea.[116] A little earlier the sheriff of Kent was directed to let twenty-four elected *jurati* levy on lands in the marsh of Romney to repair *wallas* and *watergangas contra maris impetum et periculum*.[117] In 1280, upon complaint of the abbot of Stratford, that inundation from the Thames and the sea resulted from the neglect of his neighbours to make repairs or contribute to those he made yearly, the sheriff of Essex was granted during pleasure a commission *de walliis et fossatis*.[118] In 1301, so it is reported, the earl of Warwick had granted to the king the right to have a prison in his town of Leicester, and a great sum of money had been collected accordingly from the men of the county. The sheriff was ordered to make inquest to ascertain in whose hands this was, and then cause a plot to be bought in a suitable place.[119]

Encroachments upon the public highway, such as hedges [120] or

[112] Pollock and Maitland, i. 555.
[113] In 1260 the burgesses of Gloucester were granted by the king the right to take *muragium* in order to close and repair their walls, and the sheriff was instructed not to impede them or their bailiffs in collecting this (Close Roll 75, m. 4). King Henry also granted the burgesses of Newcastle-on-Tyne a *murage* (*Rot. Hund.* ii. 19).
[114] *C.P.R.* 1281–92, p. 450.
[115] *C. Cl. R.* 1296–1302, p. 484. In *Rot. Hund.* ii. 19 is a complaint that those who collected the *murage* of Newcastle by King Henry's concession in his forty-ninth year had not yet repaired the walls.
[116] Close Roll 73, m. 15 d.   So an inquisition (*Inq. Misc.*, Chancery, i. 519) reports that the marsh of Winchelsea cannot be saved by the old wall and that the new wall cannot be kept up by the lands that kept up the old one.
[117] Close Roll 65, m. 4.
[118] *C.P.R.* 1272–81, p. 380.
[119] *C. Cl. R.* 1296–1302, p. 428.
[120] *Placita de Quo Warranto*, 391.

ditches,[121] and the diversion of waters [122] to the injury of any one, upon presentation before the justices, or sometimes in the sheriff's tourn, or upon hearing before the sheriff by the king's writ,[123] were treated as common nuisances. The usual procedure followed by the justices was to order the sheriff to remove them at the cost of the persons responsible. Often he was ordered to take the coroners with him,[124] that he might have official record of what was done and thus better assure punishment if there was resistance. Other nuisances [125] were similarly treated. Especially troublesome were weirs and other obstructions in streams.[126] Letters to the sheriff of Gloucester, in 1277, ordered him to open the banks of the Severn near certain weirs so that it might be twenty-six feet in width. Persons who resisted the sheriff in the matter were to be brought before the king in parliament on the quinzaine of Easter, to receive penalties for their contempt.[127]

The period under review marks the culmination of the sheriff's power as chief peace official of his shire. He was the officer primarily responsible for good order, and he raised hue and cry against disturbers of the peace. It was his duty to arrest lawbreakers as well as to have them indicted before him. Saving the bailiffs of franchises, he and his bailiffs were the only persons bound to make such arrest. At the beginning of the period there was a sheriff's peace. In his tourn he exercised the functions of a local judge. Both here and in the county court he received indictments of homicide and other pleas of the crown. By the time of Edward I. the power to receive such indictments was passing in some measure to the *conservator pacis*.[127a]. Against the danger of invasion from without, as well as armed disorder within the county, the sheriff might be called to act. The whole force of the county was in theory

---

[121] Salt Arch. Soc. iv. 213. The sheriff of Essex (*Lit. Claus.* i. 510) was in 1222 directed to level certain ditches made to the harm of a royal forest.

[122] *Pleas for Gloucester*, ed. Maitland, 50-51.

[123] Bracton, iii. 564 ; cf. Britton, Nichols ed., i. 81-82.

[124] Close Roll 69, m. 22.

[125] A bank raised to the harm of a neighbour (*Yorkshire Assize Rolls*, Yorks. Archæol. Soc., xliv. pp. 78-79); so other obstructions (*Oxford City Documents*, Oxford Histor. Soc., 207).

[126] *Ibid.* 207. In 1224 a sheriff was directed (*Lit. Claus.* i. 622) to take coroners and if necessary prostrate a weir which was preventing use of the stream at Shrewsbury. In 1254 the abbot of Westminster had impeded the sheriff in this matter (Close Roll 67, m. 9 d).

[127] *C.P.R.* 1272-81, p. 195.

[127a] *Assize Roll* 20, m. 6 (8 Edward I.). Here also called *custos pacis*.

at his disposal, and if occasion required he might be instructed to adopt most elaborate arrangements for local police.

The assumptions recorded in the chancery of Edward I., concerning the sheriff's responsibility for the peace, are wide and its doctrines on the subject vigorous. The death of Henry III. was followed on November 23, 1272, by letters to the sheriffs of England, notifying them that the peace had been proclaimed in the name of the new king and ordering them to cause it to be proclaimed in their bailiwicks, inhibiting all under penalty from infringing it.[128] Again, in March 1303,[129] sheriffs were ordered to make proclamation of a truce concluded with the king of France and cause it to be observed inviolably, forbidding all under pain of loss of body and goods from inflicting damage on the king of France, his subjects or his lands. A commission for fixing the metes of a certain place on the borders of three counties declares that robberies and other offences were being perpetrated here, but since it was unknown in what county that place was, sheriffs and coroners of these counties did not exercise their office in raising and pursuing hue and cry and in viewing the bodies of slain persons.[130] More than once the Statute of Winchester was sent to the sheriffs to cause it to be publicly read, proclaimed and firmly observed.[131] Sheriffs were sometimes ordered not only to exercise diligence in these matters but to direct the bailiffs of the county to do the same. If the latter failed to do so, the sheriff was to enter their liberties and assume the initiative.[132] The careless sheriff of Shropshire and Staffordshire was told, in 1286, that certain robberies and other enormities against the king's peace could not be if he displayed due diligence, as he ought, for the preservation of the peace, especially as the *posse* of the county was at his summons and distraint.[133] A letter sent, in 1306, to a sheriff declared that it was manifestly contrary to the sheriff's oath and fealty for him to pass over what the king had ordered for the preservation of the peace, and that if he were negligent and remiss the king would lay a heavy hand on him for contemning his orders and consenting to malefactors.[134] Nor was the threat by any means peculiar to this document.

[128] *C. Cl. R.* 1272–9, p. 1.   [129] *Ibid.* 1302–7, p. 80.
[130] *C.P.R.* 1272–81, p. 178.
[131] *C. Cl. R.* 1288–96, p. 330 ; *ibid.* 1302–7, p. 396.
[132] *Ibid.* 1279–88, p. 434 ; 1302–7, p. 397.
[133] *Ibid.* 1279–88, p. 434.
[134] *Ibid.* 1302–7, p. 406 ; cf. *ibid.* 1288–96, p. 330.

This guardianship of the peace had also unusual angles which are interesting.   In times of stress the sheriff was sometimes ordered to administer the oath of fealty in a county court [135] or in a community [136] where disturbance was threatened ; and he might be ordered to arrest persons who repeated false and sinister statements about the king.   Safe conduct for the king's agents upon his business was sometimes sent to sheriffs.[137]   So were lists of offenders who were to be taken [138] or bailed ; [139] and a pardoned outlaw required a mandate to the sheriff to permit him to be in peace in the county.[140] Sheriffs were sometimes ordered in the time of Henry III. to give counsel and aid to the king's castellans or other agents for the preservation of the peace, and even to aid them with horses and arms.[141]   If sheriffs made undue delay in arresting criminals they were liable before the justices.[142]   They were sometimes called to explain why they had made arrests, and they were subject to amercement for taking a man without reason.[143]   Sergeants and bailiffs exercised these powers subject to the sheriff's direction.   It was held in one case that a sheriff's clerk ought to have arrested a person who presented a writ sealed with a false seal.[144]   When malefactors took sanctuary, the sheriff sometimes received their abjuration [145] in person.[146]   The thief who publicly confessed he sometimes permitted to become an approver and accuse his associates, holding them to appear before the justices.[147]   On complaint made by the

[135] In March 1263 the sheriff of Norfolk and Suffolk was ordered to cause oath to be taken of all in the county courts and in the several bailiwicks for faithful adherence to the king, and at his death to have his son Edward for lord and prince (C.P.R. 1258–66, p. 285).
[136] In 1248 of the men of the towns of St. Edmunds and Len (C.P.R. 1247–58, p. 11).   For the sheriff's action against traducers of the king, C. Cl. R. 1272–9, pp. 568–69.
[137] C.P.R. 1232–47, p. 256 ; ibid. 1266–72, p. 306 ; Royal Letters, i: 235.
[138] C.P.R. 1232–47, p. 62 ; ibid. 1281–92, p. 520.
[139] As in C. Cl. R. 1272–9, p. 478.   [140] As in C.P.R. 1247–58, p. 119.
[141] Ibid. 1232–47, pp. 165, 228, 287.   [142] Select Pleas, no. 153.
[143] Bracton's Note Book, ed. Maitland, ii. no. 705 ; Pollock and Maitland, ii. 582-583.
[144] C. Cl. R. 1272–79, p. 487.
[145] In 1232 the sheriff of Lincolnshire was ordered to send two coroners to the church of Horncastle, where the bishop of Carlisle had liberties, that a thief might abjure the realm before them (C. Cl. R. 1231–34, p. 99).
[146] Select Pleas, no. 169.
[147] Rot. Lit. Claus. i. 361, 427 ; Pleas for Gloucester, ed. Maitland, 109.   In 1254 the sheriff of Somerset was ordered to deliver an approver from prison so that he might abjure the realm.   This was on condition that he had fought as many duels as he had promised to purge the land of thieves (Close Roll 65, m. 9) ut moris est.

Q

ecclesiastical authorities to the king, the sheriff might to a much later date be ordered to arrest excommunicated men or women who were contumacious.[148] A sheriff at Oxford might be ordered to deliver an imprisoned scholar to the chancellor of the University.[149] In 1238 sheriffs received the names of certain clerks present when insult was done the papal legate at Oxford and were ordered to imprison those responsible.[150] Sheriffs were instructed to take and deliver to the Friars Preachers an apostate from their order,[151] and similar aid was extended to the master of the order of Templars, in recovering one of the brothers who wandered about the county in secular dress.[152]

But the sheriff was called to face far sterner emergencies. The unruly conduct of the barons in the earlier years of the majority of Henry III., as well as frequent local disorders and disturbing conditions of the period of the barons' war, all added greatly to the difficulty of peace maintenance. The reign of Edward I. also witnessed some local lawlessness of armed bands with an increase in general criminal tendencies,[153] which occasioned apprehension in periods of emergency.[154] In 1233 several sheriffs were instructed to enforce a provision, recently made at Gloucester, against persons travelling armed, to have them arrested with arms and horses and, if they did not have power, to raise hue and cry and follow them from vill to vill.[155] The sheriff of Gloucester was ordered [156] to cause all passages of the Severn to be guarded, and to let no one pass until it was known of whose household he was and that he was not an enemy of the king. In that year, as part of a general mandate for conserving the peace, all the sheriffs were ordered to take armed persons who trained in the woods.[157] The sheriff of Kent and his *servientes* were instructed not only to do this, but also ordered to cause it to be cried in each hundred of the county that the men of those parts were to take certain armed men if they passed by night

[148] *C. Cl. R.* 1296–1302, p. 512 ; *ibid.* 1302–7, p. 99. Bishop Grosseteste charged sheriffs with systematic delay in carrying out royal mandates of this nature (*Ann. Monast.* i. 424).

[149] Close Roll 67, m. 8.   [150] *Cl. Rolls*, 1237–42, p. 136.

[151] *C.P.R.* 1232–47, p. 248 ; cf. *ibid.* 1301–7, p. 123.

[152] *C. Cl. R.* 1302–7, p. 339.

[153] Cf. the writer's *Frankpledge System*, 151–52.

[154] A criminal outbreak was feared in 1306 when the king departed for Scotland (*Cl. Rolls*, 1302–7, p. 396). The absence of knights removed a powerful restraining force.

[155] *Cl. Rolls*, 1231–4, p. 317.

[156] *Ibid.* 320.   [157] *Foedera*, i. 208–9.

and, if they could not, to raise the hue and cry.[158] In 1231 the sheriff of Cambridgeshire was instructed to aid the bishop, the chancellor and the masters in repressing unruly students, arresting and imprisoning or expelling certain malefactors and rebellious clerks. No student was to be permitted to remain in Cambridge who was not under the tuition and discipline of some master.[159] That the peace might not be disturbed during the king's absence the sheriff was in 1253 sent to Yarmouth with letters to the men of that town to prevent them from molesting at the fair the barons of various of the Cinque Ports.[160]

Lawlessness beyond, or almost beyond, the sheriff's power of repression continued throughout the period. In 1262 the sheriff of Essex received a mandate to have the ways leading to London watched and guarded, as robberies and murders had been committed on men going to the dedication of the church of St. Paul.[161] In the years between 1285 and 1290 sheriffs were often instructed to suppress bands that threatened the peace of the realm. In Herefordshire the problem was to prevent suspected persons from holding conventicles in the city of Hereford or elsewhere.[162] The sheriff of Staffordshire and Salop was directed to clear the passes of wood and to enlarge them.[163] In Westmoreland in 1288 magnates and others went armed, and proclamation was ordered that no one under pain of grievous forfeitures should go about with horses and arms or make assemblies.[164] The next year all the sheriffs were asked to aid in measures to break up rival bands of persons with horses and arms.[165] Letters close of September 1297, directed to two sheriffs, record the attempt to hold the town of Nottingham against those who purpose to advance thither to hold assemblies and to go thence toward London.[166]

The sheriff's work in dealing with offenders of rank seems often to be confined to the mild function of making inhibitory proclamation. This is clear in measures for the suppression of tournaments. In 1244 the king issued a prohibition of a projected tournament to be held near York, ordering the sheriff to go in person and inquire

[158] *Cl. Rolls*, 1231–4, pp. 309-10.
[159] *Royal Letters*, i. 396-97.
[160] *C.P.R.* 1247–58, p. 239.
[161] Close Roll 78, m. 3 d.
[162] Madox, *Exchequer*, ii. 143.
[163] *C. Cl. R.* 1279–88, p. 434; so the sheriff of Hereford in 1234 (*ibid.* 1234–37, p. 8).
[164] *Ibid.* 1279–88, p. 547.
[165] *Ibid.* 1288–96, p. 51.
[166] *Ibid.* 1296–1302, p. 129.

the names of those meeting to tourney, or those whose inns were taken there for tourning, so he might certify the names to the king.[167] Edward I. several times forbade such affairs during the course of the war with Scotland. The usual procedure employed was to order the sheriff to cause proclamation to be made probibiting any knight, esquire or other person from tourneying, tilting or jousting, seeking adventures or otherwise going to arms without the king's licence, and to arrest any who did so after the proclamation.[168] In 1302 sheriffs were ordered also to take into the king's hands the lands of such recalcitrants.[169] In 1306 the penalty proclaimed was forfeiture of life and limbs, goods and chattels,[170] but it is obvious that this could not *ipso facto* involve enforcement by the sheriff. More possible of execution was the king's order to the sheriff to remove lay force from a church.[171] It is sometimes recorded that the sheriff in doing this employed a *posse*.[172]

Sheriffs also called out the *posse comitatus* to demolish and prostrate castles and other strongholds fortified against the king's prohibition. All sheriffs were ordered in 1218 to comply with the provisions of the Charter of Liberty regarding adulterine castles.[173] The usage of calling out the *posse* for the purpose is evidently older than 1189.[174] As late as 1251 [175] a sheriff was ordered to take *probi homines* of the county with horses and arms and to destroy a fortified private house or seize persons there assembled. These occasions sometimes called for a systematic levy of the regular forces of the county.[176] In 1224 the sheriff of Devon, who had been ordered to raise the whole *posse* of his county and to blockade the castle of Plympton with the aid and *posse* of the bishop of Exeter, reported

---

[167] *C.P.R.* 1232–47, p. 424. In 1241 the sheriff of Nottingham was not to permit any one to sell victuals to those tourneying there (*ibid.* p. 266).

[168] *C. Cl. R.* 1296–1302, pp. 373, 408, 588.

[169] *Ibid.* 1302–7, p. 66.		[170] *Ibid.* p. 433.

[171] *Cl. Rolls*, 1234–37, p. 563 ; 1247–51, p. 40 ; Heales, *Records of Merton Priory*, app. no. lxii ; Close Roll 66, m. 20 d ; *ibid.* 67, m. 12 d.

[172] *C. Cl. R.* 1296–1302, p. 597.

[173] *Foedera*, 1816, i. 150.

[174] Above, p. 136. In 1218 (*Lit. Claus.* i. 380) the sheriff of Staffordshire and Shropshire was ordered to prevent the fortification of the castle of Dudley *cum omni posse tuo et cum viribus tocius ballie.* If necessary he was to call to his aid the sheriff of Worcestershire *cum viribus suis.*

[175] *Cl. Rolls*, 1247–51, pp. 540-1 ; cf. *Royal Letters*, i. 429. See also *Cl. Rolls,* 1231–4, pp. 311, 321. Cf. Statute of Westminster, i. cap. 17 (*Stats. of Realm,* i. 31).

[176] In *Lit. Claus.* i. 474, are writs of 1221 ordering several sheriffs of northern counties to call out the military tenants of their counties to destroy castles.

that the knights all declined to make ward of this kind, inasmuch as their lords had already answered the king's summons to serve in his army.[177]

The *posse comitatus* at the sheriff's disposal was thus assumed to consist, if necessary, of the whole force of the county.[178]   He was entitled to use distraint to enforce such service,[179] but there are complaints after about 1250 that communities are not reporting.[180] The general levy, such as that made to demolish castles, was rare, and the usual form of summons was by raising hue and cry.   If the force raised in one vill was not sufficient to take malefactors or rebels, the sheriff, in accord with ancient usage, followed them from vill to vill until they were taken *per patriam confluentem*.[181]   The local bailiffs also were expected to raise *posses* [182] in the same way. The Statute of Winchester shows that sheriffs as well as bailiffs were expected to keep horses and armour to follow the hue with the country.[183]

Determined opposition to a sheriff was by no means unknown,[184] and in 1253 the sheriffs were instructed that if, in the execution of the king's mandates, they found hereafter any resistance, they were then to take with them as much of the *posse* of their counties as should be necessary.[185]   Use of a *posse* was sometimes made in levying upon property by judicial order,[186] as well as in making arrests [187] or guarding an important prisoner.[188]   According to the order of 1253 all persons were commanded that, so often as they were warned by the sheriff to come to him and give their aid and counsel, they were to hold themselves so prepared, that the king might not have to go against them.   In the king's writs frequent injunction is given

---

[177] *Royal Letters*, i. 232.

[178] In *C.P.R.* 1281–92, p. 317, a sheriff is enjoined to take the *posse* of two counties if necessary.   In Gross (*Coroners' Rolls*, 29) occurs an instance in which the sheriffs and men of two counties pursue and decapitate a murderer.

[179] *C. Cl. R.* 1279–88, p. 434.   [180] *Ibid.*.1272–79, p. 19 ; cf. p. 24.

[181] *Royal Letters*, i. 418, 428.   [182] *C. Cl. R.* 1279–88, p. 434.

[183] In or just before 1253 the under-sheriff of Dorset and Somerset was wounded and beaten at Shaftesbury with his bailiffs and men (*C.P.R.* 1247–58, p. 225).

[184] *Stats. of Realm*, i. 98 ; Stubbs, *Sel. Charters*, 472, 474.

[185] *C.P.R.* 1247–58, pp. 223-34.

[186] Coram Rege Roll 77, m. 5 (11 Edward I.).   A sheriff is earlier enjoined by court order to take with him a sufficient *posse comitatus* in making a levy upon goods within a liberty (*ibid.*).   He also took similar action at times in releasing beasts held contrary to gage and pledge.   Cf. *Placitor. Abbreviatio*, 261.

[187] As in Salt Arch. Soc. v., part 1, 142.

[188] The sheriff was to be at Devizes *cum coronatore et toto comitatu* to guard Hubert de Burgh in the church (*Cl. Rolls*, 1231–4, p. 309).

to the men of the counties to use due diligence to assist the sheriff in preserving the peace.

The sheriff's other duties were often too numerous and too pressing for him to give much systematic attention to peace enforcement. The difficulty of restraining criminals increased, and in the time of Edward I. king and council showed an anxiety for the peace which reminds one of the days before the Norman Conquest.[189] Special aids to the sheriff were sometimes employed to maintain the peace. In 1242 one Peter Turald was associated with the sheriff at Oxford, so that one of them might be able to attend constantly to the peace there and use such diligence as the king commanded. The reason assigned for so unusual an act was that the sheriff was too much employed to attend constantly to the matter.[190] The special *custodes pacis*, employed largely, one for each county, in the period of the barons' war, and occasionally mentioned in the time of Edward I.,[191] were officials of much the same type.

A more important innovation dates from 1233, when the magnates of the realm discussed the threatening state of affairs and agreed that watch and ward should be kept at night by four men in each vill. All the sheriffs were ordered to proclaim this in full county, in hundreds and in markets, and all sheriffs and bailiffs to use due diligence in guarding the peace according to this form. No strangers were to be allowed to pass through any vill at night, and if the watch could not arrest them, they were to raise the hue and cry upon them.[192] This same procedure was invoked later in the year against enemies of the king and the next year against certain armed persons.[193] It was revived again from Ascension day to Michaelmas, 1242, at which time was ordered the well-known appointment of a chief constable in each hundred, to be under the direction of the sheriff, and two knights in each county, who were to have power to convene all the *jurati ad arma* of the hundred.[194] This provided a

---

[189] For the statute of 1293, making the men of the community responsible for damages incurred by robbery when they did not produce the criminals, *C. Cl. R.* 1288–96, p. 330.

[190] *C.P.R.* 1232–47, p. 301. Cf. the special custodians of the Jews at Lincoln (*Rot. Lit. Claus.* i. 357). [191] Above, pp. 175, 200, 221.

[192] *Cl. Rolls*, 1231–4, p. 309.

[193] *Royal Letters*, i. 428 ; *Cl. Rolls*, 1231–4, p. 544.

[194] *Ibid.* 1237–42, pp. 482–3. In the following March, the chancery complains of negligence on the part of sheriffs and bailiffs in preserving the peace and of the prevalence of homicide and rapine. Sheriffs are ordered to have watch kept at night by the *jurata ad arma* of each vill that malefactors may be taken (Close Roll 66, m. 18 d). In 1260 the sheriff of Kent was ordered to make proclamation

device for effective pursuit of malefactors when the local *posse* made ineffective response.   In 1251 all the sheriffs were again enjoined to establish watch and ward by several men of each vill in order to deal with malefactors, both footmen and horsemen, who passed by pillaging and killing.[195]   On both of these latter occasions sheriffs were enjoined in case of neglect of this duty on the part of the men of a vill to see that it was performed at their expense.[196]   This arrangement tended to place the whole system under control of the sheriffs.[197]   They were also ordered occasionally to keep watch and ward in regions in danger of attack from the king's enemies by land [198] or sea.[199]   Such special methods were no doubt necessitated by a decline in effectiveness of the old local constitution, such as one sees in the gradual disintegration of the frankpledge system.   The main change in the position of the sheriff of the fourteenth century was to lie in the transfer to others of much of this responsibility for the observance of the peace.

The custody of the king's prisoners regularly rested with the sheriff.   Castellans who were not sheriffs received and guarded prisoners by special order.[200]   Towns [201] and private persons [202] kept the prisons required for the exercise of their special jurisdictional privileges.   But a prominent duty of local bailiffs was to deliver offenders to sheriffs that they might be committed to jail. Indeed the common complaint of both local and private bailiffs was that money was demanded of them by sheriffs before they would receive prisoners.[203]   In the later decades of the thirteenth century,

---

calling on the *jurata ad arma* to be sworn and ready to take malefactors (*ibid.* no. 77, m. 12 d).   In Northampton in 1283 the sheriff was ordered to swear to arms for the arrest of malefactors all those whom it had been usual to summon (*C.P.R.* 1281–92, p. 66).

[195] *Cl. Rolls*, 1247–51, p. 540.

[196] A writ to the sheriff of Shropshire and Staffordshire, whose men have been negligent in this matter, directs that the cost is to be placed by view of good men, and to be levied and rendered at the expense of those to whom this *custodia* and *tuicio* belongs (*ibid.* 528).

[197] In March 1252 the sheriff of Wiltshire was commanded to cause men to be sworn to arms to keep the peace (Close Roll 66, m. 18 d).

[198] As in *Royal Letters*, ii. 300.

[199] In 1296 sheriffs were ordered to place a light guard along the sea coast to give warning to wardens (*C. Cl. R.* 1296–1302, p. 74).

[200] As in *C. Cl. R.* 1247–51, p. 37 ; *ibid.* 1288–96, p. 425.

[201] *E.g.* Worcester and Winchester ; see below.

[202] *Northumberland A.R.* 74 ; *C. Cl. R.* 1227–31, p. 367 ; *Lit. Claus.* i. 490 ; *Cl. Rolls*, 1234–37, p. 302 (a jail for women).

[203] Stubbs, *Sel. Charters*, 385 (sect. 20) ; *Rot. Hund.* i. 10.

moreover, the arrest and safe keeping of debtors was already being enjoined upon sheriffs by judicial process.[204] The power of these officials to take and imprison persons stands out in a popular griev-ance which recurs over and over again. It is charged that they wrongfully incarcerate persons to extort money from them. To prevent this an enactment of 1285 gave action for false imprisonment to persons imprisoned by sheriffs or bailiffs of franchises for felony without indictment.[205] The well-known writ of 1252, in directing that strangers be taken at night, specifies that this is not to be made the occasion of exaction on the part of sheriffs or bailiffs.[206]

The prisoners in charge of the sheriff were confined in the king's prison.[207] In the time of Edward I. it was broadly assumed that there was a chief prison in each county.[208] At one period, however, the sheriff of Essex had no prison and made use of Newgate. Leicestershire had none until 1301, the jail at Warwick serving for both [209] counties held by the one sheriff. The new one then erected at Leicester was an exception in that it was not in a castle. One reason assigned for the committing of the *corpus* of a castle to a sheriff was that he might keep his prisoners in the jail there.[210] In 1285 the outer bailey of Sherburne castle was assigned the sheriff for this purpose.[211] In 1231 the bailiffs of Worcester were ordered to allow the sheriff the use of their jail until the one within the castle was repaired.[212] In 1228, however, a sheriff was directed not to keep his prisoners at the castle of Winchester but at the jail of the same city.[213] In 1260 the only jail which the king had in Kent was in Rochester castle, and the constable was directed to receive all the prisoners delivered to him by the sheriff according to the custom hitherto prevailing.[214] In the time of Edward I., however, the sheriff had other prisons in this county,[215] and it is clear that in addition to the chief prison of a county there were some-times others which he used.[216]

---

[204] As in De Banco Roll 158, m. 43 ; 110, m. 238. Cf. Statute of Merchants.
[205] *Stats. of Realm*, i. 81. [206] Stubbs, *Sel. Charters*, 371.
[207] Cf. *Cl. Rolls*, 1247–51, pp. 38, 39. [208] *Stats. of Realm*, i. 132 ; cf. i. 40.
[209] *C. Cl. R.* 1296–1302, p. 428 ; as to Essex, Close Roll 74, m. 10.
[210] *C.P.R.* 1272–81, p. 127.
[211] *Ibid.* 156. [212] *C. Cl. R.* 1227–31, p. 567.
[213] *Cl. Rolls*, 1227–31, p. 31. [214] Close Roll 75, m. 10, P.R.O.
[215] At Canterbury (*Rot. Hund.* i. 49) and at Maidstone (*C. Cl. R.* 1296–1302, p. 327). As to local prisons see *The Eyre of Kent*, Selden Soc. ii. pp. lxv-lxvi.
[216] In Warwickshire, at Kenilworth (*Lit. Claus.* i. 420), as well as at Warwick (*Select Pleas*, 100).

Ultimate responsibility for the keeping of prisoners might sometimes be transferred from the sheriff to others. The castellan at Rochester was presumably responsible for escapes, but at Newcastle-on-Tyne, although there was a castellan, the sheriff was nevertheless in charge of the prison and responsible for those who absconded.[217] That this burden was not light is shown by a pardon granted the sheriff of York in 1298 for the escape of twenty-six prisoners from the jail of York, the consideration being his many cares and occupations at the time as well as his diligent pursuit of the fugitives.[218] Had the matter come before the itinerant justices, the amercement of a hundred shillings, which according to Britton [219] would have been levied upon the unfortunate official for each escape, would have sadly depleted his resources. Some jailers were also brought to account for their acts or negligence. The one at Ilchester, when some of his prisoners escaped and fled to the churches, himself took sanctuary with his family.[220] Jailers are often seen to be the agents of sheriffs. Under-sheriffs and bailiffs are mentioned in 1285 as keepers of prisons.[221] Despite the fact that the keepership of a prison is sometimes a sergeanty [222] and that the king appoints keepers [223] and removes them from office for negligence,[224] a *custos prisonae* is also mentioned as having served under this or that sheriff.[225] Newgate prison certainly had its own keeper, yet it was the custom for new sheriffs to ride thither each year on the vigil of St. Michael to receive the prisoners.[226] The jailer of the sheriff of Norfolk in the Hundred Rolls [227] is called his constable. Many of the amercements levied upon sheriffs [228] for the escape of prisoners

[217] *Northumberland A.R.* 83, 96.
[218] *C.P.R.* 1292–1301, p. 364.
[219] Nichols ed. i. 87.
[220] *Rot. Claus.* ii. 13-14.
[221] *Stats. of Realm,* i. 30.
[222] As at Winchester in 1219 (*Book of Fees,* part i. 259). The keeper of Fleet prison held a serjeanty which also included the custodianship of the king's manor of Westminster and the repair of the bridge of the Fleet (*Cal. Inq.* ii. 201, 8 Edward I.). At Exeter the serjeanty of the keepership of the jail carried with it the holding of the king in chief of two carrucates of land (*ibid.* i. 117, 45 Henry III.). In 1212 it was combined with the post of doorkeeper of the castle and was held by grant of Henry I. (*Book of Fees,* part i. 96).
[223] *Lit. Claus.* ii. 168.
[224] *C. Cl. R.* 1247–51, pp. 19-20.
[225] *Rot. Hund.* i. 49; cf. Britton, Nichols ed., i. 75.
[226] Fitz Thedmar, *De Antiquis Legibus,* 24. For the keeper see *Lit. Claus.* i. 450; *C. Cl. R.* 1247–51, p. 9.
[227] *Rot. Hund.* ii. 525.
[228] As in *Select Pleas,* no. 154; *Cl. Rolls,* 1237–42, p. 42; Fitz Thedmar, *De Antiquis Legibus,* 22; *Northumberland A.R.* 98.

arose of course from the fact that they were the keepers of the jails.[229] This is accentuated also by another fact. They received special instructions concerning the rigour of the treatment to be accorded a prisoner,[230] and in some instances were directed to be more lenient.[231] The sheriff of Southampton in 1306 was not only ordered to keep the rebel bishop of St. Andrews in irons in a strong place in the tower of the castle, but also to appoint good and faithful men under security to assume his custody.[232]

The custody of prisoners involved the obligation to produce them before the justices, the council, or the exchequer [233] when required, to release them when specially ordered, and to exercise care in admitting them to bail. Here there was long a conflict of legal opinion concerning the scope of the sheriff's discretion. If he thus liberated a person suspected of felony [234] he was sometimes held liable for his escape. If without authority he admitted to mainprise a person taken at the special command of the king or a person accused of homicide he was liable, but he had a wide power. Sometimes he was ordered by the writ *replegiare* to release prisoners to mainpernors, usually twelve in number,[235] who assumed the obligation to produce them before the justices when required.

King Henry once ordered the sheriff of Somerset to admit to bail the bishop of Bath and Wells, who had been cited to appear before the king.[236] In the time of Edward I., as earlier, there was complaint that sheriffs took money to admit to bail indicted and

[229] Cf. *C.P.R.* 1301–7, p. 112.

[230] Thus Hubert de Burgh was to see no one without the king's special permission except a man in religion and was to have food but once a day, then a halfpenny loaf and one great *cafata* of beer (*Cl. Rolls*, 1231–4, p. 161).

[231] In two instances, in 1248 and 1250, prisoners are to be removed from the dungeon to the upper prison, although one of them is still to be in chains (*ibid.* 1247–51, pp. 56, 254–5). According to Britton, Nichols ed. i. 44, only prisoners apprehended for felony or imprisoned for trespasses committed in parks or vivaries or detained for arrears of accounts might be put in irons. These may be the prisoners who (Bracton, ii. 402) appear before the justices in leg irons.

[232] *C. Cl. R.* 1302–7, p. 410.

[233] A writ of *duci facias*: Madox, *Exchequer*, ii. 63–4.

[234] As in *Oxford City Documents*, Oxford Histor. Soc., 201 ; *Northumberland A.R.* 98. On this whole subject see Pollock and Maitland, ii. 584–6.

[235] See letters on behalf of various prisoners to the sheriff *quod ponantur per ballium* (*Cl. Rolls*, 1247–51, p. 494). Cf. Britton, i. 57, which mentions the admission to mainprise of the thief until the county court or jail delivery ; also the writ (Bracton, ii. 292, 500) for delivering to twelve pledges a man accused of a death.

[236] Close Roll 75, m. 11 d.

imprisoned persons who were not replevyable,[237] and that they and others were holding and grieving persons entitled to bail in order to gain of them.[238] The result was the enactment of a section of the first statute of Westminster, specifying which prisoners were and which were not henceforth entitled to be admitted to mainprise.[239] The sheriff or other keeper of a prison who released prisoners not bailable, upon attaint thereof was to lose forever his fee or office ; he who withheld prisoners entitled to bail after they offered sufficient surety was to pay to the king a grievous amercement ; and, if he took reward for delivering such, he was to pay double to the prisoner and also to be at the great mercy of the king. Sheriffs and keepers of jails in 1285 were also made liable in the amount of damages done the king, if they released a receiver of moneys bound to yield account at the exchequer. Such persons were not replevyable, and these officials were to take heed that they were not released by the common writ *replegiare*.[240] The interest of the sheriff as well as that of the prisoner made against observance of those statutes. In 1299 a second enactment complains that sheriffs are releasing by replevin prisoners not replevyable by the form of the statute recently made, even persons openly defamed and incarcerated for murder and felony. Upon the justices of jail delivery was now imposed the duty of inquiring whether sheriffs or others had released by replevin persons not bailable and of punishing those whom they found guilty.[241]

In general the hanging of the thief in a franchise seems to have presupposed the presence of sheriff or coroner.[242] The power to execute criminals after trial was a regal one, and a private person

[237] *Rot. Hund.* ii. 194. [238] *Stats. of Realm,* i. 30. Cf. *Rot. Hund.* i. 15.
[239] Those taken for the death of a man, or by commandment of the king or the justices of the forest were not at that time bailable. Henceforth persons who had been outlawed or had abjured the realm, approvers, those taken with the mainour, those who had broken prison, those not of good name who were appealed by approvers so long as the latter were living, or those taken for felonious house-burning or for uttering false money or counterfeiting the king's seal, or excommunicate persons taken at the request of the bishop, or persons taken for manifest offences or treason touching the king himself, were to be in no wise replevyable whether by common writ or without writ. Persons designated as bailable were those accused by an approver now deceased and those indicted for larceny by inquests before sheriffs or bailiffs by their offices, such persons not being previously guilty of larceny nor more serious felony nor of aiding felons and being of light suspicion, or else accused of petty larceny amounting to less than twelve pence (cap. 15).
[240] *Stats. of Realm,* i. 80. [241] *Stats. of Realm,* i. 129 (cap. 3).
[242] As in *V.C.H. Rutland,* i. 172

was permitted to have a gallows only by prescription or special grant.[243] In 1234 the men of Gretton in Northampton were pardoned for raising in their town a gallows upon which two outlaws were hanged, since this was done by order of the king's constable of Rockingham.[244] It is obvious that the erection of a gallows upon which to execute the king's prisoners might be the function of persons other than the sheriff. Nor does it appear that he was as yet personally expected to attend all executions.[245] In 1280 the sheriff of Northumberland was ordered [246] to take coroners and go to the church of St. Andrews, Newcastle-on-Tyne, and, if he found that a certain man condemned and hanged for a death was alive after he was carried to the churchyard for burial, he was to cause his peace to be proclaimed that he might go safely to his house and there remain at the king's peace. While it is hardly safe to press too vigorously the inference to be drawn from one instance, it seems clear that the sheriff had not attended, or at any rate had not been presumed to attend, this particular hanging.[247]

The purely military functions of the sheriff, which, by the Magna Charta period had reached their highest importance, in the reign of Henry III., declined, and before 1307 practically disappeared. Stubbs has shown that the sheriff was leader of the minor tenants-in-chief and the body of freemen sworn under the Assize of Arms.[248] But he had no power outside his county, and it is possible to find him directing military operations only in King Henry's earlier years, when engaged in the duty of taking and prostrating a castle held by one of the king's unruly subjects. For this service he might be ordered to summon those who regularly owed the king military service, the *jurati ad arma*,[249] or men of the county with necessary implements.[250] But after the forces were assembled others were

[243] Cf. *Oxford City Docs.*, Oxford Histor. Soc., 211. In *Rot. Hund.* ii. 31 there is mention of a gallows where bailiffs of the bishop of Lincoln condemn and hang those taken with the mainour.

[244] *Pat. Rolls of Henry III.* 1232–47, p. 65.

[245] According to *Oxford City Documents*, 217-18, 219, clerks on two occasions lead to the gallows a man adjudged by the justices to be hanged, and in so doing place the condemned in a church where he may abjure the realm.

[246] *C. Cl. R.* 1279–88, p. 34.

[247] But there were in the time of Henry II. hangings *visu vicecomitis*.

[248] *Const. Hist.* ii. 220.

[249] *Lit. Claus.* i. 474 (Feb. 1221); *Royal Letters*, ii. 300. Against Fawkes de Breaute, who was then besieging Bedford castle, the sheriff of Salop and Stafford was directed in 1224 to levy a force from vill to vill of his county with horn and hue (*Rot. Lit. Claus.* i. 632).

[250] *Lit. Claus.* i. 632; *C. Cl. R.* 1272–79, p. 410.

in charge of the operations [251] which were not purely local.   The most important exclusive function of the sheriff touching military affairs was soon confined largely to the delivery or publication of the king's summons, or in a few cases revocation of the summons, to those owing him military service.[252]   In 1264 the sheriff was instructed to take the *custos pacis* of the county with him to make summons.[253]   In Oct. 1299 the king's writs touching military service in Scotland were handed over at the exchequer to the sheriffs, who were to deliver them to the barons residing in their respective counties.[254]   The writ compelling knighthood, as Stubbs has shown,[255] was also directed to the sheriff, and it was he who was commanded to take into the king's hand the lands of those who disregarded the order.[256]   At the end of the reign of Edward I. the sheriffs of London were ordered to seize the goods and chattels, and to arrest the persons, of certain knights and men-at-arms,[257] who were declared to be acting in manifest contempt of the king in crossing to foreign ports while the Scottish war continued.

The sheriff also performed important functions as mustering officer, although in this his powers in the thirteenth century were no longer exclusive.   At the end of the century the knights of the shire might still be ordered to come before the sheriff to be mustered.[258] But the old form of order to bring to a specified place the force of the shire owing knight service is mentioned only in the earlier years of Henry III.[259]   In 1232 all sheriffs were enjoined to notify both

[251] In Oxfordshire in 1198 one Robert fitz Alan held land as a sergeanty *per servicium portandi baneram populi prosequentis per marinam* (*Book of Fees*, part i. p. 11).   In Sussex there was a similar sergeanty (*ibid.* p. 71 ; part ii. p. 1413). A thirteenth-century law writer (Liebermann, *Gesetze*, i. 656, 657) assumes the existence of an *heretochus ad conducendum exercitum comitatus sui iuxta praeceptum regis.*   This is clearly not the sheriff.

[252] Stubbs, *Const. Hist.* ii. 221.   For specific examples, Madox, *Exchequer*, i. 653, note x (ordering the usual proclamation) ; *Cl. Rolls*, 1231–4, p. 313 (ordering notice to be cried to all *servientes* who wish to enter the king's service for pay) ; *Cl. Roll* 75, m. 9 (1260) ; *C. Cl. R.* 1272–9, p. 410 ; *ibid.* 1296–1302, p. 209.   Cf. Stubbs ii. 291. In Sept. 1260 sheriffs were ordered to proclaim revocation of the summons to go against Llewellyn (Close Roll 75, m. 7 d).

[253] Madox, *Exchequer*, i. 654.   The *custos pacis* was at this time in charge of some of the county *posses* (*C.P.R.* 1258–66, p. 361) mustered to ward off alien invasion.

[254] *Ibid.* i. 654.   [255] *Const. Hist.* ii. 221, 295-6.

[256] As in *Cal. Fine Rolls*, 1272–1307, p. 317.

[257] *Fine Roll* 104, m. 2, 34 Edward I.   [258] Madox, *Exchequer*, ii. 104.

[259] In 1223, when Lewellyn besieged the castle of Builth (*Foedera*, i. 170), sheriffs were notified merely to summon those who owed military service to join the king at Gloucester.

the regular force and all men of the county sworn under the assize of arms to arm and be ready for the king's service.[260]   In the preceding year, when a third of this latter force was actually called out, the sheriff of Gloucester was required, if necessary, to exceed the quota in sending two hundred men with axes and with victuals for forty days.[261]   He was to send all the carpenters of the county, who were to come at the king's pay, and with these and the others a trustworthy person to certify their number.   Several of the sheriffs were ordered to be at Bridgenorth with the military tenants and to have there also a third of the *jurati* equipped with good axes.[262]   In 1254 sheriffs were notified that all knights and *servientes* should proceed at the king's pay to Gascony, and that public proclamation should be made, summoning these to be at Westminster three weeks after Easter.   Sheriffs were to be present at that time to inform the council how they had carried out this mandate.[263]   But in the reign of Edward I. sheriffs could not generally have been required to accompany their men, and the business of mustering was passing to others.[264]

Throughout the period they retained various other functions in military affairs.   In 1230 sheriffs were employed with three or more other persons to assess and swear to arms [265] men of their respective shires, and such was the arrangement under the famous writ of 1252.[266]   In 1253 sheriffs were ordered to hire as many knights and sergeants as possible to go with the barons to Gascony,[267] and in 1263 the king called on the sheriffs to summon by proclamation the tenants-in-chief of a knight's fee or more to be at Worcester on Aug. 1, to proceed against the Welsh Llewellyn and his rebel accomplices.[268]   In 1263 the sheriff of Kent was ordered to guard the sea coast,[269] and in 1264 the sheriff of Norfolk and Suffolk, in mustering the *jurati ad arma* to ward off alien invasion, was ordered to proclaim that this should not be made a precedent.[270]   In September of that year the sheriffs of Norfolk, Suffolk and Essex

---

[260] *Cl. Rolls*, 1231-4, p. 60.

[261] Stubbs, *Const. Hist.* iii. 297 ; *Cl. Rolls*, 1227-31, p. 595.   The victuals the sheriff was to cause to be found *per homines comitatus sui juratos ad alia minuta arma* whom the king wished to remain in those parts.

[262] *Ibid.*   [263] Close Roll 67, m. 12 d.

[264] Stubbs, *Const. Hist.* ii. 297-8 ; Madox, *Exchequer*, ii. 105-6 ; Cf. *C. Cl. R.* 1288-96, p. 456 ; *ibid.* 1296-1302, p. 379 ; *C.P.R.* 1301-7, pp. 103, 146-7.

[265] *Cl. Rolls*, 1227-31, pp. 398-402 ; *Royal Letters*, i. 371.

[266] Stubbs, *Sel. Charters*, 371.

[267] Mitchell, 262.   [268] Close Roll 80, m. 10 d.

[269] *Ibid.* 78, m. 19.   [270] *C.P.R.* 1258-66, p. 360.

were commanded, along with some discreet knights, to go with all speed to where the commonalties of the county were assembled to guard the seashore in defence of the realm against aliens. They were to order these to stay where they were, and not to retire at the end of forty days as proposed. For those who did not have expense money sheriffs were to levy it. But they were to seize the lands and goods of those who had not come nor sent a substitute.[271]   In 1303 the sheriff of Nottingham acted with the steward of Sherwood forest to select a hundred and twenty woodmen for the expedition against the Scots, and a special agent was named who was to pay their wages and conduct them to the king at Roxburgh.[272] By the fourteenth century sheriffs sent up the required number of archers, who were selected by specially appointed persons.[273]   In 1295, when providing for the defence of the north, sheriffs merely assembled before the king's special agents the knights of the shires and two men from each township to hear and execute the order of these officials.[274]   Sheriffs were ordered to provide messengers to carry military messages,[275] and they still secured provisions [276] and transport,[277] or even in time of hostilities broke down bridges,[278] or prevented knights, merchants with horses and arms and other armed persons from entering the realm without the king's license.[279] In at least one instance they prepared beacons to give the alarm in the event of a Scottish invasion which was threatened, and were instructed to warn out all those with horses and arms to meet such an emergency.[280]   They were of aid in war in countless instances, but their duties are purely auxiliary.[281]

---

[271] C.P.R. 1258–66, pp. 367-8.   [272] Ibid. 1301-7, p. 136.

[273] In 1295 sheriffs were directed to arrest the footmen thus selected (C. Cl. R. 1296–1302, p. 379).

[274] Stubbs, Const. Hist. ii. 222. At this time the sheriffs were ordered to proclaim that all those who had horses and arms, regardless of the amount of lands held, were to enter the king's service at his wages (L.T.R. Mem. Roll 67, m. 10 d).

[275] As in Madox, Exchequer, ii. 106.

[276] In 1231 the sheriff of Gloucester was ordered not to permit any fair or market during the muster and to cause all wines to follow the army (Cl. Rolls, 1227–31, p. 595).

[277] The sheriff of Dorset in 1226 was ordered to take security of all ships to go into the king's service (Royal Letters, i. 356). A writ of aid in 1303 directed various sheriffs to assist Walter Bacun, the king's clerk, in procuring fifty ships (C.P.R. 1301-7, p. 121).

[278] Letters of Henry III., ii. 253 ; Close Roll 81, m. 7 d (1263.)

[279] Ibid. 76, m. 2, Apr. 1260.   [280] L.T.R. Mem. Roll 69, m. 91 b (1298.)

[281] An inquest of 1276 (Inq. Misc., Chancery, i. 309) well illustrates this. A sheriff of Kent with horses and arms started toward the siege of Rochester castle. but was ordered back by the constable of Dover castle.

This same observation may be made as to the sheriff's position in general at the death of Edward I. His initiative indeed remains so far as peace preservation is concerned, but special conservators are already being appointed, and their successors will in a generation take precedence in this respect. In the shiremoot he still does justice, but its importance is much dimmed by the rise of the king's court. Much of what he does in the county court, including his conduct of elections, he does by special order. His position in relation to the king's courts is ministerial, and he is in addition called to execute many mandates of various kinds. Writs of aid require him to assist special agents of the king, whether in the capture of the king's enemies,[282] in the prevention of counterfeiting,[283] in the conduct of his treasure,[284] or in the maintenance of his other interests.[285] Gradually other officials have entered the field he once occupied. Custodians of the peace, commissioners of array, special collectors of various revenues are each pre-empting ground. Except for some years subsequent to 1274,[286] the guardianship of the king's wards and escheats in the county is given over to especial sub-escheators.[287] Moreover, the elected knights who serve as coroners now stand next to the sheriff in importance. When there is no sheriff, or when he is officially called to answer,[288] or fails to act,[289] writs are addressed to them. To a certain extent he is a check upon them, for he is now required to have counter rolls with them.[290] But they are also checks upon him. In criminal matters their rolls are more authoritative than his, and in judicial as well as

[282] C.P.R. 1232–47, p. 26.   [283] Ibid. 1247–58, p. 114.
[284] Ibid. 1232–47, pp. 60, 301.
[285] As matters pertaining to a Jewry (ibid. 1247–58, p. 70).
[286] On Nov. 3, 1274, it was ordered that the office of escheator on both sides of the Trent be filled by the respective sheriffs except in the counties held as of fee. The sheriff of Gloucester was to act also for Worcestershire, the sheriff of Northampton for Rutlandshire, the sheriff of Devon for Cornwall, the sheriff of Cumberland for Lancashire and Westmoreland (Fine Roll 73, m. 5). But in the Close Rolls after 1277 sheriffs are not addressed as escheators in their respective counties.
[287] Cf. Pat. Rolls, 1216–25, p. 70 ; C.P.R. 1232–47, pp. 7, 486.
[288] As in De Banco Roll, 158, m. 130. An exchequer writ of the Michaelmas term 1264 directs the custos pacis of the county of Hereford with one other person to make inquest in pleno comitatu whether a certain sum had been paid to a former sheriff (L.T.R. Mem. Roll 38, m. 1 d). Cf. Bracton's Note Book, no. 1010 ; Sel. Coroner's Rolls, p. xxvii.
[289] Thus in 31 Edward I. coroners are ordered to levy upon the goods of certain collectors of the aid after the sheriff has failed to collect (K.R. Mem. Roll 76, m. 84 d).
[290] Stats. of Realm, i. 29 (cap. 10).

administrative business they sometimes assist and possibly supersede the sheriff.[291]   The remaining discretionary power of the latter is too often attested by various kinds of peculation and petty extortion which still exist at the end of the period.   To check these was probably the chief motive a generation later in placing the sheriff on an annual tenure.   But by 1307 the powerful official of the Norman and Angevin periods, almost dictator in his county except for the rare intervals when the king's justices were there, has made way for a sheriff of circumscribed authority, who in most things is ruled by writs.   This subordination appears very distinctly also in the realm of finance.

[291] Gross, *Sel. Coroners' Rolls*, pp. xxvi-xxvii.   But a passage in Britton (i. 89) seems to show that sheriffs procured the writs by which unfit coroners were removed from office.   According to this passage sheriffs had sometimes procured writs of false suggestion for removing coroners and were amerciable for this offence.

IX

THE FISCAL FUNCTION

OF the ties which bound the sheriff to the central government in the thirteenth century the firmest was the financial one. Important as were his other duties, there are many indications that in official estimation they yielded to those touching the royal income. It was an indispensable part of the sheriff's duty to account for the farm of the county, to collect the other crown debts, and to maintain the king in seisin of his fiscal rights. Royal mandates might issue either to the sheriff or to the exchequer, directing the collection of sums due the king, the expenditure of funds on the king's behalf, the respite of claims, and of days set for account or the pardon of arrearages. But these, except in the matter of disbursement, were quite the exception. To his fiscal obligations the sheriff was held by the exchequer, which enforced the rules and usage affecting crown debtors, and before which he made regular appearances at periodic intervals.[1]

This strong administrative and judicial body announced and often made the appointment of the sheriff, held him to his fiscal duties, and exercised control over him in various ways. Its officials had power to take cognizance of his misdeeds,[2] trespasses,[3] or negligence,[4] to impose due penalties, and even to remove him from

---

[1] In 1236 was established a similar responsibility of sheriffs and bailiffs in Ireland to answer at the exchequer in Dublin (*C.P.R. 1232–47*, p. 153).

[2] An exceptional case occurs in 31-2 Edward I., when it was decided in the exchequer before the treasurer and barons, the chancellor, the chief justices of both benches and others of the king's council that a sheriff had acted unjustly, maliciously and in derogation of the liberties of Shrewsbury in seizing a citizen of that town. The release of the prisoner was ordered and the sheriff incarcerated in the Fleet (*L.T.R. Mem. Roll* 74, m. 27 d).

[3] Madox, *Exchequer*, ii. 281.

[4] *Ibid.* ii. 65, note (g). Barons of the exchequer were sometimes ordered to place a sheriff under arrest for failure to carry out the king's orders (*Cl. Rolls, 1247–51*, p. 41). The sheriff of Cambridgeshire in Mar. 1262 was pardoned 100s.,

office. In the performance of fiscal duties sheriffs, when local officials failed to act, were given authority at the exchequer, as by the central law courts in other cases, to enter liberties from which they were supposedly excluded and within which writs were regularly served and executed only by seigniorial or municipal bailiffs.[5] So advantageous was the method of delivering precepts and writs to sheriffs in person at the exchequer that other departments sometimes employed it. Writs were sent for delivery, particularly at the opening of the semi-annual sessions when all sheriffs were expected to attend.[6] Sheriffs sometimes received at the exchequer important orders relative to military affairs,[7] certified to it their proceedings in certain matters of peace preservation,[8] or reported through messengers Scottish inroads.[9] Thither on at least one occasion came King Henry with his council to issue commands to sheriffs by word of mouth.[10] It was the only place in England where the sheriffs as a body might be reached at one time.

The sheriff's responsibility for the revenue derived from his county of course explains his marked subordination to the exchequer. Each year he was expected to appear there twice, at the opening of the Easter and of the Michaelmas session, to make a proffer or advance payment upon the debts collected during the fiscal year, and also to come on one or more occasions especially designated to close up his accounts for the preceding fiscal year.[11] The sums for

---

which he was fined for not executing the royal precept to levy damages recovered by a litigant (Close Roll 77, m. 15). In *Parliamentary Writs*, i. 192, a sheriff is ordered to appear *coram Rege* for such contempt and also for making a false return. In *L.T.R. Mem. Roll* 74, m. 19, the sheriff of Wilts is amerced 100*s.* for not producing at the exchequer the heir of a certain debtor of the king.

[5] As in *K.R. Mem. Roll* 26, m. 3 d (Worcester); *L.T.R. Mem. Roll* 18, m. 6 d (Winchester); *ibid.* 27, m. 12 (liberty of Windsor and of the abbot of Reading); *ibid.* 33, m. 2 (liberty of bishop of Carlisle); *ibid.* 74, m. 19 (liberty of St. Edmunds).

[6] On Oct. 4, 1298, certain writs and letters were delivered to practically every sheriff in England (*K.R. Mem. Roll* 72, m. 2 d). On Oct. 2, 1301, a baron of the exchequer, in the presence of the treasurer and other barons, delivered under his seal to the sheriff of Cornwall estreats with writs and other records pertaining to jail delivery for the itinerant justices in the county (*L.T.R. Mem. Roll* 73, m. 17 d).

[7] Above, p. 235.

[8] *C. Cl. R.* 1279–88, p. 547.

[9] *L.T.R. Mem. Roll* 69, m. 91 d. In case of a Scottish inroad the sheriff of Lancaster was ordered in 1298 to send news by a mounted man.

[10] Madox, *Exchequer*, ii. 102-3; *L.T.R. Mem. Roll* 25, m. 2.

[11] Upon this whole subject of thirteenth-century exchequer procedure the writer gratefully acknowledges his indebtedness to Miss Mabel H. Hills, whose notable contributions, embodying the results of diligent research on this subject,

which he was thus ordinarily accountable arose from three distinct sources : the *ferm* of the shire with kindred payments, the fixed annual rents, and the oblates, fines and amercements arising in the course of administration.

The *ferm* of the various shires rarely shows any variation in amount from the ninth year of Richard I.,[12] when the sum is first recorded on the Pipe Roll, to the end of the period under discussion. In eight of the ten instances in which a single sheriff regularly held two counties there was but one *ferm* ; in two sets of linked counties [13] *ferms* were reckoned separately for each county. To the annual *ferm* required of the sheriff of the year 1200 was usually added from John's time a fixed sum each year from his *proficua*. Late in the century this was being designated as the *firma pro proficuo comitatus*. The amount for any county was fixed at the time of the sheriff's appointment and varied from time to time.[14] The Pipe Roll by the later years of Edward I. had become the official record of this, the roll of the year when a change was first recorded being cited as authority for the amount.[15] Occasionally, though not often, the *ferm* was still further augmented by an *incrementum*. From the sheriff who held his shire as *custos* and did not farm it, such a payment could not be demanded.[16] After the year 1284 the amounts of the annual *ferm* recorded upon the Pipe Roll for the various shires are much smaller, but this is a change in form rather than fact, the new figure merely representing the net *ferm* after fixed alms and allowances were met, the so-called *remanens firma post terras datas*.[17] The new method of computation was not

---

appear in *E.H.R.* xxxvi. 481-96 and xxxviii. 351-4. As to certain counties that did not follow the normal rules see *ibid.* xxxvi. 483.

[12] In most cases the amount in that year was the same as that shown by the computation of Mr. G. J. Turner (*Trans. R.H.S.*, N.S., xii. 142-9) for 10 Henry II. In Cornwall there was an increase after 10 Henry II., and in Essex and Hertfordshire, Kent, Norfolk and Suffolk, and in Nottinghamshire and Derbyshire a decrease.

[13] Dorset and Somerset ; Leicestershire and Warwickshire.

[14] Generally speaking, it was highest about 1258. In 1305 it ranged downwards from the 200 marks paid respectively by the sheriffs of Kent, of Norfolk and Suffolk, and of Yorkshire, to the 40 marks paid by the sheriff of Warwick. In some counties the amount is but seldom mentioned.

[15] Thus in the Pipe Roll of 1304 the sheriff of Oxfordshire and Berkshire owes £120 *de firma pro proficuo comitatus sic contentum in rotulo xxiii* (*i.e.* the Pipe Roll of 1296).

[16] Madox, *Exchequer*, ii. 226. For the *incrementa* of the various counties in 1240 see *Red Book*, ii. 771.

[17] Cf. *Stats. of Realm*, i. 69. *The Great Roll of the Pipe for 26 Henry III.* (ed. H. L. Cannon) shows well how the nominal *ferm* had been thus reduced.

thenceforth always employed, for there was a tendency to slip again temporarily into the old ways.[18]

There was in practice no distinction between the sources of the revenues designated as *ferm* and as *proficua*. In keeping the account of money which belonged under these heads sheriffs give no heed to exchequer classifications, listing on their own rolls under the general head *proficua* or *exitus* the amounts received. From the sum total the requisite amount was taken to pay the *ferm* of the shire, the *corpus comitatus* as it was called, the remainder being applicable toward the *proficua* of the shire.[19] The *ferm* and *proficua*, which stand as the two earliest items of account in the Pipe Roll of a given year and county in the reign of Henry III. have only historic significance. They are the names respectively of the older and the newer annual exactions imposed upon sheriffs for the concession of their shires. So far had the amount of the former become conventionalized, that a sheriff who held his shire as *custos*, and who in theory did not farm it, paid exactly the same amount as the sheriff who was *firmarius* of the same county.[20]

So far as the revenues listed under these two heads are concerned, the sheriffs' problem of collection was merely that of continuing to levy the customary amounts from whatever source derived and by whatever title locally designated. Minute sums were collected in many communities, and might be listed in the sheriff's roll either under the names of persons or of villages without more exact designation. Other entries represented the *ferms* paid by bailiffs for hundreds still in the king's hands, for lathes in Kent,[21] and for ridings in Yorkshire and Lincolnshire [22] also let at farm by

[18] The record for the year 1304–5 (as in *Chancellor's Roll* 98) shows a complete return to the older form of statement.

[19] Thus Thomas de la Hyde, sheriff of Cornwall, holding as *custos* in 31 Edward I., renders an account (*Exchequer L.T.R. Misc. Rolls*, 5/12) *de proficuo comitatus et hundredorum* which totals £160 : 5 : 3 *de quibus subtrahuntur* £60 : 19 : 10 *numero ad perficiendum corpus Cornubiae post terras datas de anno xxxi. De quibus idem vicecomes respondet in rotulo de eodem anno per corpus comitatus . . . et remanens* £91 : 5 : 5 *numero in quibus idem Thomas reddit in praedicto rotulo xxxi. in firma proficui comitatus.* The *proficua* exacted for the shire in this period was 50 marks. This usage appears in the period 1235–40 (*L.T.R. Misc. Rolls*, 5/43, 5/74, 5/2, 2/29, 6/30), when sheriffs are *custodes*, but also in rolls 5/17 and 5/34 of this series kept by sheriffs who were *firmarii*.

[20] See note 19. The sum exacted from the sheriff of Buckingham and Bedford as *firmarius* for the year 1259–60 was £369 : 19 : 11 *bl.* and £108 *numero.* His predecessor, who was *custos*, accounted for the same sum the preceding year (*Chancellor's Roll* 53).   [21] *Rot. Hund.* i. 226.

[22] *Ibid.* i. 130, 337. Cf. Magna Charta, sect. 25. The sheriff also lets at *ferm* a ward in Northumberland (*Rot. Hund.* i. 20).

sheriffs.[23] In some instances whole groups of hundreds were farmed together.[24] The attempt of 1258 to prevent the farming of baili-wicks was unsuccessful, and the Hundred Rolls complain of much extortion occasioned by the high *ferms* which sheriffs exact from bailiffs.[25]

The plan of accounting whereby the amount of the *ferm* of the shire was taken from the general issues which appear on the sheriff's roll makes it difficult to gain information as to the constituent portions of this important element of the royal revenue before the exaction of annual *proficua* in John's time. But it is clear that income from the king's lands continued to be included in the time of Edward I.[26] The judicial perquisites of county and hundred courts are still to be traced on the sheriff's rolls, those of the hundred presumably forming the chief basis of the bailiff's *ferm*. Rents of assize, which included in some localities such renders as grain and fowls,[27] seem to be survivals of an ancient *feorm*, and in some instances were clearly regarded as belonging to the *ferm*.[28]

The classification of certain other local payments, which are listed upon the sheriff's roll of the issues of his county, is a more difficult matter. There is fairly conclusive evidence that some amounts collected for view of frankpledge in the twelfth century formed a part of the *ferm* of the shire,[29] and thus could hardly have

[23] In Shropshire nine hundreds were in the king's hands, a tenth let at fee farm (*Rot. Hund.* ii. 63). In Gloucestershire seven were in private hands, yielding the king nothing. But the abbot of Cirencester held four others at fee farm, while the sheriff had seven at farm (*L.T.R. Misc. Rolls*, 1/8). The fact that perquisites of the hundred courts sometimes appear upon the sheriff's *Particule Proficuorum* (*e.g. ibid.* 6/13) leaves a doubt as to whether they were always farmed.

[24] Three Buckinghamshire hundreds for 7 marks (*Rot. Hund.* i. 23).

[25] That of the ward of Tyndale in Northumberland, regarded as a quasi-hundred, had been increased *ad gravamen populi* from 6 marks a year to 12 (*Rot. Hund.* ii. 22); that of the wapentake of Lovedon in Lincolnshire *per extorsionem* from 8 marks to 26 (*ibid.* i. 331).

[26] Above, p. 243.

[27] A roll of the sheriff's *proficua* for Berkshire, 22-23 Henry III. (*L.T.R. Misc. Rolls*, 5/12) amounted to £86 : 6 : 6. Part of this was from hidage (£30 : 17 : 10), view of frankpledge (108*s.* 10*d.*), wardpenny (£6 : 11 : 4), grain of assize (£4 : 9 : 8) and hens (21*s.*).

[28] In the *Chancellor's Roll* for 1258–59 (no. 51, m. 1, Essex and Herts) *proficua comitatuum*, *res assisae* and *placita et perquisita comitatus* are once mentioned as component parts of the *corpus comitatus* for the year. Rents assessed since John's time could not have belonged to the ancient *ferm* of the county.

[29] In an exchequer record of John's time attributed to Robert Mantel, sheriff from 1170 to 1181, view of frankpledge is reckoned as part of the *ferm* of the county (*Red Book*, ii. 775). This payment was collected and turned over by the bailiff to the hundred (*L.T.R. Misc. Roll*, 5/25; *Inq. Misc.*, Chancery, i. 173).

been appropriated by the treasury in John's time under the title of *proficua*.   Sheriff's aid, which appears in the sheriffs' accounts along with hidage and wardpenny, had been claimed successfully by Henry II.,[30] and now belonged to the crown.[31]   In John's time it was being held at the exchequer that sheriff's aid should not go to the sheriff as one of the sources of revenue farmed by him, since he had formerly paid a full *ferm* without it.[32]   This payment thus seems to be the principal basis of the sheriff's *proficuum*.   The sheriffscot,[33] rendered annually at the hundred,[34] was presumably the same thing.   Collections for view of frankpledge and sheriff's aid undoubtedly went hand in hand, both growing out of the sheriff's tourn of the hundreds.[35]   Land was geldable for sheriff's aid just as it was for *murdrum* and other communal impositions.[36]   The original form of the levy was thus the same [37] as that of the hidage, which in some places still lingered as a customary obligation.

The second class of revenues for which sheriffs responded annually was derived from minor *ferms* or rents, notably those for assarts, purprestures and sergeanties.   Annual income from some manors and occasionally from other sources also fell under this head.[38]

The third source of income collected by the sheriff, that arising chiefly from *oblata* or fines and from the perquisites of justice in the king's courts, was known as the summonses because the writs

[30] See Robert Mantel's roll, cited n. 29.

[31] The king makes grants (*Cl. Rolls*, 1227–31, p. 14), respites (*ibid.* p. 15) and pardons (*C. Cl. R.* 1279–88, p. 39) of sheriff's aid, and loss of income from this source is loss to the crown (*Rot. Hund.* i. 355).

[32] *Red Book*, ii. 769.

[33] *Scirreweschot, chiryveschot, Rot. Hund.* i. 454, 484.

[34] *Ibid.* i. 484.

[35] The sheriff of Wiltshire had ingress twice a year into a certain hundred and received thence *per annum* 6 marks *ad opus domini regis* and 20*s. ad auxilium vicecomitis* (*Rot. Hund.* ii. 232).   So at his tourn in the hundred the sheriff took 10*s.* 6*d.* sheriff's aid (*ibid.* i. 67).   He was not supposed to collect at his tourn money from geldable vills and hundreds contrary to custom (Madox, *Exchequer*, i. 446).   Bailiffs of hundreds are mentioned in sheriffs' accounts as turning over sheriff's aid (*L.T.R. Misc. Rolls*, 5/25).

[36] *Rot. Hund.* i. 239 : 110 carucates.   Cf. *Cl. Rolls*, 1247–51, p. 477-8.   The two forms of geldability mentioned (*Feudal Aids*, ii. 14, 23) are that to amercements of itinerant justices and that to the sheriff's tourn.

[37] Above, p. 114.

[38] But *ferms* of manors held by sheriffs as temporary *custodes* were accounted for as foreign estreats (see *Red Book*, iii. 879).   For various minor *ferms* others than sheriffs were responsible.   Those rendered by sheriffs are shown in *Great Roll of the Pipe*, 26 Henry III., ed. by H. L. Cannon, pp. 3, 62, 71, 87, 145.   Cf. *Dialogus*, 80.

ordering the sheriffs to collect these debts were so called.[39]   Payments for writs issued at the Chancery illustrate one type.   Amercements before the itinerant justices are one of the most familiar and prolific sources of such income.   Of the latter, as well as amercements levied in the two benches and in the exchequer, lists known as estreats were preserved in the exchequer and copies sent to sheriffs as part of their summons.[40]   The sheriff is mentioned as sending his estreats on to the bailiffs of hundreds, who responded to him for such debts.[41]   The difference between the two classes of summonses sent to sheriffs, the summonses of the Pipe and those of the Green Wax, was technical, growing out of the origin and method of listing the various sums due.[42]

The opening days of the semi-annual sessions of the exchequer were occupied with the business of the sheriffs' proffer.   The sheriffs assembled for this, and all were supposed to be present on the first day.[43]   Various matters requiring the attention of all the sheriffs were brought up in the brief period while they remained in attendance.[44]   As they paid instalments of their *ferm* and had it blanched, they were assigned days for an audit of the account of the preceding year.   Ordinarily they were then responsible for further appearance only on the days allotted them individually.   These were set in the thirteenth century at any time exclusive of exchequer vacations,[45] between Michaelmas and the next

[39] See *E.H.R.* xxxviii. 345.

[40] Such lists appear, for instance, in *L.T.R. Misc. Roll*, 1/6 and in *Sheriffs' Accounts, Exchequer*, K.R. 8/5.   In 1226 certain sheriffs were ordered to receive these directly from the justices (*Pat. Rolls*, 1225–32, p. 87).

[41] *Feudal Aids*, ii. 20, 21, 23.   Special provision has to be made to carry this out in hundreds privately held.   Thus a bailiff of a franchise is *juratum vicecomiti*. In 1236 the Exchequer ordered the sheriff to return a sum which he had taken as murder fine *per ballivum hundredi de Sutton* (*K.R. Mem. Roll* 14, m. 10).

[42] Miss Mills (*E.H.R.* xxxviii. 345) shows that both classes were in the later thirteenth century probably listed upon one piece of parchment when the summons was sent out.   But the list headed *De Pipa* was copied directly upon the Pipe Roll for the current year.   It included such sums due and not paid on the last Pipe Roll.   After 1284 it included only debts arising in the chancery and appearing on the Originalia Roll.   The summonses of the Green Wax or foreign estreats, on the other hand, were listed according to the court and year to which the estreat roll belonged.   The *oblata* of the Pipe Rolls represent sums collected upon summonses of the Green Wax.

[43] A penalty for absence, as in other instances when appearance was required, was incurred the first day, and was levied again for the second and third days. For the fourth day the maximum penalty was pronounced.   See below, p. 249.

[44] Above, p. 242.

[45] For exchequer term times, Tout, *Chapters in Administrative History*, ii. 97, n. 3.   Cf. Madox, *Exchequer*, ii. 5–6, and *Red Book*, iii. 973.

July.[46] Such a scheme varies considerably from the one outlined in the *Dialogus*, which presupposes that only view of accounts will be made at the Easter session, the actual accounting for the year being reserved to the Michaelmas session.

Sheriffs in the time of Henry II. clearly attended as a body at the opening of both Michaelmas and the Easter sessions,[47] but it becomes apparent only in the thirteenth century that they were expected on both occasions to make a proffer. The fact, though understood by Madox, is only now gaining recognition.[48] No other attested by the double series of Memoranda Rolls is better established. The proffer was highly important in administration, for prior to the sheriff's audit it supplied funds to meet current needs. The order that the sheriff hasten certain special collections sometimes specified that these were not to reduce the amount of the usual proffer,[49] and occasionally enjoined them to strengthen and increase

---

[46] The days set (*L.T.R. Mem. Roll* 53, m. 25 d) in the Michaelmas term, 1279, for sheriffs to account are these :

> Gloucester. Morrow of St. Michael (Sept. 30).
> Oxford. Octave of St. Michael (Oct. 5).
> Lancaster. Quindene of St. Michael (Oct. 12).
> Essex. Three weeks from Michaelmas (Oct. 19).
> Norfolk. Morrow of All Souls (Nov. 3).
> Southampton. Morrow of St. Edward the Confessor (Jan. 6).
> York. Wednesday before feast of St. Andrew (Nov. 29).
> Hereford. Morrow of St. Hilary (Jan. 14).
> Devon. Octave of St. Hilary (Jan. 20).
> Warwick and Leicester. Morrow Purification of the Blessed Virgin (Feb. 3).
> Northampton. Octave of the Purification (Feb. 9).
> Wilts. Quindene of the Purification (Feb. 16).
> Surrey and Sussex. Monday after St. Peter *in Cathedra* (Feb. 20).
> Cambridge. Monday after Sts. Perpetua and Felicitas (Mar. 13).

The days set after Easter 1280 are these (*ibid.* m. 26) :

> Bedford and Bucks. Morrow of the Close of Easter (May 3).
> Somerset. Thursday after the feast of St. John before the *Porta Latina* (May 9).
> Nottingham and Derby. Friday after the feast of St. Dunstan (May 24).
> Cumberland. Monday after feast of St. Augustine (May 27).
> Northumberland. Morrow of Ascension (June 5).
> Salop and Stafford. Morrow of St. Trinity (June 21).
> Kent. Morrow of St. John (June 25).
> Worcester. Wednesday after feast of Sts. Peter and Paul (June 30).
> London and Middlesex. Morrow of the Translation of St. Thomas the Martyr (July 8).
> Lincoln. Saturday after feast of the same translation (July 13).

[47] See *Dialogus*, 112, 113.
[48] Miss Mills in *E.H.R.* xxxvi. 483.
[49] As in *L.T.R. Mem. Roll* 60, m. 24.

it.[50]   As a rule the proffer of a sheriff is said to be derived both from *ferms* and from summonses.[51]   When outlays had been required by mandate, writs were presented instead of cash.   A fixed proportion of the *ferm* was required on these occasions,[52] and exact amounts in arrears are mentioned.   Proffers were sometimes paid at the wardrobe instead of the exchequer of receipt.[53]   Miss Mills has shown that after 1298 they decreased very much in amount through this process, for in a period of warfare there was an increasing demand that sheriffs pay amounts required in advance of the time of the proffer, and the wardrobe organization was best suited to the management of these affairs.[54]

The ancient rule of the exchequer required both sheriffs and bailiffs of franchises to be present in person on the day assigned to make their proffer on penalty of amercement according to the heavy scale of penalties [55] traceable as far back as the *Dialogus*.   Such amercements are recorded.[56]   But the *Dialogus* provided a form of excuse by which sheriffs unable to come to the exchequer put other persons in their places.[57]   So the *adventus vicecomitum* of the Memoranda Rolls in a great many instances show the sheriff's son, his clerk or his undersheriff making the proffer.[58]

The period required to complete the audit of a sheriff's account in ordinary cases was estimated in advance at from one to twelve days, varying with the volume of business for the different counties. The average allowance in 1299–1300, according to the Michaelmas

---

[50] *C. Cl. R.* 1279–88, p. 529.

[51] As in Madox, *Exchequer*, i. 134, note (s) ; ii. 154, note (f).   At Michaelmas, 26 Edward I., the sheriff of Northumberland sent £100, which included £14 cornage, £20 *proficua* of the county, £26 minute *ferms* and sergeanties and £20 disafforestation and summonses.   No more was sent because of the war with Scotland.

[52] Apparently half at each term.   According to the *Dialogus* (Oxford ed. 116) half was due at the Easter term.   Arrears of proffer appear on the schedule for Easter term, 10 John, attached to *Pipe Roll*, no. 54.   In *L.T.R. Mem. Roll* 62, m. 3, the sheriff of Norfolk owes £19 : 13 : 2 of his proffer for the Michaelmas term, for payment of which a day is set.   So the sheriff of Surrey and Sussex, *ibid.* 65, m. 19.

[53] Madox, *Exchequer*, i. 268.

[54] *E.H.R.* xxxviii. 337-8, 351-2.

[55] Madox, *Exchequer*, ii. 153-4, 155 ; *Dialogus*, 121.

[56] Eleven sheriffs, 5 marks each (*L.T.R. Mem. Roll* 61, m. 8, 17-18 Edward I.).

[57] Pp. 123, 213.

[58] In 1220 the clerk of Geoffrey de Neville, sheriff of Yorkshire, appeared at the proffer and reported that the under-sheriff was *en route* with the money (*K.R. Mem. Roll*, no. 3, m. 2).   At this time nearly all the sheriffs sent substitutes. (*L.T.R. Mem. Roll*, no. 2, m. 8, *adventus*).

list, was about seven days.[59] It was the rule that there could be no accounting until cash had been paid toward the proffer. Only then might a sheriff be assigned a day to sit at the exchequer.[60] Even the preliminary statement or view (*visus*) of account, which according to the writer of the *Dialogus* belonged in the Easter rather than the Michaelmas term, when settlement was expected, was a statement of debits and payments applying upon them. The essence of such a view, and of all computation from the exchequer standpoint, is the production of evidence of credits to apply upon obligations.

Exchequer procedure in the thirteenth century no longer presupposes a general view of accounts of the current year after Easter, when it is half over, and a general audit after Michaelmas, when it ends. The Michaelmas and Hilary terms suffice for the accounts of approximately half of the counties, the remainder going over to the Easter and Trinity terms. Thus some accounts, including occasionally those of sheriffs retiring from office, are not heard until the summer following the end of the fiscal year when they are supposedly due. Moreover the *visus* is now relegated to a comparatively unimportant position. Such a preliminary survey of the annual account may be ordered at Michaelmas,[61] although it still occurs at

[59] The *dies dati* of this session (*K.R. Mem. Roll* 73. *L.T.R. Mem. Roll* 71, m. 2, for Michaelmas 1298 corresponds in all details) are these :

> Wilts. Morrow of St. Michael (Sept. 30), 8 days.
> Cornwall. Same day.
> Lincoln. Friday in the feast of St. Dionysius (Oct. 9), 8 days.
> Hereford. Wednesday after the feast of St. Luke (Oct. 21), 5 days.
> Gloucester. Vigil of St. Simon and St. Jude (Oct. 27), 5 days.
> Somerset and Dorset. Morrow of All Souls (Nov. 3), 8 days.
> Essex and Herts. Morrow of St. Martin (Nov. 12), 10 days.
> Northumberland. Vigil of St. Katherine (Nov. 24), 6 days.
> York. Morrow of St. Andrew (Dec. 1), 12 days.
> Kent. Morrow of St. Hilary (Jan. 14), 8 days.
> Notts and Derby. Morrow of St. Vincent (Jan. 23), 8 days.
> Warwick and Leicester. Morrow of Purification (Feb. 3), 6 days.
> Northampton. Thursday after St. Valentine (Feb. 19), 6 days.
> Salop and Staffs. Thursday after St. Peter *in Cathedra* (Feb. 26), 8 days.

A complete list (*Red Book*, iii. 838), presumably of later date, in nearly all instances reduces somewhat this time allotment. From Madox *Exchequer*, ii. 170, note (b), and *Red Book*, iii. 851, it appears that sheriffs attended but part of the day, and that the length of the period was probably estimated by the amount of writing the clerks would have to do.

[60] *Dialogus*, 48. This payment involved blanching also.
[61] As in *K.R. Mem. Roll* 73, m. 70.

Easter as a view for the half-year.[62]   But assignments are set for
it in only a few cases whether on the Michaelmas or the Easter list
of *dies dati*.   It is the *compotus* or account proper which engrosses
attention.

The sheriff or other collector, unless especially excused, had to
appear on the day appointed for his accounting under penalty of a
hundred shillings a day for each of the first three days and after that
liability at the king's pleasure.[63]   Nor was the exchequer back-
ward in imposing these penalties.[64]   As a rule sheriffs were subsequently pardoned upon showing good reason for their absence.
Those engaged in necessary business sometimes obtained a royal
mandate respiting the accounting to a later date.[65]   In case illness
or other urgent cause detained any of them, it was possible to
arrange, by the king's leave or otherwise, for some person to act
as his attorney.[66]   It appears that a fair measure of lenience was
extended [67] in this respect.

When the sheriff accounted the presence of the bailiffs of the
various liberties within his county was essential,[68] for the fiscal
business of the county was treated as a unit.   By the thirteenth
century the officials of various municipalities were permitted to
answer at the exchequer *per manum suum*, appearing at the sheriff's

[62] As in *L.T.R. Mem. Roll* 61, m. 12.   Sometimes it seems to deal merely
with the sheriff's proffer of the Easter term (*Great Roll of the Pipe*, 26 Henry III.,
ed. Cannon, p. 350).

[63] Madox, *Exchequer*, ii. 235-7.   The rule is expressly stated in *L.T.R. Mem. Roll*
73, m. 30 (30-31 Edward I.).   Cf. *Dialogus*, 121.

[64] As in the case of the sheriff of York (*L.T.R. Mem. Roll* 48, m. 2, 3 Edward I.).

[65] *Cl. Rolls*, 1237-42, pp. 13, 58.

[66] *Ibid.* 1247-51, p. 558; Madox, *Exchequer*, ii. 177-9; *Dialogus*, 214.   *Magister
Rogerus de Seyton vicecomes Northampton ponit in loco suo Nicholaum de Saundon
clericum suum ad reddendum pro eo compotum suum nondum redditum et ad ipsum
onerandum et exonerandum et ad firmam suam faciendam* (*L.T.R. Mem. Roll* 47,
m. 2).   In *ibid.* 66, m. 18, the sheriff of Wilts, who was ill, was assigned a later
day with the privilege of designating some one *per litteras suas patentes qui potestatem
habeat ad reddendum compotum*.   The account was rendered by his son.   A former
sheriff who is going to foreign parts is in 1258 permitted to send an attorney to
render his account (Close Roll 73, m. 6).   In this year the *compotus* of the sheriff
of Oxford and Berkshire was postponed by letters close on account of his many
duties in providing by precept for the parliament of Oxford (*ibid.* m. 8).

[67] The new rule laid down in 1323 is that sheriffs are not henceforth to account
by attorney without the king's mandate, unless the treasurer and barons are well
aware that they who ought to account are incapable of travelling or otherwise
hindered (*Red Book*, iii. 893).

[68] The sheriff of Yorkshire was once ordered to have at the exchequer all his
bailiffs and all the bailiffs of liberties in his county to affeer certain issues and
amercements (*L.T.R. Mem. Roll* 69, m. 121).

compotus.[69]   If the representatives of the town did not appear at the time designated, the sheriff might be directed to ignore town liberties and to enter and distrain for all debts for which he had summons.[70] Liberties were placed in jeopardy by such default and might be seized into the king's hand.[71] The presence of a former sheriff was sometimes necessary also, especially if he had given up office before the end of the fiscal year.[72] In the later part of the century the sheriff was expected to warn the bailiffs of liberties within the county of the time of his accounting, and to make proclamation of it in his county court.[73]

On the appointed day the sheriff sat at the exchequer to render his account. Public notice of this, given in the county, enabled not only the requisite bailiffs of liberties to be present, but also any of the county who held exchequer tallies or sheriffs' tallies might come also to assure themselves that they received allowance.[74]   In the time of Edward I. there was much complaint that sheriffs received moneys from debtors of the king, issuing their own tallies by way of receipt,[75] and then neglecting to acquit the debts in question at the exchequer.   In King Edward's reign the sheriff, on the day set for the beginning of his account, was required to take oath in full exchequer that he would faithfully charge himself with the debts of the king which he had received and which he could have received.[76] At his accounting the sheriff must have been attended regularly by

[69] Thus the sheriff was not to distrain the bailiffs of Southampton to render to him the *ferm* of the vill and debts and tallages, but to accept security of them to respond at the exchequer (*Cl. Rolls*, 1231-4, pp. 459-60). · Again, according to the *Hundred Rolls* (ii. 1) the vill of Northampton was held in chief of the king *in feodo* for £120 payable annually at the exchequer.

[70] Madox, *Exchequer*, ii. 244-7 ; so also in the case of free hundreds (*ibid.* i. 622).

[71] *Ibid.* ii. 244-5.

[72] The separate audit of a former sheriff's account for a fraction of a year appears as a special privilege (*Cl. Rolls*, 1231-4, p. 278).

[73] Madox, *Exchequer*, ii. 197.

[74] *Ibid.* ii. 197.

[75] Lists of accounts for which were issued tallies *contra vicecomitem* occur in *Sheriffs' Accounts, Exchequer K.R.* 44/1 and 3/2.

[76] According to *L.T.R. Mem. Roll* 66, m. 16, the sheriff of Devonshire, in making his accounting of the year 22-23 Edward I., was involved in difficulty because he objected to the portion of this customary oath which called for more than his burdening himself with what he had received. Just a little earlier, in 1292, the sheriff of the same county failed to answer for certain sums and was said to conceal these *contra sacramentum suum quod fuerat de debitis regis per ipsum receptis regi fideliter persolvendis* (*L.T.R. Mem. Roll* 63, m. 38).   Whatever may have been true at an earlier period (*Dialogus*, 73) Madox has justification in this one when he places an oath (*Exchequer*, ii. 182) before the account. Cf. Dalton, *Office and Authority of Sheriffs* (London, 1700), p. 475.

the clerk who served as his receiver of moneys, for an accounting might be respited because a sheriff's *receptor* had died.[77] The rolls and writs kept by this clerk are often mentioned as essential to the exchequer process of accounting. In the absence of the sheriff, his clerk was usually the most obvious individual to act as his attorney.

The business of the occasion was that of completing the sheriff's account of the issues of his county or counties [78] as well as that of the summonses. The latter he was directed to have with him at his accounting. It was the duty of the accountant to discharge certain obligations with which he was automatically charged, as well as others which he formally admitted. The *ferm* of the county could of course easily be charged upon the Pipe Roll in advance along with unpaid remnants of the *ferm* of the former year. In 1284 it was directed in the Statute of Rhuddlan that the *corpus comitatus* be written in a certain roll, to be read each year at the sheriff's account. Unpaid remnants of former years were to be entered upon the Pipe Roll and sheriffs charged therewith. At the same time it was provided that *ferms* of sergeanties, assarts and all other *ferms* for which annual response was made,[79] should similarly be listed on the great roll in advance. Upon these matters the sheriff was opposed and accounts rendered in full exchequer before the treasurer and barons. Concerning amounts in the summonses the sheriff could be charged only *per suas responsiones* [80] as he confessed collection of individual items. Upon these accounts he was opposed by a single baron of the exchequer,[81] attended by a clerk to record the result. After 1284 only gross debts of which there was hope of payment were entered upon the great roll ; but a special roll of desperate debts and dead *ferms* was read at the sheriff's account to suggest the possibility of realizing something upon them.

A further stage in the process was that known as the *summa*. Strictly speaking, this was a computation of the total combined arrearage of *ferm*, other fixed annual payments and summonses.

[77] *K.R. Mem. Roll* 78, m. 1 d, 33 Edward I. The *generalis eius procurator* of *Dialogus*, p. 123.

[78] *Ad reddendum et perficiendum compotum pro eodem vicecomite unde debet computare ad scaccarium de exitibus predictorum comitatuum* (*Cl. Rolls*, 1247–51, p. 558).

[79] *Stats. of Realm*, i. 69.

[80] *Red Book*, iii. 843 ; Madox, *Exchequer*, ii. 170, note (b).

[81] As in *L.T.R. Mem. Roll* 70, m. 38 d. The case is noted also in Madox, *Exchequer*, i. 354. Cf. *Red Book*, iii. 853.

Separate days were often set for this portion of the accounting.[82] That there was often, perhaps usually, a break in the proceedings before it was reached, is shown by the list of *dies dati* or assignments of days for sheriffs' accounts, which bear many notations that sheriffs accounted as far as the *summa*.[83] When the amount of arrearage was calculated the sheriff might then claim allowance for sums disbursed. In practice this was regarded as a part of the *summa*. Allowance of fixed alms and other fixed payments by custom of the exchequer automatically followed the usage of the preceding year until a change was ordered. Allowance by oath of the viewers of public works called for definite action. It was only subsequent to 1284 that the exchequer made view of all allowances claimed by virtue of the writ authorizing the expenditure.[84] Upon certificate,[85] issued to the chancellor by the treasurer and barons, writs of allocation were issued. Many of these matters were settled in the exchequer by putting them in a view at the end of the account.[86] Allocation was sometimes craved by sheriffs on equitable grounds, as when certain issues of the county could not be collected because of war, or when official or even personal losses were sustained by sheriffs in war time by reason of the king's service.[87] Allowance upon an exchequer tally lost by the sheriff[88] or the genuineness of a tally presented,[89] were also questions to be settled before a sheriff's business could be concluded. Concerning the loss

---

[82] *Vicecomes habet diem ad arreragia sua reddenda et ad summam faciendam in xv Sancti Hillarii. Postea habet diem ad faciendam summam super vicum (K.R. Mem. Roll* 26, m. 1).

[83] *Computavit usque summam* is the form.

[84] *Stats. of Realm,* i. 69.   [85] *E.H.R.* xxxviii. 349.

[86] Only later was a special day set for the sheriff to crave allocation on account (*Red Book,* iii. 874-6).

[87] Michael de Harcla, sheriff of Cumberland in 1305, received reasonable allowance *propter incendia, depredaciones et alia multimoda dampna et gravamina . . . per Scotos,* especially an arrearage of £37 for wool taken *ad opus regis (L.T.R. Mem. Roll* 76, m. 4 d). Madox (*Exchequer,* ii. 174) cites a case growing out of the barons' war. By special mandate of 1269 the barons of the exchequer exonerated Robert de Nevill, former sheriff of York, from 13*s.* 4*d.* of the issues of his county between Michaelmas 1263 and the battle of Lewes because he was impeded by the barons and received no more (Close Roll 87, m. 13). John de Oketon, former sheriff of York, received a similar allowance because he was hindered by the barons in the exercise of his office between Michaelmas 1265 and the Easter following (*ibid.*).

[88] Or, as in *K.R. Mem. Roll* 51, m. 3 d, misplaced by the chamberlain. Below, p. 258. Procedure when exchequer tallies were lost or destroyed is shown in *C.Cl.R.* 1272-9, p. 433.

[89] As in the case of William Trussel, 25 Edward I. (*K.R. Mem. Roll* 70, m. 54). Cf. Madox, *Exchequer,* ii. 160, 161.

in the value of coins known as pollards and crockards, levied by sheriffs just before they were depreciated in value according to the king's orders, at Christmas 1296, sheriffs had to petition the king for allocation.[90]

The form of account recorded in the Memoranda Rolls under the heading *facta summa*, like that of the *visus compotus* at an earlier stage of the proceedings, is a statement of arrearages due and allocations upon them. Obviously this was not the end of the matter, for even after the computation there were payments and often collections to be made. The ultimate stage of the account, so it has been shown, was represented by what in the time of Edward I. was termed the final view.[91] After the *summa* was made orders were issued to sheriffs concerning amounts still to be collected. These *Praecepta super compotum vicecomitis* in the earlier years of Henry III. appear merely as marginal notes in the Memoranda Rolls, but later become a distinct section of the record. Care was exercised that funds collected between the *summa* and the sheriff's acquittal for the year be reported,[92] so the final condition of the account might be known. In case the combined payments and outlays exceeded the indebtedness, this *superplusagium*, as in earlier periods, became a credit on the next year's account.[93]

Before the sheriff was free to go he was required to certify his account by oath[94] and to pay his arrearage. Often the same day was set for both acts.[95] Delays in rendering arrearage were sometimes due to difficulty in securing returns from bailiffs.[96] In default of payment at the time set, the sheriff might be required to find a manucaptor,[97] or committed to the custody of the marshal until he found[98] manucaptors or in some other way gave satisfaction.[99] In

---

[90] *K.R. Mem. Roll* 74, m. 26.

[91] Miss Mills in *E.H.R.* xxxviii. 347, n. 1. Cf. *Red Book*, iii. 874-5.

[92] The coroners of Devon are ordered (*L.T.R. Mem. Roll* 63, m. 38, 19-20 Edward I.) to attach a sheriff for concealing *contra sacramentum* a fine collected after the *summa*.

[93] Such a credit due the sheriff of Shropshire in 1260-61 for some reason was not placed upon the great roll until 5 Edward I. (*Chancellor's Roll*, 54).

[94] See *Dialogus*, 51, 73.

[95] *Vicecomites habent diem ad arreragia sua reddenda in Octabis Sancti Martini et ad fidem suam faciendam* (*L.T.R. Mem. Roll* 14, m. 14; so also m. 15 d). The *fides* or *veredictum*, by which the sheriff is said to make his view of account and which Madox believed to be a declaration, was apparently an oath (*Dialogus*, 213).

[96] Madox, *Exchequer*, ii. 244-5.

[97] As the sheriffs of Lancashire and Westmoreland (*L.T.R. Mem. Roll* 68, m. 14).

[98] As the sheriff of Dorset and Somerset, 17-18 Edward I. (*ibid.* 61, m. 20 d).

[99] Cf. Madox, *Exchequer*, ii. 161, 288.

S

1257 the king commanded the treasurer and barons in full exchequer that no sheriff or bailiff who owed any crown debt should depart from the exchequer until he had fully paid it.[100] The debtor who failed to answer for the sums with which he was charged was often committed either to the Fleet prison or the Tower.[101] If a sheriff departed from the exchequer without leave before his accounting was completed, his arrest was ordered.[102] After his accounting he might be held for other service.[103] For giving false responses at the exchequer a sheriff might also be imprisoned.[104] If he failed to appear in response to summons, the coroners of the county were ordered to produce him.[105]

For some accounts sheriffs were assigned special days. Former sheriffs were in the same way called to account for the king's debts arising in their time,[106] or for property which had been in their custody.[107] As in the past, great persistence was shown in prosecuting such claims. By writ of *venire facias* a sheriff was often enjoined to have at the exchequer to answer on a certain day a former sheriff or his agent.[108] The lands and chattels of a deceased sheriff, like those of many another debtor of the king, were quite regularly taken over by order of the exchequer to assure payment of the sums he owed.[109] His clerk, his wife, his son or his executors are

[100] Madox, *Exchequer*, ii. 231.

[101] *Ibid.* ii. 234-5; *Cl. Rolls*, 1247-51, pp. 397-8.

[102] *Ibid.* ii. 243-4. In 25 Edward I. the sheriff of Norfolk and Suffolk was given permission to return to Norfolk to dispatch ships on the king's business, his clerk rendering his account as far as the *summa* (*L.T.R. Mem. Roll* 68, m. 46).

[103] In 1306 (*ibid.* 76, m. 75) a sheriff was forbidden to leave until he had had the names of those holding fees of the honour of Albemarle.

[104] Madox, *Exchequer*, ii. 244.

[105] As in the case of the sheriff of Westmoreland, 19-20 Edward I. (*L.T.R. Mem. Roll* 63, m. 33 b), and the former sheriff of Devon (*ibid.* m. 38).

[106] *Ibid.* 43, m. d (53 Henry III.); the sheriff of Northampton is to cause the former sheriff to come to respond to Ralph Pippard for 40s. which he received from him when he was sheriff. So Henry Perot, late sheriff of Kent (*ibid.* 52, m. 8 d). In *ibid.* 54, m. 3, the former sheriff of York is given day to render account of divers escheats of the crown, felon's chattels and other debts as well. Cf. Madox, *Exchequer*, ii. 234, note (p).

[107] A former sheriff of York is called to account (27-28 Edward I.) for 30 pairs of *fergiae* which he showed had been delivered with the castle of York.

[108] In 1265 William fitz Herbert is to be made to appear with all rolls, writs and tallies to render account for the counties of Nottingham and Derby (*K.R. Mem. Roll* 40, m. 1). William de Costello, late sheriff of Warwick, whom his successor has order to distrain *per terras et catalla* and to produce his body on a specified day, is detained by illness. He is excused the default, but his son and heir is to bring the money which is due (*ibid.* 73, m. 23).

[109] As those of William de Rothingge (*L.T.R. Mem. Roll* 61). Cf. Madox, *Exchequer*, ii. 233.

to be found struggling to clear up such accounts.[110] Nor was it always easy to bring to a reckoning a former sheriff who was alive. If, after the goods and chattels of such a debtor were seized, the sheriff reported that he could not find him in the county, a writ issued to the sheriff of another county to attach him by his body.[111] About 1302 Robert de Vere, former sheriff of Northampton, after incurring the usual penalties for default on the fourth day of non-appearance, took his family and all his moveable chattels into Cambridgeshire. When a local bailiff finally delivered him to the doorkeeper of the exchequer at Peterborough, the latter reported that he was sick and unable to ride or walk. But the treasurer and barons ordered that he be brought if this were possible without peril of death. At a day appointed he came and was ordered to prison, two manucaptors being designated. Subsequently he was released to be present from day to day, and ultimately fined with the king in the amount of twenty pounds for relaxation of the amercement he had incurred.[112]

Of the king's debts, aside from those for which the sheriff was personally liable, he was still regarded as the collector whenever one was necessary. The matter occasioned no difficulty when prominent personages or their bailiffs, after affording security [113] to the sheriff, appeared at the exchequer to acquit their own obligations ; nor when the bailiffs of municipalities and other franchises appeared at the sheriff's annual account to acquit the *ferm* or other debt due from their respective communities ; nor when escheators accounted for the income from estates in the king's custody. But the sheriff was constantly ordered in his summons to collect arrears, and had many special precepts to levy by distress, even in cases wherein he had no desire to collect.[114] His office was treated by the exchequer as a power constantly in reserve, guaranteeing the ultimate success of all special devices for collecting public debts.

[110] As in *C.P.R.* 1247–58, p. 481. In *L.T.R. Mem. Roll* 58, m. 7 d, the clerk and former *receptor* of the deceased sheriff of Dorset and Somerset delivers to the remembrancer of the exchequer two rolls of *proficua* and debts of the king which he has received since the sheriff's decease.

[111] As in the case of the late sheriff of Westmoreland, 23-24 Edward I. (*ibid.* 67, m. 57—an order to levy).

[112] *L.T.R. Mem. Roll* 73, m. 39, 1302–3.

[113] *Fidem.* In *ibid.* 18, m. 12, appears the *fides* of a manorial seneschal.

[114] Ordered to levy on goods and chattels of the Archbishop of York to the value of 4000 marks, the sheriff reported goods to the value of 100 marks for which he found no buyer. He was ordered to levy without delay (*ibid.* 67, m. 22).

To the hundredors, as to the other bailiffs, he passed on estreats for collection.[115] The special collectors of taxes no doubt often lacked the means of enforcing the collection from the king's debtors.[116] Moreover, many of the more usual types of debt due the crown could be handled through no other collector. Most persons who owed sums at the exchequer would hardly have been regarded as financially responsible, even had they been able to make the trip to Westminster to settle their accounts. In these cases the sheriff collected the debts and himself paid in the amounts due.

  To the debtors he gave his own tallies of receipt,[117] which, unfortunately, did not always insure his clearing the account at the exchequer. In 1238 sheriffs were directed to make proclamation, that such persons come before them and the coroners with their tallies of payment to receive due credit.[118] There were similar inquests in the next reign. The first Statute of Westminster [119] required that the parcels of common amercements, levied on the whole county before the justices, should for the same reason be estreated into the exchequer, and not merely the whole sum. Furthermore, it ordered an investigation concerning those who had paid the king's debts, and required that they be acquitted whether the sheriffs were living or dead ; and it provided that in future sheriffs were not only to make tallies to all who paid their debt to the king, but, if they failed to acquit these debtors at the next account, were upon conviction to pay the plaintiff three times the amount received and in addition make fine to the king.[120] Moreover, some exchequer tallies were lost and some sheriffs died in office, leaving their accounts in poor condition. In 1284 a series of remedial measures were adopted by the council.[121] All who had tallies of the exchequer for debts of themselves or ancestors, usually paid there, but which came in summons of the exchequer, were required to deliver these to the sheriff to gain credit. But at the same time debts which came in summons of the exchequer whereof

---

[115] *Feudal Aids*, i. 87 ; ii. 21.   [116] Cf. Madox, *Exchequer*, i. 355.

[117] Above, p. 252, note. The sheriff's bailiffs and beadles also issued tallies (*Feudal Aids*, i. 87).

[118] Madox, *Exchequer*, ii. 194-6 ; cf. *C.Cl.R.* 1272-9, p. 433.

[119] *Stats. of Realm*, i. 31, cap. 18.

[120] *Ibid.* i. 32. A statute of 1299 (*ibid.* i. 129) provided that a baron and a clerk of the exchequer were annually to be sent through each shire to enroll payments made by those of whom they were exacted through the Green Wax, and to hear and determine complaints against the sheriffs.

[121] *Stats. of Realm*, i. 69-70.

debtors proffered tallies against sheriffs were to be acquitted. Commissions were to visit the shires to make inquiry concerning such debts. For the future the chamberlains of the exchequer were ordered to make no receipt tallies to sheriffs, until they first received of them the particular sums of the various actions of debt involved and the names of the persons who paid them. The sheriff henceforth produced at the exchequer receipt for all debts which fell within its summons, but it was possible through the chamberlain's receipt tallies to trace his payment of the debts of individuals, and through the sheriff's own tallies to individuals to acquit them and make him liable.

An emergency device of the thirteenth century is seen in an order to sheriffs to collect and pay at unusual times certain sums already in the summonses, or even *ferms* as well. This meant the acceleration of the work of individual sheriffs who had been ordered by writ to collect specified accounts. The process usually concerned the clearer debts already entered as due. The writs issued in such cases often ordered sheriffs, on penalty of their bodies and goods and of the king's wrath, to cause to be levied specified sums, which they were to have at the exchequer by a certain date, usually set within a few weeks.[122] Thus a mandate to the sheriff of Lincoln, in 1284, ordered him to levy at once the amercements imposed by the itinerant justices and have the same at Westminster fifteen days from St. Hilary.[123] Sometimes these levies were general. A writ of December 26, 1291, ordered the sheriffs to have at the exchequer at the Feast of the Purification all the money in their hands from the issues of the county, to exceed in each case a specified sum.[124] The occasion for such levies was usually the king's visit to transmarine parts, although the purpose of such a levy in 1266 was to aid the king in besieging Kenilworth,[125] and in 1286 to pay debts to foreign merchants.[126] The amounts a sheriff was thus peremptorily ordered to collect varied with circumstances from ten to some four hundred pounds.

[122] Madox, *Exchequer*, i. 355-6; *K.R. Mem. Roll* 35, m. 5 d; *L.T.R. Mem. Roll* 59, m. 26 d; *ibid.* 60, mm. 21, 24; *ibid.* 61, m. 5 d.
[123] *Ibid.* 58, m. 29; cf. Madox, *Exchequer*, i. 356, and *C. Cl. R.*, 1279–88, pp. 38, 91.
[124] *E.g.* £200 (*L.T.R. Mem. Roll* 62, m. 39) for the sheriff of York, smaller amounts for other sheriffs from *exitus* of their counties.
[125] Madox, *Exchequer*, i. 356, note (c); *L.T.R. Mem. Roll* 41, m. 3 d: *de fermis et clarioribus debitis quae veniunt in summonitionibus.*
[126] Madox, *Exchequer*, i. 357, note (g).

In the levy of taxation the sheriff of the thirteenth century was probably assistant more often than he was collector. Certainly this was more and more true as the century advanced. Yet the latter position he also held in the case of some classes of levy, and his general relation to the assessment and collection of every form of secular tax was a very important one. In the levy of taxes on moveables his position has been shown to be that of a subordinate co-operating with royal commissions for the county, who supervised the work, and with a committee of the locality, on whom fell the burden of assessment and taxation.[127] He summoned men of the various communities to elect assessors, and he distrained men who had failed to pay their tax and collectors who had not paid at the exchequer. He was often ordered in general terms to aid the taxors in their work,[128] and in some instances he himself served with two or three others as one of the chief taxors of the county.[129] The carucage and hidage, authorized by the great council of the realm in the earlier years of Henry III., was also levied and collected in practically the same way. Both in 1217 and 1220 the money was collected by knights chosen for the work,[130] in the latter instance at a county court convened by the sheriff for this purpose.[131]

The collection of scutages, feudal aids and tallages brought the sheriff into the more immediate foreground. The tenant-in-chief who owed scutage seems to have accounted usually to the sheriff,[132] although it is clear that sometimes he paid to the exchequer.[133] When the tenant obtained a royal writ authorizing him to collect from his tenants, the sheriff aided him with distraint. When he did not have such a writ, it was the business of the sheriff to enter the lands of the rear vassals and collect from them.[134] Before collection could be made, the sheriff had to have a list of fees, which

[127] Mitchell, 352. For details concerning specific levies, cf. 164-6, 204, 216, 298 ; and for the reign of Edward I. Rot. Parl. i. 224, 227-8, 239, 240, 241-2, 269-70.
[128] C.P.R., 1232–47, p. 231 ; ibid., 1292-1301, p. 299 ; ibid., 1301-7, p. 15 ; Rot. Parl. i. 228.
[129] As in collecting the ninth of 1297 (L.T.R. Mem. Roll 69, m. 38).
[130] Mitchell, 123, 135-6. It is possible that in 1220 the sheriff was the chief receiver. Notices of remission or respite of the levy of that year were sent to him (Lit. Claus. i. 428, 442), and in 1218 bailiffs of a liberty were directed to pay him the amount they had collected (ibid. i. 348).
[131] Lit. Claus. i. 437.
[132] Mitchell, pp. 227, 229, 231. For directions to sheriffs for collecting scutage in 1218, see Lit. Claus. i. 371.
[133] Mitchell, 231.
[134] Ibid. 188, 229, 231-2, 335-7. In 1231 the sheriff was directed to distrain only those who held in chief (Cl. Rolls, 1231-4, p. 10).

was until after 1300 prepared by inquest. In 1306 two sheriffs were amerced at the exchequer,[135] when they made return that they had failed to levy as ordered because they had no writ authorizing inquest of fees, the exchequer ruling that they could have sought writs. The inadequacy of the levy through the sheriff is shown by the employment in 1314 of special collectors who completed the levy of two scutages of Edward I.[136]

The collection of the feudal aid in the reign of Henry III. was by the same process.[137] For this purpose inquests of fees were also held, sometimes by the sheriff.[138] The levy of the aid of forty shillings a knight's fee for marrying the daughter of Edward I. was made by six men of each hundred and liberty, summoned by the sheriff before a special collector entrusted with the work. The sheriff was directed to be present and to have all summonses of the exchequer and writs of the king, as well as rolls, inquests and verdicts in his possession.[139] Both sheriffs and collectors accounted at the exchequer for the sums collected.[140]

The levy of tallage repeated the same details. Assessment was made by justices or other officials assigned to this duty,[141] and often aided by the sheriff.[142] In 1251 the statement is made that no sheriff may enter the king's demesnes or *custodiae* to assess and levy tallage without being deputed or having special mandate from the king.[143] The sheriff is mentioned as collector,[144] and appears to be the only one mentioned in the time of Henry III. It is stated that in 1260 the assessment roll was sent to the sheriff for collection.[145]

[135] *K.R. Mem. Roll* 79, 34 Edward I., m. 31.
[136] *E.H.R.* xxxviii. 19. In *L.T.R. Mem. Roll* 77, m. 80 (34-5 Edward I.), are precepts of the exchequer *pluries* ordering sheriffs to levy arrears of scutages.
[137] Mitchell, 349-50. For detail, pp. 209-10, 257-8.
[138] As in *Feudal Aids*, i. 195.
[139] *K.R. Mem. Roll* 76, m. 49; cf. *Rot. Parl.* i. 266. The sheriffs were sometimes directed to assist the collectors in the levy (*L.T.R. Mem. Roll* 75, m. 75 d).
[140] The sheriff of Cumberland is directed to deliver the money in his hands (*K.R. Mem. Roll* 76, m. 70; *Feudal Aids*, v. 299). Both sheriffs and collectors were given day at the exchequer to account (*L.T.R. Mem. Roll* 75, mm. 50 d, 60; *ibid.* 77 m. 27). The sheriff of Norfolk and Suffolk was ordered to appear at the exchequer with his rolls, writs, tallies and all other accounts of the collection of the aid, and to produce the two collectors for these counties (*ibid.* no. 74, m. 65 d).
[141] Cf. *Cl. Rolls*, 1227-31, p. 23.
[142] Mitchell, 342. For detail as to specific levies, cf. 172, 207, 251, 283, and *Cl. Rolls*, 1231-4, p. 368; *Rot. Parl.* i. 266. Mitchell finds no mention of his participation in 1230, 1260, 1268 (pp. 190-1, 289, 294), but the *Foedera* (1816, i. 478) shows sheriffs making the levy in 1268 according to estreats furnished by the escheator *citra Trentam.* [143] *Cl. Rolls*, 1247-51, p. 526.
[144] Mitchell, 190-1, 289, 294. [145] *Ibid.* 289.

In the reign of Edward I. are mentioned special collectors of the tallage.[146] To tenants-in-chief concession was made to tallage portions of the ancient demesne which they held,[147] the sheriff being ordered by the king's letters to cause such persons to have a reasonable tallage. The same official was sometimes ordered to tallage a borough and turn over the proceeds by a certain time,[148] and directions for the tallage of the Jews in 1255 were transmitted to the various sheriffs.[149]

Royal claims relating to prisage and the customs were enforced, at least occasionally, through the sheriff. For a period beginning in 1288 the sheriff of Devon was ordered to levy four shillings on each tun of the wine of Bergerac.[150] In 1278 the sheriff of the same county was ordered to collect the king's new custom on wool, woolfells and leather, and had authority to arrest persons carrying away and concealing the custom.[151] In the seizures of wool, made by Edward I., the sheriff appears as the king's agent, also in arranging for the carriage of this commodity. After the king in 1297 remitted the maltolt on wool, he directed that the custom be collected henceforth at certain ports ;[152] and in 1299 all the sheriffs were ordered to obey and be intendant to the knight and collectors of the custom, appointed at each of these ports.[153]

Custodianship of the king's fiscal rights in the county incidentally added much to the sheriff's fiscal responsibility. He was expected to preserve the king's fiscal rights and keep him in seisin. He was frequently directed to take into the king's hand lands or liberties and to account for their issues. The king's right of primer seisin [154] was pressed still further when tenants-in-chief died indebted to the crown. Thus sheriffs were even ordered to take into the king's hand all the goods and chattels of the countess of Gloucester, daughter of Edward I.,[155] and of Edmund his brother.[156] Just

[146] In *L.T.R. Mem. Roll* 77, mm. 84-5 (35 Edward I.) sheriffs are ordered to distrain them to render account.

[147] *Cl. Rolls*, 1234-7, p. 218; *ibid.*, 1237-42, p. 109; *ibid.*, 1247-51, p. 9; *ibid.*, 1302-7, p. 305 ; Close Roll 65, m. 17 ; *ibid.* 75, m. 5).

[148] Madox, *Exchequer*, i. 356, note (z).

[149] *Pat. Rolls*, 1247-58, pp. 439, 442-4. This occurred also in 1259 (*K.R. Mem. Roll* 32, m. 12 d).

[150] *Cal. Fine Rolls*, 1272-1307, p. 275.   [151] *C. Cl. R.*, 1272-9, p. 442.

[152] *C. Cl. R.*, 1296-1302, p. 187.   [153] *Ibid.* p. 316.

[154] Pollock and Maitland, i. 311.

[155] *L.T.R. Mem. Roll* 77, m. 84 (34-5 Edward I.) ; *per cyrographum inter te et ballivos locorum in quibus bona et catalla illa existant.*

[156] *Ibid.* 67, m. 65 (23-4 Edward I.).

prior to 1294 the sheriff held the town of Worcester, which had temporarily lost its privileges.[157] In May 1292 the city of York was taken over by the sheriff on account of arrearage.[158] When the lands of Hubert de Burgh were occupied in 1232, the men of the hundred of Wormelow and Archenfield were directed to respond to the sheriff of Hereford *ut suo custodi*.[159] Similarly other hundreds,[160] as well as various manors, which did not belong to the escheator's province, fell to the sheriff's custodianship. In 1256 the sheriff of Norfolk with one of the king's clerks held in custody during a voidance the abbey of St. Benet Hulme,[161] and in 1300 the sheriff of Nottingham was accounting for the issues of the priory of Lenton, an alien priory seized into the king's hands some three years earlier.[162] It also fell to the lot of the same official to have appraised and delivered into custody the goods of a person accused of crime, so in case of conviction, they might not be lost to the king.[163] In periods of conflict with France sheriffs seized for the king goods and chattels of merchants who were French subjects.[164]

Other forms of fiscal obligation also arose through the sheriff's duties of custodianship. The sheriff of Devon in 1262 was directed to permit no person to appropriate within the county gold, silver or copper mines, which ought to belong only to the king.[165] The sheriff of Cumberland in 1223 responded for the king's mines of silver and lead, as his predecessors had done in the time of John and Henry II.[166] Sheriffs are still mentioned in the thirteenth century who stock lands in the king's possession,[167] attend to their tillage,[168] and assume charge of their crops.[169] A sheriff was in 1253 ordered to dam up the king's fishpond at Newcastle under Lyme, which had been broken through to effect sale of the fish.[170]

---

[157] The citizens elected a coroner without warrant (*L.T.R. Mem. Roll* 65, m. 41 d).

[158] *Ibid.* 63, m. 38.   [159] *Pat. Rolls*, 1225–32, p. 500.

[160] As the Yorkshire wapentake of Staincliffe and Ewcross (*L.T.R. Mem. Roll* 40, m. 2 d); *manerium de Crichelawe* (*ibid.* 69, m. 21 b (25-6 Edward I.); the lands of John Balliol (Sheriff's Accts. 1/3).

[161] *C.P.R.*, 1247–58, p. 470.

[162] The monks retained their sustenance, *P.R.* 145, Nottingham.

[163] *Inq. Misc.* i. 405.

[164] *L.T.R. Mem Roll* 77, m. 10 d (1297); *Cl. Rolls*, 1227–31, p. 483.

[165] *Cl. Roll* 80, m. 15.   [166] *Lit. Claus.* i. 534.

[167] *Rot. Claus.* ii. 34, 95; *Cl. Rolls*, 1227–31, p. 359; Madox, *Exchequer*, ii. 151-2.

[168] The sheriff of Norfolk (*K.R. Mem. Roll* 61, m. 3, 1287) is ordered to cultivate and sow lands; so *Lit. Claus.* i. 502.

[169] *C.P.R.*, 1232–47, p. 226.   [170] Close Roll 66, m. 12.

They sometimes distrain bakers to do suit at the king's mill,[171] and exact his rights of multure there.[172] At times they appear also as keepers of royal parks or vivaria,[173] protectors of the king's fishing rights in streams,[174] and as temporary custodians of the dyes of the coiners.[175] The Statute of Westminster [176] of 1275 requires that goods saved from a ship, not adjudged a wreck, shall be kept by view of the sheriff, coroner or king's bailiff and held in the locality where found a year and a day before the sheriff or coroner seizes them into the king's hand.

As of old the sheriff was often commissioned to receive and entertain distinguished guests. When in 1194 the king of Scots paid a visit at the royal court, sheriffs and bishops were directed to conduct him from county to county until he returned to the Tweed.[177] In 1253 the sheriff of Northumberland received the monarch of the same realm in the castle of Newcastle-on-Tyne and found him wine and provisions.[178] In 1260 eight sheriffs had orders to meet the king of Scotland and his wife, the king's daughter, when they came to their bailiwicks, to offer them the king's castles, manors and game in his parks and warrens, and to procure them honours and courtesy.[179] Some time prior to this visit, in May 1258, the duties of the sheriff of York were extended somewhat further. He received a mandate to go frequently to the king's daughter, the queen of Scotland, and to afford her solace and succour when necessary.[180] In 1257 the sheriff of Kent was to afford to Master Richard Rostand, papal subdeacon, a speedy return passage at Dover.[181] Local arrangements for the reception of the papal nuncio, as he travelled through England in 1307, were in charge of sheriffs, whose duty it was to meet him with noblemen, knights and others of their counties warned out for the purpose, to

---

[171] At Canterbury, *Cl. Rolls*, 1231–4, p. 205.

[172] Wilts, *Inq. Post Mortem*, Henry III.-Edward I., Index Library, 118. In *Inq. Misc.* Chancery, i. 328, it is shown that the sheriff drew to the king's mill the multure of malt belonging to the mill under the castle of Sarum, because the lord made there a fuller's mill and a fishery.

[173] *Cl. Rolls*, 1231–4, p. 10.

[174] Thus a sheriff is to remove certain nets from the Severn and prevent the maintenance of boats to the detriment of the king's fisheries (Close Roll 66, m. 18 d).

[175] As the sheriff of York, 33 Henry III. (*L.T.R. Mem. Roll* 22, m. 2 d); the mayor and sheriffs of London about this time had a *custos cuneorum* chosen and presented him to the exchequer (Madox, *Exchequer*, ii. 88).

[176] *Stats. of Realm*, i. 28.   [177] Hoveden, *Chronica*, iii. 245.
[178] Close Roll 65, m. 28.   [179] *C.P.R.*, 1258–66, p. 95.
[180] Close Roll 73, m. 8.   [181] *Ibid.* 71, m. 9.

conduct him through their bailiwicks as securely, befittingly and honourably as possible, and to make proclamation for procuring him victuals and carriage.[182] The sheriff's powers of purveyance, authorized or implied in some of these instances, were largely exercised at London in the time of Edward I. in providing hospitality to the counts of St. Pol and Boulogne.[183]

The functions of the shrievalty in making disbursements on behalf of the crown are no less important in the thirteenth century than in the twelfth. The last Pipe Rolls of Edward I. still show the customary payments to cover fixed alms, repairs of buildings, wages of workmen and serving men, expenses of approvers and justice, and similar items just as at the beginning of the period. Much of the business of the sheriff's annual account had to do with the allocations which he sought to cover these and other expenditures. The usage which regulated these matters is not often disclosed. Sometimes the sheriff was ordered to ascertain and pay sums due workmen.[184] Many repairs and estimates upon them are mentioned as made under the sheriff's direction by view of responsible persons.[185] The sheriff himself appears occasionally among those who make such view.[186] In 1274 there was complaint of fraud, and an inquest was ordered as to what sheriffs or custodians of manors or castles had caused those who were viewers of works according to the king's precept to compute more than they reasonably expended ; and also as to what timber or stone had been provided at a loss to the king.[187]

The items paid by the sheriff are in part the familiar ones of the Angevin period. A few may be mentioned. There are still the wages of local officials, fixed and special alms,[188] the third penny of

[182] *C. Cl. R.*, 1302–7, p. 521.

[183] *Rot. Hund.* i. 421.

[184] Close Roll 66, m. 20 d : *per rotulos suos et rotulos visorum operacionum, castri Oxon.*

[185] *E.g.* the walls of Hereford (*K.R. Mem. Roll* 62, m. 9 d) ; the houses of Sherburne castle (*C. Cl. R.*, 1272–9, p. 33) ; the king's fisheries and weirs on the Severn (*Rot. Claus.* ii. 95) ; the *camera* of the castle of Rockingham, the sheriff to take experts who know the work (*ibid.* ii. 35).

[186] The constable of the new castle under Lyme to expend funds in repairs by view of the sheriff and other good men (*C.P.R.*, 1247–58, p. 110) ; the sheriff to see what was required to finish the castle of York and to provide funds from the fines of the eyre (Close Roll 72, m. 2 d) ; the sheriff to appoint four men to overlook the works of Porchester Tower (*ibid.* 66, m. 21 d).

[187] *Foedera* (1816), i. pars. ii. 518 ; *Rot. Hund.* i. introd. 14.

[188] The lepers of St. Leonard, Lancaster, through the sheriff are guaranteed animals from the king's forest, wood to burn and timber for building (*Lit. Claus.* i. 414) ; so the wood allowed the monks of Lenton (*ibid.* i. 421).

the shire,[189] money or supplies for King Edward's daughter Marie, who was a nun,[190] supplies for the king's or the queen's [191] household,[192] victuals, delicacies for his table,[193] including the preserving and transportation of his game,[194] robes for his men,[195] passage money for his messengers,[196] and sustenance for his men, his horses, his falcons and his hounds.[197]   There are also galleys and barges for the king's visit to Wales,[198] necessaries for King Henry's white bear and elephant in the Tower of London,[199] three hundred besants offered at St. Edmund's shrine for the king and queen and their children,[200] stockings and tunics for the king's alms,[201] frequent sums expended on behalf of King Edward's Scottish prisoners, and, at least on one occasion, the wages of knights and burgesses in parliament.[202]

The thirteenth-century sheriff, though not the only royal purveyor,[203] in this capacity often fulfilled important and extensive functions.   Incidentally he was creating a grievance familiar in later days, in exercising the right of prisage without paying for goods taken.   Abuse of this right by sheriffs occurred before Magna Charta, and it seems to have continued throughout the century. The first statute of Westminster forbids the taking of victuals from men of religion and the taking of animals, ships and barges for purposes of transport without the consent of the owner.   At Easter 1307 the men of the soke of Peterborough complained before the

---

[189] Madox, *Exchequer*, ii. 164-5.

[190] *L.T.R. Mem. Roll* 71, m. 56 d.   [191] *C. Cl. R.*, 1302-7, p. 55.

[192] Wax and other purchases for the king's wardrobe (*Exchequer, K.R. Accounts*, 552/3); three ships of wood for the Tower of London (*Lit. Claus.* i. 419); oaks for making charcoal for the king's hearth at Acton Burnel (*L.T.R. Mem. Roll* 61, m. 13, 17-18 Edward I.); stumps for charcoal for the king's children (*C. Cl. R.*, 1302-7, 507).

[193] As a half *carca amigdalarum* and two pounds *croci* (*Lit. Claus.* i. 401).

[194] At Hilary, 1291, a former sheriff of Nottingham is allowed expenses in salting thirty deer taken in Sherwood forest, and in carrying them from Nottingham to London (*L.T.R. Mem. Rolls* 62, m. 10); cf. *Cl. Rolls*, 1247-51, p. 493.

[195] As in *K.R. Mem. Roll* 73, m. 7; *Rot. Lit. Claus.* i. 459.

[196] Close Roll 65, m. 13.

[197] *Lit. Claus.* i. 305; *L.T.R. Mem. Roll* 70, m. 79, 26-7 Edward I. (14 *pullani regis*); *Cl. Roll* 65, mm. 16, 18.

[198] Close Roll 72, m. 2 d, 1257.   [199] Madox, *Exchequer*, i. 376-7.

[200] *Ibid.* i. 378.   [201] Close Roll 78, m. 12.

[202] *C. Cl. R.*, 1302-7, p. 330; *Parliamentary Writs*, ed. Palgrave, i. 192.

[203] In 1271 the king ordered the sheriff to take the liberty of Boston into the king's hands because his purveyors sent thither to make purchases at the fair were impeded (Close Roll 89, schedule).   In 1298 writs were issued ordering sheriffs to aid the king's agents sent to purvey oats, flesh, fish, etc. (*C.P.R.*, 1292-1301, p. 344).

king and his council that the sheriff of Northampton took from them for the use of the king a great part of their goods and chattels and gave them only tallies in return.[204] The response reveals a distinction of the period, important also in later law. The treasurer and barons of the exchequer demand to know whether the goods taken were for the use of the king or others. In 1267 the sheriff of Salop was ordered to provide wheat, oats, oxen and sheep for the king's coming.[205] Just a little earlier a sheriff failed to pay for cattle taken at the king's instance, although he obtained allocation for the expenditure.[206] Edward I. issued writs of privy seal to sheriffs commanding them to supply provisions for his household, due allowance to be made for the costs.[207] Sometimes, as in 1306, the king's letters patent ordered various sheriffs to pay the expenses of a special purveyor sent to supervise, prosecute and cause provision to be made of victuals in their counties for the king's use.[208]

Frequently the needs of the king's household required fairly large purchases. In 1225 a sheriff was ordered to cause to be sought in his county five hundred lambs and twenty kids and to have them at Westminster by the Wednesday before the feast of the Purification.[209] In preparation for the king's observance of Easter in 1231 the sheriff of Southampton was required to deliver to the seneschal at Clarendon fifty hens and twenty kids.[210] In Feb. 1248 the sheriff of Gloucester was ordered to purchase and pay for all the lampreys he could find for the king's use and to send two parts of them to Norfolk, the other part to Windsor for the queen's use.[211] In 1232 the sheriff of Somerset was commanded to find in his bailiwick against Easter for the king's use a hundred and fifty congers, and one, two or three porpoises if they could be found.[212] In 1304 the sheriff of Cumberland by order took cattle to stock the

---

[204] *L.T.R. Mem. Roll* 77, m. 41 d.   Cf. Cam, *Studies*, 174-5.
[205] Close Roll 84, m. 3.   [206] *Rolls of Parl.* i. 164.
[207] As in *Exchequer, K.R. Accounts*, 554/32, 32-35 Edward I. In 1256 various sheriffs of the home counties still procured provisions for the king's Christmas festivities at Westminster (Close Roll 72, m. 14 d) and bread for his observance of the feast of St. Edward (m. 13 d).
[208] *C. Cl. R.* 1302-7, p. 370.   [209] *Rot. Claus.* ii. 13.
[210] *Cl. Rolls*, 1227-31, p. 483; the requisition upon several sheriffs prior to Christmas 1253 was very heavy (Close Roll 66, m. 27 d).
[211] *Cl. Rolls*, 1247-51, p. 33.
[212] These were to be delivered at Gloucester to the bailiffs of the town, who were to send them to Worcester with the sheriff's men, through whom the cost was to be signified to the king that payment might be made (*Cl. Rolls*, 1231-4, p. 172).

king's larder prior to the advent of the king of Scotland.   In the following year some of the sheriffs provided a considerable amount of wheat and a supply of grain for the king's hostel at Westminster during the parliament.[213]

Purveyance was upon a far larger scale when required to meet the needs, not of the king in person but of his service.   An indenture of moneys paid by the sheriff of Cumberland in Oct. and Nov. 1291, and witnessed by the treasurer and the keeper of the king's wardrobe, shows expenditures for beer, meat, fish, hay and oats.[214]   In Nov. 1259 the sheriff of Lancaster was ordered to sell the meat which he had provided against the coming of the king and his army to Chester.[215]   Fairly large sums were paid in 1277 by the sheriff of Southampton to supply hay and oats for numbers of horses [216] at Winchester.   Again, a mandate to the sheriff of Dorset and Somerset in 1303 requires him to purvey large amounts of wheat, oats and beans under supervision of one of the king's clerks, and as soon as possible to have them at Berwick-on-Tweed.[217]   Purchases of grain for the army in Wales in 1277 [218] were made by a special agent in whose behalf writs of aid were issued to various sheriffs.

The Gascon war occasioned the issuance for the fiscal year 1296–7 of a *forma capcionis bladorum* designed to protect the poorer classes.   In each county grain was to be taken by the sheriff and a clerk, who were to give receipt tallies for the number of quarters taken.   To these tallies their seals were appended.   Payment was to be made at the exchequer a month from Easter.[219]   The various sheriffs were ordered to take specific amounts [220] and to have them transported to Plymouth.   The purveyance of grain in 1307 was still a grievance which the crown was attempting to mitigate.[221]   Another commodity purchased on a large scale for the king's use at this time was wool.   Fifteen hundred sacks were required in Lincolnshire, the sheriff acting with four others to procure it,[222]

[213] *Rot. Parl.* i. 164, 407-8.

[214] *Exchequer, K.R. Accounts,* 554/9.

[215] Close Roll 74, m. 15.   The sheriff of Cumberland was to do similarly concerning grain taken (*ibid.*).

[216] Fifty-two for 22 days, 66 others for 13 days (*L.T.R. Mem. Roll* 50, m. 1).

[217] *C.P.R.* 1301–7, p. 126.   [218] *Ibid.* 1272–81, p. 219.

[219] *L.T.R. Mem. Roll* 68, m. 20.

[220] The sheriff of Cambridge and Huntingdon took 3000 quarters *frumenti,* 1000 quarters *ordo,* other sheriffs large amounts (*ibid.*).

[221] *C. Cl. R.* 1302–7, p. 506.

[222] *L.T.R. Mem. Roll* 68, m. 63 ; cf. *C.P.R.* 1292–1301, p. 299.

and large purchases were also ordered in many other counties.
This was to be carried to St. Botulph and delivered to the collectors
of the customs for transportation to foreign parts.[223]

The need supplied by the sheriff was often urgent. In Oct.
1298 the sheriff of Northumberland was ordered to provide victuals
and other necessary supplies for the garrison at Newcastle-on-
Tyne,[224] the cost to be allowed the sheriff on his account. The
negligence or inefficiency of a sheriff was here a risky matter. A
few months later a writ of privy seal, sent to the barons of the
exchequer, recites that several sheriffs have made defaults in not
sending to Berwick according to the king's commandment corn,
oats and other kinds of provision for the sustenance of his host.
The barons are accordingly ordered to punish these sheriffs and
other *ministri* severely and to use diligence that the king be supplied
with provisions and money.[225]

Repairs upon the public buildings and the provision of building
materials called for further purchases and outlay. In 1222 the
sheriff of Nottingham was ordered to buy at the market fifty *carrate*
of lead for the king's use and to cause it to be taken to Bristol.[226]
Allocation to sheriffs for expenses in providing building stone
continue throughout the period.[227] The work on Westminster
Abbey is attested by an order of King Henry to the sheriff of Kent,[228]
directing him under great pain and forfeiture to buy one hundred
ship-loads of grey stone and convey it to Westminster. The same
king directed that the stone to be used by the dean and canons of
Lincoln in building their church and other buildings, as well as that
of the citizens for buildings, should be quarried by counsel of the
sheriff of Lincoln.[229] Again the sheriff is often ordered to provide
timber for building purposes. Usually an order to the keeper of
the king's forest authorizes him to let a sheriff have a certain number
of oaks [230] or timber enough for a certain purpose. A tally between
the sheriff and the forester was made to show the number of trees

---

[223] *L.T.R. Mem. Roll* 69, m. 107. The order was superseded in the following
year before it had been completely filled.

[224] *L.T.R. Mem. Roll* 69, 26.   [225] Madox, *Exchequer*, i. 380.

[226] *Lit. Claus.* i. 501.

[227] Thus 2500 *lapides* brought from a quarry near Grimshill to the castle of
Shrewsbury, 28-29 Edward I. (*L.T.R. Mem. Roll* 72, m. 29 d), together with expenses
of carriage.

[228] Madox, *Exchequer*, i. 377-8.

[229] *Cl. Rolls*, 1227-31, p. 472.

[230] Ten from the forest of Clarendon (*C. Cl. R.* 1272-9, p. 309).

taken,[231] the sheriff holding the counter tally.[232] Sometimes oaks were purchased.[233] Provision for oaken shingles [234] and for beams and planks [235] was made in this way. In 1223 the sheriff of Dorset was ordered to cause Master Stephen, carpenter in the hall of the king's castle at Sherburne, to have a cable for drawing thither the timber for the hall.[236]

Often the sheriff is ordered to have specific repairs made. He is to provide a strong rope for the king's well within the manor of Clarendon, and to cause the houses of the manor to be roofed and repaired when necessary.[237] He is to find and fit up ships for the king's service,[238] to have rebuilt the king's mills at Carlisle lately burned,[239] to have repaired the hall, chamber and cellar of the castle of Winchester,[240] or the tower and hall of the castle of Bamburgh,[241] to have constructed a jail at Ipswich or at Norwich,[242] or to repair the one in the castle of Lincoln or Wallingford.[243] He is to enclose a park and make necessary repairs,[244] to build a good oven in the castle of Winchester,[245] to begin a butlery and dispensary of stores at the king's hall of Oxford and to cause the chambers to be repaired.[246] He is to provide and fit up quarters for holding the exchequer at York,[247] to cause the image of St. Christopher with Christ in his arms and the image of St. Edward the king to be painted in the queen's chapel at Winchester, or the queen's chamber at Nottingham to be painted with the history of Alexander.[248]

Carriage for the materials, supplies and equipment required by royal order, was a second form of purveyance usually arranged through the sheriff. From the time of King John there is ample evidence that transport under the sheriff's direction was a very important feature of the royal service. In the period, especially of the Scottish and Gascon Wars of Edward I., large outlays, made by sheriffs in carrying grain or wool, appear among the items for which

[231] *Cl. Rolls*, 1247–51, p. 19. So in case of the timber required in 1262 for repairing the great tower of the castle of Lancaster (Close Roll 77, m. 9).
[232] *Cl. Rolls*, 1231–34, p. 270. In this case made by *visores*.
[233] The sheriff of Salop and Staffs paid 34s. for two (*L.T.R. Mem. Roll* 61, m. 13, 17-18 Edward I.).

[234] *C. Cl. R.* 1272-9, p. 309.
[236] *Lit. Claus.* i. 529.
[238] *Lit. Claus.* ii. 10 (1224).
[240] *Lit. Claus.* i. 367 (1218).
[242] *Ibid.* i. 410 (1219).
[244] Close Roll 72, m. 13.
[246] *Rot. Lit. Claus.* i. 42 (1225).
[248] *Ibid.* i. 377.

[235] *Ibid.* 1231–4, p. 270.
[237] *C. Cl. R.* 1272-9, p. 144.
[239] *L.T.R. Mem. Roll* 68, m. 99 (1297).
[241] *Ibid.* i. 545 (1223).
[243] *Ibid.* i. 466, 469 (1221).
[245] *Ibid.* 66, m. 25.
[247] Madox, *Exchequer*, ii. 9.

they ask allowance at the annual account.[249] When the king's treasure [250] or official records were sent, it was necessary that they provide safe conduct. In the same way a special escort had to accompany prisoners and approvers transferred from one prison to another.[251] For horses and carts taken for transport a fixed payment was established by the Magna Charta of 1217. The sheriff of London is seen enforcing by distress the right to requisition carts for removing the king's wardrobe.[252] Provision of sumpter horses to carry the king's chapel and the rolls of chancery was made in June 1257 by the sheriff of Lincoln and certain abbots.[253] The impressment of ships for carriage from port to port, or for transporting armies, special officials, horses and equipment to the Continent, was regularly ordered by writ to the sheriff. Sometimes a special official was sent with the sheriff to arrest ships found in the harbours and to take the usual surety from masters to be at a port named by a certain date.[254] In the time of Henry III. the writ directed sheriffs to impress ships able to bear from fourteen to sixteen horses or more.[255] In 1217 the sheriff of Devon provided the ships to take the queen mother to France,[256] and in 1254 ships for the queen's passage were provided in the same way,[257] and in 1257 the sheriff of Essex was ordered to provide bridges and hurdles for the equipment of a hundred ships for the passage of Richard, earl of Cornwall, from Yarmouth to Germany.[258]

The carriage required for the king's household supplies was provided in a similar manner. Sheriffs, for instance, were often ordered to take venison in the royal forests and to have it carried to some designated place ; [259] or, to take a specific instance, the sheriff of Lincoln arranged concerning the carriage of wines bought for the

[249] Good examples may be seen in the accounts in 1297 of the sheriff of Oxfordshire and Berkshire (*L.T.R. Mem. Roll* 69, m. 20 b) and the sheriff of Southampton (*ibid.* m. 74). According to this roll (m. 25) a sheriff placed £9 : 11 : 2 in the carriage of 150 quarters of wheat and 184 quarters of oats from divers places in his county to Gascony for the sustenance of the king's lieges.

[250] As in Mitchell, 204, 352.

[251] In 1296 sheriffs are directed to aid in conducting Scottish prisoners to their prisons (*Foedera*, i. part ii. 841). A sheriff of Staffs expends 6s. 8d. in the carriage and conduct of an approver to the jail of Newgate (*L.T.R. Mem. Roll* 61, m. 13).

[252] Letter Books, City of London, Bk. A, 192.

[253] Close Roll 72, m. 7 d.   [254] *Pat. Rolls*, 1225–32, p. 259.

[255] *C.P.R.* 1225–32, p. 326 ; *ibid.* 1232–47, pp. 270, 273.

[256] *Rot. Claus.* i. 315.

[257] Close Roll 67, m. 10.   [258] *Ibid.* 72, m. 9.

[259] London, *L.T.R. Mem. Roll* 69, m. 26 (25-26 Edward I.) ; York, *L.T.R. Mem. Roll* 71, m. 55 (27-28 Edward I.).

T ·

king's use in Boston fairs.[260]   For supplies thus being transported
the custodian was legally responsible.   In 1282 the king's letters
close notified the sheriff of Lincoln that he was not to molest Roger
Bogard, who in carrying wine to Chester for the king's use had lost
a tun through leakage occasioned by the breaking of the cart.   Since
Robert was not culpable, the king had pardoned him the trespass.[261]
An interesting entry shows the sheriff of Southampton seeking
allocation for carrying wine by water from Portsmouth to Fishburn,
thence to Chichester by water, and also for outlay in mending the
casks and in guarding, unloading and in carrying and placing them
in the cellar.[262]

The king's building materials, particularly timber,[263] stone,[264]
and lead, were naturally among the commodities for which sheriffs
were required to provide carriage.   Lead was likely to be taken
considerable distances.   In 1224 the sheriff of Lincoln was ordered
to send to Bawtry and receive there from Robert de Lexinton for
the king's use thirty *carrate* of lead, and from the sheriff of Nottingham fifty *marcate* of lead, and cause all to be carried to Dover.[265]   In
war time the effectiveness of the sheriff's transport was of course
quite as vital as his provision of supplies.[266]   No less urgent under
such circumstances, and always highly important, was the carriage
of the king's money, for which sheriffs were as a rule required to
provide safe conduct,[267] and even when they were not the custodians
to aid in transport.[268]   The sheriff's facilities in this respect often
made him custodian and forwarder of moneys collected in his county
by special collectors of the taxes.[269]   In the transfer of public
records he played the same part.[270]   The transfer of the exchequer

---

[260] Close Roll 72, m. 6 d.   The sheriff of London is to provide at once three
strong carts to carry wine to St. Albans (*ibid.* 65, m. 6).

[261] *C. Cl. R.* 1279–88, p. 169.

[262] *L.T.R. Mem. Roll* 69, m. 74.   [263] As in *Lit. Claus.* i. 459.

[264] In Apr. 1261 the sheriff of Surrey was to find carriage to Lambeth of stone
from the quarries of Ryegate destined for the works at the Tower of London
(Close Roll 77, m. 14 d).

[265] *Lit. Claus.* ii. 6.   [266] Above, p. 269.

[267] *C. Cl. R.* 1227–31, p. 535.

[268] So the order to sheriffs, bailiffs and *ministri*, when the king in 1298 sends
treasure to Scotland (*L.T.R. Mem. Roll* 70, m. 3 d); also the arrangements for
sending whited silver from the south-western counties to the Tower of London to
be coined (*ibid.* m. 3).

[269] As in *C.P.R.* 1232–47, p. 60; cf. Mitchell, p. 352.

[270] The sheriff of Bedford in 1224 was to send to London at the king's cost
per bonas et fortes carectas et sub salvo et securo conductu the king's tents and those
of the justices (*Lit. Claus.* i. 605); the sheriff of Cumberland sought allowance

to York, in 26 Edward I., occasioned orders to various sheriffs for the safe conveying of rolls and writs as well as treasure.[271] In 1304, when the exchequer was to be returned to Westminster, the sheriff of Yorkshire was ordered to cause its rolls, tallies and memoranda, as well as the rolls of the bench, to be carried thither according to directions given by the treasurer.[272]

In conclusion, it may be observed that the sheriff's work as collector of the King's revenue was a vital concern to thirteenth-century monarchy, which supported a strong exchequer to supervise and enforce the completion of these labours. This official, in addition to his usual activity as collector year by year, was an aid in the collection, and sometimes a collector, of extraordinary impositions laid upon the realm by the king and the great council. Furthermore, until the period of King Edward's wars he was one of the chief disbursing agents of the crown, and the powers of prisage and purveyance which he exercised were of great importance to administration as well as to popular liberty. Finally, his place in providing carriage and directing transport of supplies and materials, both in time of peace and in time of war, made him an official who was indispensable.

(*Exchequer, K.R. Accounts*, 554/10) for taking from Carlisle to Newcastle by writ of privy seal 16 wagons carrying the king's tents and for 10 *valetti* guarding them.

[271] Madox, *Exchequer*, ii. 9. Directions were given sheriffs each to be at a specified place at a set time (*cum sufficiente posse*) to carry out the order and to follow a certain route (*L.T.R. Mem. Roll* 69, m. 79).

[272] *C.P.R.* 1301-7, pp. 190-1.

X

THE REWARDS AND ABUSES OF OFFICE ABOUT 1300

AN inquiry concerning the profit which accrued to the sheriff from the revenues he collected may well form the conclusion to a study of his work. This question of the material advantage which the sheriff derived from his office is not an easy one. Although much is said of his duties, one finds but little concerning his rewards, and this largely by process of inference.

Some sheriffs because of faithful service received special gifts to mark the king's favour.[1] But for every one of these, it seems clear that there was one who underwent imprisonment, at least for a brief period, and several who incurred the king's displeasure to the extent of a fairly heavy amercement for trespass upon the crown,[2] for the escape of a prisoner in custody, for technical error in receiving inquests in criminal matters, for remissness at the exchequer, or for delay in executing some peremptory order. The ordinary routine of the office was often burdensome as well as exacting. The justices in eyre required all sheriffs since the last eyre, sometimes a period of twenty years, to answer concerning the pleas of the crown for their respective terms. Furthermore, the possibility of dying a king's debtor, of the consequent seizure of all one's goods and chattels, of the clearing up of the accounts of the office only through strenuous exertion on the part of heirs, executors,[3] or official aids,[4]

---

[1] A cask of wine or two bucks *de bono regis* (*Cl. Rolls*, 1234–7, pp. 78, 98) ; an annual grant of a hundred marks (Madox, *Exchequer*, ii. 207).

[2] *C.P.R.* 1247–58, pp. 111, 120, 125, 128 ; Close Roll 65, m. 30 d. Thomas de Divelston, when sheriff of Northumberland, was amerced by the justices of the bench several times, thus incurring an indebtedness of £50 (*L.T.R. Mem. Roll* 61, m. 2 d).

[3] *C.P.R.* 1247–58, p. 481. As to the plight of a minor son of a sheriff, *C. Cl. R.* 1272–9, p. 547.

[4] Above, p. 257. At Michaelmas 1279 John de Acton, husband of one of the daughters of John de Ancre, former sheriff of Somerset and Dorset, along with

275

was not pleasant to contemplate. No wonder that some men sought of the king formal lifelong exemption from the obligation to hold the office.

Some hope of gain must have balanced this risk and inconvenience. At intervals throughout the reigns of Henry III. and Edward I. the assignment of a shrievalty to a son or a brother of the king, and in one case to a noble lady as part of her dower,[5] is a sufficient reminder that the office was regarded as a source of income. Even the reformers of 1258, who wished to make an end of the sheriff's exactions, were compelled to acknowledge, at least tacitly, that the conduct of his office involved an outlay. Their attempt to abolish the system of *ferms* called for reasonable allowance to the sheriff, to meet the expense of keeping his bailiwick and rewarding his bailiffs.[6]

But the sources of a regular income from the office are not obvious. The fiscal demands made upon the sheriff of John's time probably did much to reduce the profit of the *ferm* of the shire. More than once when the sheriffs developed a profitable financial device the exchequer had appropriated the proceeds. Such was the fate of sheriff's aid. Moreover, the amount from the *proficua*, added from John's time to the annual *ferm* paid by the sheriff, must have absorbed much of his income from miscellaneous sources. Apparently a certain income from the chattels of thieves tried before the sheriff [7] had gone the same way. The extension of the jurisdiction of the king's courts naturally brought this more and more to the crown.[8] The words of Britton [9] imply that a sheriff might not levy money of the chattels of felons except upon writs of the Green Wax through estreats of the exchequer. There remained some exactions of a questionable nature, which in the latter half of the thirteenth century were being placed under the ban. About these centres almost the whole story of the abuse of the sheriff's

---

the other daughter and Hugo de Fif, former under-sheriff and *receptor* as well as executor of the estate of the deceased, rendered account for the last quarter of the fiscal year 1254–55 (*L.T.R. Mem. Roll* 52, m. 3).

    [5] Above, pp. 181, 182.
    [6] Above, p. 170.
    [7] Above, p. 114. The *misericordia*, arising from pleas concluded in the county court, which, according to Glanville (ix. cap. 10), was due the sheriff, increased his income only when it increased the amount collected as *ferm* of the shire.
    [8] The Assize of Clarendon claims for the crown the chattels of felons tried upon jury indictment.
    [9] Nichols ed., i. 88.

office. Of the many irregularities recorded in the Hundred Rolls
a very few were due to malice, some to carelessness and inefficiency,
but nearly all were associated with schemes which brought in money.
Yet the petty and tyrannical demands of the shrievalty, which were
inherent in the methods of Angevin monarchy and which constitute
the chief blot on the shrievalty of that period, were gradually
ceasing to be regarded with indulgence.

Of these older exactions some are occasionally mentioned. The
payment of a hundred shillings, begun by the men of Nottingham
during the darkest days of John's reign to secure the good will of
the sheriff and induce him not to invade their liberty, continued to
be made as late as 1260.[10] The men of Newcastle-on-Tyne accused
the sheriff of exacting a much heavier payment for his good will.[11]
The exaction of scotale was a grievance which even John's
government had felt it necessary to suppress. It was regarded as
unwarrantable practically from the accession of Henry III. The
king's council in 1220 ordered all sheriffs to prevent the observance
of scotale and to make proclamation that henceforth no one make
either greater or lesser scotale.[12] The ecclesiastical authorities in
this period sometimes intervened to prevent breaches of sobriety
and good order by impleading in the court Christian those who
attended scotales.[13] A writ addressed in 1234 to the notorious
Engelard de Cigogné, as sheriff, informs him that the king has
conceded by his charter to the men of the realm that no bailiff of his
make scotale, and orders that none be made *per vos vel ballivos
nostros* nor permitted in any royal manor of his bailiwick.[14]

The sheriff's arbitrary and unjust use of his usual fiscal powers
gave rise to further exaction. He required high *ferms* of bailiffs,
thus forcing them under various pretexts to extort sums from their
communities. From Bracton's time on sheriffs were under the
imputation of taking money twice for the same amercement, of dis-
training several persons of the same name to collect when only one
amercement had been levied, of compelling persons to pay or do

[10] *Inq. Misc.* i. 90.
[11] Ten marks *ne eos per eandem inquisitionem gravaret nec . . . quadam
contumelia que facta fuit inter eos burgenses et homines ipsius vicecomitis* (*Rot. Hund.*
ii. 19).
[12] *Lit. Claus.* i. 436. A charter of Richard I. terms it the king's scotale
(*Charter Rolls*, i. 382).
[13] *Somerset Pleas*, Somerset Record Soc., 63 (1225).
[14] *Cl. Rolls*, 1231–34, p. 518. Lesser bailiffs seem to have exacted it from
suitors of the hundred as late as 1254 (*Ann. Monast.* i. 332).

more than was required by summons of the exchequer.[15]   Justices
were also instructed to inquire concerning sheriffs who took money
for watches appointed but not observed.[16]   In the Hundred Rolls
there is also complaint that the sheriff makes a local levy, such as
pontage, and then fails to apply it to its proper object.[17]   The monks
of Barnwell claimed that sheriffs made wrongful distraint to collect
sheriff's aid and wardpenny.[18]   One sheriff especially stood accused
of overtaxing both men and vills in levying the twentieth of 1270.[19]
Sheriffs were also charged with abusing the right of prisage in the
king's name,[20] and with obtaining allocation at the exchequer for
purchases made when the purchase price had not been tendered.[21]
So also in the matter of debts in summons which the sheriff collected,
and for which he even issued a receipt tally, but which he neglected
to discharge at the exchequer.   According to Britton, the justices
in eyre were expected to have inquest made concerning forbidden
demands of sheriffs for hospitality, their borrowing horses and carts
or money, and their begging timber and wood.[22]   Here there
is a question of the observance of statute law [23] as well as the Great
Charter.   Finally, some sheriffs were guilty of accepting money to
remit [24] or respite [25] distraint of knighthood or debts of the king,[26]
a conspiracy against the rights of the crown rather than the liberties
of the realm, the detection of which could not fail to bring
punishment.

Sharp practice in the exercise of his judicial powers for much
of the century operated to the sheriff's profit.   Sheriffs or their
agents are accused of making wrongful distraint to enforce suit of
shire and hundred, of amercing vills on questionable pretexts,[27] of
mulcting for default those summoned to inquests when a sufficient
number came,[28] even of taking money *ad opus vicecomitis* for failure

[15] Bracton, ii. 250; Britton, i. 89; *Rot. Hund.* ii. 306; *Feudal Aids*, i. 87.
[16] Bracton, ii. 252; Cam, *Studies*, 94-95.
[17] Above, p. 219.
[18] *Memoranda de Barnewell*, ed. J. W. Clarke, 238.
[19] Cam, *Studies*, 172.   [20] *Ibid.* 174-75.
[21] Above, p. 267.   [22] Britton, Nichols ed., i. 93.
[23] Violation of the statute (below, p. 282) regulating the sheriff's right to
entertainment is stated in 1313 to be punishable by fine and imprisonment (*Eyre
of Kent*, Selden Soc., i. 41).
[24] *Ann. Monast.* i. 331; *Rot. Hund.* i. 3; *Chartulary St. Peter of Gloucester*,
Rolls Ser., ii. 278.
[25] *Rot. Hund.* i. 3, 69.   [26] Cam, *Studies*, 169.
[27] *Rot. Hund.* ii. 306.
[28] *Foedera* (1816), i. part ii. 517: cf. *Ann. Monast.* i. 331; Bracton, ii. 251.

to come to jail delivery when there was no one in jail.[29]   There is also accusation of downright acceptance of bribes.[30]   This and other evils were being attacked on the eve of the barons' war.   The justices were instructed in the new articles of the eyre, first used in 1254, to inquire concerning sheriffs and other ambidextrous bailiffs who took from both parties.[31]   The king's council is said to have declared, in 1257, that justices and bailiffs who received gifts should be removed from the service of the king,[32] leaving it to the council to make provision concerning sheriffs.   The articles of the barons in 1258 required the latter to do justice fairly between parties, and restricted their rights to receive presents to food and drink for the day and presents not exceeding twelve pence in value.   This rule was apparently enforced in the time of Edward I., for it appears in effect in the oaths of councillors and judges.[33]   The judges were also charged to inquire in and after 1254 whether sheriffs were maintaining any cause in such a way as to stifle justice.[34]

Other types of judicial irregularity mentioned are the holding of tourns by sheriffs oftener than twice a year [35] and their levy of fines for beaupleader,[36] otherwise amendment of pleas, an imposition forbidden by the great council in 1259 and again in 1266, but obviously continued to a later time.[37]   There was also much exaction and trickery in summoning jurors.   The bailiffs were certainly the worst offenders in this respect,[38] but sheriffs also took gifts to remove recognitors from juries and assizes.[39]   The practice required amend-

---

[29] *Rot. Hund.* ii. 306.

[30] *Rot. Hund.* i. 53, 441 : taking money from one indicted at the tourn and doing nothing.   There were apparently many such cases in 1274 (*Foedera* (1816), i. part ii. 517 ; *Rot. Hund.* i. introd. 14.

[31] *Ann. Monast.* i. 332.   See Cam, *Studies*, pp. 92-94, for the new articles of the eyre dating from this period.

[32] *Ann. Monast.* i. 396.

[33] Above, p. 171.   Stubbs, *Const. Hist.* ii. 271.

[34] Cam, *Studies*, 94 (i. 43).

[35] *Foedera* (1816), i. part ii. 517 ; *Rot. Hund.* i. introd. 14.   The under-sheriff of Gloucester was accused (*Rot. Hund.* i. 167) of holding his four times a year.

[36] *Rot. Hund.* i. 4.

[37] Stubbs, *Sel. Charters*, 402 ; *L.T.R. Mem. Roll* 43, m. 5.   It subsequently appears as a customary imposition, as in *Feudal Aids*, i. 88.

[38] *Ann. Monast.* i. 331.   In 9 Henry III. an assize could not be held because the bailiffs of the hundred had for reward removed all the knights first put on (*Salt. Archaeol. Soc.* iv. 37).   The farmers of wapentakes and their *subbedelli* are again accused in the *Hundred Rolls* (i. 4, 436 ; ii. 305).   Cf. Cam, *Studies*, 160-61.

[39] See below, note 42.   Miss Cam's conclusion that the empanelling of juries was left to underlings (*Studies*, p. 161) must not be pressed too far.   See Tout and Johnstone, *State Trials*, R.H.S., p. 19.

U

ment because of the unsatisfactory quality of those retained.[40]   A sheriff is even accused of causing inquisition to be taken by the friends of one party to the detriment of the other.[41]   The Statute of Westminster [42] complains that sheriffs and bailiffs had for purposes of extortion summoned unwarrantable numbers of jurors, including persons who could not possibly serve, with the result that the poor served and the rich remained at home by reason of bribes.   The result was the establishment of a property qualification for jurors and a limitation upon the numbers to be summoned to any one assize.

Sheriffs exacted money from prisoners in their charge, and sometimes from persons under indictment or accusation [43] on condition that they should not arrest them.   Arrests, moreover, were often associated with the taking of fines for replevin, the practice finally requiring a very sharp delimitation between offenders who were and those who were not repleviable.[44]   Sheriffs and their agents knew well how to exploit the desire of the accused to escape imprisonment.[45]   These officials were aided by the fact that until the time of Edward I. nearly all offences were bailable.   In the time of Henry III. they were also taking money for admitting to plevin persons accused of homicide.[46]   In the next reign there was still complaint,[47] and some amercement of sheriffs [48] for such offences.   They were accused also of imprisoning persons *per potestatem officii* until money was given them,[49] even making arrests on feigned indictments or the accusations, real or pretended, of approvers whom they held in

[40] In *Rot. Hund.* ii. 306, a sheriff's clerk is charged with retaining only jurors who are willing, contrary to their oath, to amerce their neighbours *ad voluntatem.*

[41] *C.P.R.* 1272–81, p. 151.

[42] *Stats. of Realm,* i. 89.   It is clearly implied in Britton (Nichols ed. i. 90) that sheriffs and bailiffs are accountable to the justices not only for taking bribes to remove jurors, but also for placing on juries or inquests persons sick, disabled by gout, sick or maimed persons past seventy years of age, or persons not resident in the county.

[43] Cam, *Studies,* 158.   Cf. note 31 and *Feudal Aids,* i. 89.   In 1276 appears a new article of the eyre requiring inquiry concerning sheriffs who have taken money from excommunicates *ne caperentur* (Cam, *Studies,* 94).   There is complaint that a sheriff takes money *pro pace habenda* from a man placed on the exigent by the justices (*Rot. Hund.* i. 155).

[44] Above, p. 233.   The law presupposed the letting of repleviable prisoners to pledge without payment (Bracton, ii. 251).

[45] Thus the *servientes* of a sheriff take of a man 6s. *ut non iret ad prisonam* (*Rot. Hund.* i. 436).

[46] *Ann. Monast.* i. 332.

[47] *Rot. Hund.* i. 363 ; Unpub'd. Hund. Rolls, Tower Series, Hunts, no. 6.   Cf. *C. Cl. R.* 1279–88, p. 109 ; Cam, *Studies,* 156-57, 159.

[48] *Assize Roll* 131, m. 15.

[49] *Rot. Hund.* i. 65.

custody.[50] The Statute of Westminster [51] declares that extortion of this kind has been practised, and forbids arrest except upon indictment of twelve jurors, extending to injured persons remedy by action for false imprisonment. A similar evil is manifest also in certain irregular jail deliveries held by sheriffs. A striking expression regarding these was made by the king's council in 1280, forbidding sheriffs to make delivery of jails without special writ, and declaring that neither they nor their sergeants should take fine for delivering prisoners, since by reason of such fines loyal people were aggrieved and thieves encouraged.[52]

There is every indication that in the thirteenth century sheriffs were establishing a series of charges in the form of fees. The barons strove in the Provisions of Oxford [53] to put an end to all these, with what ill success the numerous complaints in the Hundred Rolls but too well testify. The local exactions known as sheriff's palfrey and sheriff's welcome,[54] which were presumably collected in connexion with the tourn, may have been payments of this kind. Far less justifiable were the payments taken from individuals, to excuse them from suits of shire and hundred,[55] and the sums extorted from men of a vill before the sheriff would take over the prisoner they had in pursuance of duty followed and captured.[56] Sheriffs also demanded sums for estreats and return of writs, the latter service being rightfully claimed [57] against them *sine dono et precio*. It is also charged that the sheriff will not execute the king's writ without money,[58] that he will not deliver the wool of certain merchants *per breve domini regis* except for a redemption,[59] that he will not execute a writ of jail delivery until he receives money from the prisoner.[60] The sheriffs who refused upon writ to deliver oxen held *contra vadium et plegios* [61] were apparently demanding expense money [62]

[50] *Rot. Hund.* i. 17, 50, 94 ; cf. Cam, *Studies*, 340-41.

[51] *Stats. of Realm*, i. 81 (cap. 13).

[52] *C. Cl. R.* 1279–88, p. 109. The article of the eyre inquiring concerning jails delivered without warrant dates from 1227 (Cam, *Studies*, 92-93).

[53] Stubbs, *Sel. Charters*, 391.

[54] *Cartae Antiquae*, 1 (A), P.R.O., no. 19 ; *Placita de Quo Warranto*, 147 ; *Rot. Hund.* i. 157.   [55] *Ann. Monast.* i. 331.   [56] Britton, Nichols ed. i. 46.

[57] By the burgesses of Lincoln (*Rot. Hund.* i. 323). At Bristol the under-sheriff took ten shillings of a chaplain for making return of the king's writ (*ibid.* i. 177).

[58] *Ibid.* i. 15 ; cf. Cam, *Studies*, 184.   [59] *Rot. Hund.* i. 178.

[60] *Ibid.* i. 19. The sheriff takes money of a prisoner confined at Aylesbury jail *pro officio exercendo per praeceptum domini regis* (*ibid.* i. 44).

[61] *Ibid.* i. 3, 67, 71.

[62] That such a charge was made is clear from Britton, i. 89. Owners were

for their keep while in their custody. Some of the payments exacted of prisoners before their release seem to be of similar character.[63] Sums required by sheriffs to remove prisoners from the dungeon, to release them from fetters,[64] or to place them at mainprise, were certainly in the nature of fees. The first Statute of Westminster forbade any sheriff or king's officer to take reward to do his office.[65] But this clearly was not intended to put an end to all fees. A statute of 1285 specifies what the sheriff's fee shall be in an assize case,[66] and Britton admits that a fee of fourpence, though not more, may be taken for keeping a prisoner.[67]

Aside from fees and capricious exactions, usually illegal, the perquisite of entertainment and the *ferm* of the shire remained. The letter of the law still permitted the sheriff to lodge occasionally, with five or six horses, at the houses or manors of religious establishments and private persons.[68] From 1280 the justices in the eyre were charged with the enforcement of this by indictment of those who violated it.[69] The sheriff of Cambridge for years found hospitality at the priory of Barnwell with his wife and a large establishment, and as late as 1287 came for three days and two nights with his family, his household and twenty-two horses, even taking offence because the priory would not lend him a cart to carry a cask of wine to his manor.[70] But this he clearly classed as an extension of voluntary hospitality.

The remaining source of the sheriff's official income was the

supposed to have permission to feed such beasts. The sheriff's servants sometime drove distresses away (*Rot. Hund.* ii. 123) to make these charges inevitable. Cf. the charge for releasing a horse (*ibid.* ii. 306).

[63] As in *ibid.* i. 1, 71 ; cf. i. 50. In this relation it is to be observed that the king in 1257 restored certain lands to William de Insula to support him, his wife and his children, while he was in prison (Close Roll 73, m. 4).

[64] *Pro leni imprisonamento* (Unpub'd Hundred Rolls, Tower Series, Hunts, no. 6) ; to relieve prisoners of chains until jail delivery (*Rot. Hund.* i. 5). Cf. *Rot. Hund.* i. 68. So apparently in *ibid.* i. 67.

[65] *Stats. of Realm*, i. 33. The penalty was twice the amount taken and punishment at the king's pleasure.

[66] One ox not above 5s. in value from the disseiser only, not from the disseisee also ; only one ox to be taken when several disseisors are named in one writ (*ibid.* i. 85, cap. 25).

[67] Nichols ed. i. 46. Cf. Greenwood, *Bouletherion* (1668), p. 217-19, where the fees of the seventeenth-century sheriff are listed in detail. They range from 4d. to 6s. 8d.

[68] *Stats. of Realm*, i. 27-28.   [69] Cam, *Studies*, 98 (iii. 8).

[70] *Memoranda de Barnewell*, ed. J. W. Clark, 180. *Ibid.* 181-82 tells of his violence and of proceedings which sought to bring him to account. In the end he made terms with the monks.

ferm of the shire. The letting of counties and their subdivisions, other than the king's demesne manors, at a *ferm* in excess of the customary one was forbidden by the Magna Charta [71] of 1215, though not by its successors. Although the total *ferm* of the county paid by the sheriff was rarely changed down to 1307, the added *proficua* operated in practice as such an increase. The petition of the barons in 1258 complained of sheriffs who had their counties at so high a *ferm* that they could not collect it,[72] and the attempt to abolish the farming of bailiwicks followed, sheriffs being forbidden to let counties, hundreds or wapentakes at *ferm*.[73] This plan, it has been shown, soon failed. The Hundred Rolls complain that sheriffs are frequently letting hundreds at such high *ferms* that no one can pay them without oppressing the people.[74] It is unnecessary to believe that high *ferms* represent an increase in manorial rents in so far as they were founded upon these. Increase in judicial impositions and miscellaneous exactions was sufficient for the purpose. The complaint of the barons in 1258, that sheriffs amerce men above the measure of their offence,[75] accompanies the claim that sheriffs are paying too much *ferm* for their counties. The bailiff whose payment for his hundred was much increased, naturally resorted to all sorts of exactions of a vexatious nature. Attendance at court sessions, even though fraudulently demanded, was clearly yielding the sheriff a revenue in the time of Edward I. The *exitus* of courts, one must believe, formed a main basis of the sheriff's profit. A century was to elapse after the death of Edward I. before a statute finally forbade the sheriffs to *ferm* any of their bailiwicks.

The *ferm* of the shire, then, with a greater or smaller amount in fees, and possibly some other miscellaneous sources of income, at the end of the thirteenth century still afforded the sheriff financial remuneration for his work. It is well known that in the Tudor or Stuart period the king had the sheriff's services at practically no cost to himself.[76] The statement holds true for any earlier age. The

[71] Cap. 25. An inquiry concerning this is made at the eyre from 1280 on (Cam, *Studies*, 96).   [72] *Ann. Monast.* i. 441.

[73] Above, p. 171. As early as 1250 the king had forbidden the letting of hundreds and wapentakes at rack-rents (Madox, *Exchequer*, ii. 102).

[74] *Rot. Hund.* i. 251, 281 (Lincoln); ii. 27 (Nottingham); i. 40-1 (Bucks); i. 153 (Essex); i. 115 (Yorkshire). A clerk who farmed the Lincolnshire wapentake of Skirbeck for 8 marks 10s. let it at once to an under-sheriff for 12 marks, to the loss of the *patria* (*ibid.* i. 348).

[75] *Ann. Monast.* i. 441.

[76] Cheyney, *European Background of American History*, 264-65.

burden was in various ways shifted to the men of the vills and hundreds. But the tyrannical exactions of the past were being vigorously attacked, even before the barons' war. In the next reign the sheriff's exactions of too high *ferms* from the bailiffs of his hundreds was regarded as questionable if not illegal, a fact which brought it under investigation and made it a matter of record in the Hundred Rolls. It has, indeed, been regarded as doubtful whether the inquiries into the conduct of local officials, the results of which appear in these records, represent a practicable ideal of government more than an attempt of a new reign to gain popularity.[77]   However that may be, the concession of Edward I. that the men of the county might elect their own sheriff is a striking proof of the king's desire at the end of the century for just administration.

Furthermore, before that time numerous statutory enactments for the protection of popular rights speak for themselves. Surely it was not in vain that the sheriff was required to have a separate record at the exchequer of each separate debt collected on summons ; that he was made punishable for failure to acquit at the exchequer the king's debtor from whom he had collected ; that his right to official entertainment and his right of prisage were limited ; that he was made liable for proper service of writs ; that his power of arrest for felony was confined to cases wherein indictment was made ; that the number and quality of jurors whom he summoned, as well as the offences for which he might replevy offenders, were carefully defined ; that he was forbidden to take reward to do his office ; that the exaction of attendance at the tourn was confined within strict limits ; and that fines for beaupleaders were abolished.

The itinerant justices of the period as well as the barons of exchequer and the judges of the other central courts were perfectly willing to punish unjust exaction or other forbidden practices. Britton, a decade before the end of the century, certainly understood that the sheriff was punishable for many of them. Statutory provisions laid specific penalties on sheriffs who did not acquit debts collected, and who admitted to bail persons who were not repleviable. The itinerant justices were even ready to decide that the customary number of bailiffs in a hundred should not be increased. Before the reign of Edward I. there were some ten points affecting the conduct of the sheriff concerning which these justices were regularly required to have presentments made, and nearly as many

[77] Cam, *Studies*, 192.

were added in this reign.[78] In the latter category was one concerning sheriffs who let hundreds to extortioners. Furthermore, from the year 1278 a person might make complaint by bill before the justices and demand satisfaction. Complaint against a sheriff in this form is recorded.[79] The eyre, to be sure, was seldom held, and much injustice doubtless continued. Complaints concerning sheriffs may be found even as late as the Tudor period. But, according to its light, the age was progressing. Certainly not every sheriff was guilty of the excesses enumerated in the Hundred Rolls. Indeed the worst extortion mentioned is that of the bailiffs, not that of the sheriffs. The half-barbarous fiscal devices employed by sheriffs of the Norman and Angevin periods were being assailed with unwonted .energy, although unremunerated service still had its temptations, and present-day standards of efficiency and uprightness in local government were still far in the future when King Edward passed away.

[78] Cam, *Studies*, pp. 92-101.   There were also special inquests like that of 1276 and another of 1284–86 (cf. *Feudal Aids*).

[79] Cam, *Studies*, 136-37 ; *Select Bills in Eyre*, Selden Soc., no. 50.

INDEX

Abjuration of the realm, 44, n. 27, 150
Account, the sheriff's, 147, 249-255, 256-257 ; within the year, 170
Accounting, at Exchequer, 95, 97-98, 125, 132-135, 161
Aid, 148 ; feudal, 261
Aid, sheriff's, 99, 114, 147, 157, 246, 276
Aiulf, the chamberlain, 47, n. 48, 76, 80
Alderman, the Saxon, 2-3, 5, 6, 10, 16, 21, 24 ; his *gingra* or deputy, 5, 21
Alfred, and his judicial officials, 9
Amercement, of county, 120, 155 ; of sheriff, 120, 212, 275, 280 ; of hundred, 122-123 ; in county court, 199
Appointment, of reeve, 21 ; of sheriff, 23, 38, 176, 177-178; by Henry II., 111; by Richard I., 135, 136-137, 138 ; by John, 144-145 ; by barons, 172, 177 ; by Earl Richard, 173 ; by letters patent, 173 ; by Henry III., 174-175 ; in Ireland, 175
Approver, 150, 223, 280
Archers, 237
Arrest, 28, 60, 150, 158, 223, 224, 225, 227, 230, 256, 280
Assisa, an aid, 133
Assize of cloth, 216
Assumption of Shrievalty, compulsory, 178
Attorney, sheriff's, 125, n. 125, 253
Aumale, sheriffs of count of, 108

Bail, 232-233. See also Mainprise
Bailiffs, of commotes, 176 ; of hundreds, see Hundred ; of lathes, 188-189, 244 ; of ridings, 189, 244 ; seignorial, 146, 211, 252 ; of sheriff, 24, 54, 88, 115, 158, 171, 188, 191, 209, 210, 221, 222, 223, 224 - 225, 228, 279 ; of towns, 146
Baldwin of Exeter, 47, n. 47, 48, n. 49, 51, 59
Banishment, 150

Barons, as sheriffs, 41, 47-53, 75-77, 79-80
Basset, Ralph, 80-81, 87, 101 ; Richard, 86-87
Beacons, 237
Beadle, 9, 190, 191
Beauchamp, William de, Earl of Warwick and hereditary sheriff, 181 ; Guy de, 181
Beaupleader, 198, 279
Belmeis, Richard de, 77-78
Bigod, Hugh, 112 ; Roger, 47, n. 47, 52, n. 80, 76, 85 ; William, 79
Bishop, supervision of reeve, 11 ; of church dues, 15
Bishops, as sheriffs, 112, 167, 169
Breauté, Fawkes de, 162, 168
Brewer, William, 145, 161, 165
Bribery, 11, 15, 157, 160, 279, 280
Bridges, repair of, 219
Burgh, Hubert de, 160, n. 170, 161, 168, 169, 180
Burghal district, 6, 7, 10, 21

Carlisle, conquest of, 45
Carriage, 29, 157, 158-159, 265, 270-273
Carucage, 133, 139, 260
Castellan, 152, 159, 223, 229, 231 ; sheriff as, 117, 151, 174
Castles, royal, 51, 59 and n. 148, 60, 117, 126, 151, 174, 186-187
Castle guard, 159
Chattels, of criminals, 114, 131, 150, 276
Clerks, the sheriff's, 88, 115, 146, 189, 209, 223
Coinage, 118, 153, 215
Collector, sheriff as, 148, 257, 259, 261
Commerce, regulation of, 154, 215-217
Commission, the sheriff's, 170, 176
Conservator of the peace. See *Custos pacis.*
Constable, of castle, see Castellan ; of hundred, 228

Control of sheriff, by Henry I., 103; by Henry II., 111-113; by Richard I., 140. See Exchequer.

Cornwall, Richard, Earl of, as sheriff, 181; Edmund, his son, 181-182; Margaret, wife of Edmund, as sheriff, 182

Coroners, 209, 238-239, 256, 258; election of, 200-201; oath of office, 199

Council, the, 168, 213, 232, 241, n. 2, 267

Courts (for earlier period see *Gemot*), of shire, 88, 90-91, n. 143, 93, 119-120, 121, 122-123, 135, 150, 155, 156-157, 171, 175, 187, 192-202, 205, 207, 208, 218, 221, 223, 238, 278, 281; of hundred, 56, 74, 88, 89, 90, n. 136, 92, 175, 188, 194, 195, 202, 278, 281

Crementum, of county, 64, 69, 128, 144, 243

Curials, as sheriffs, 47-49, 52, 73, 77-80, 83, 84-87, 113, 114, 169

Custodianship, sheriff's, 62-63, 185-186, 262-264; of lands, 29, 61, 62, 63, 65-66, 69, 149, 158; of sea coast, 172, 186; of jewry, 186; of stannary, 186

Custody, of prisoners, 150-151, 229-232. See Prisoners

Customs, 262

Custos, of county, 86, 129, 130, 144, 145, 155, 171, 173, 177, 185, 243, 244; *custos pacis*, 174, 175, 200, 221, 228, 235

Danegeld, 23, 31, 32-33, 53, 66, 89, 96, 133

De Vere, Aubrey, 81, 85, 86-87, 100

Disbursements by sheriff, 30, 96, 125-126, 133, 149, 254, 265-270, 273

Distraint, 88, 134, 148, 208, 217, 219, 257, 260, 271, 277, 278

Duel, trial by, 119, 150, 196

Earl, Anglo-Saxon, 25, 38; relations with sheriff, 24, 25, 37, 38, 44, 45

East Riding of Yorkshire, sheriff of, 185

Edward of Salisbury, 47, notes and 47, 48, note, 51, 53, 76

Election, of abbot, 153; of conservator of the peace, 200; of coroner, 199; of justice, 103; of knights, 201; of reguarders, 200; of sheriff, 17, 100, 170, 175, 182-183, 184-185; of verderers, 200

Encroachments, 220-221

Engelard de Cigogné, 160, 163, 168

Entertainment, of king's guests, 68, 264; of sheriff, 157, 171, 189, 278, 282

Escheators, county, 199, 238

Escheats, 131

Esgar the staller, 27, 35, 38, n. 166, 42

Essoins, 208

Estreats, 258, 281

Exactor, 3, 4

Exchequer, 94-95, 105, 124, 132, 147, 148, 175, 177-178, 184, 233, 235, 253, 256, 257, 267, 270, 272-273; control over sheriff, 241-242, 269

Excommunicates, 118, 224

Excommunication, of sheriff, 119, 158

Execution, of criminals, 116, 233-234

Expense money, military, 237

Fairs, 154, 216

Ferm, of boroughs, 65, 95, 129-130; of hundreds, 155, 188, 244; of lathes, 244; sheriff's *ferm* (of county), 28, 32, 55-56, 63-66, 95, 99, 105, 107-108, 125, 127-129, 133-134, 155, 172, 243-246, 253, 276, 282-283; of ridings, 155, 189, 244

Feorm, the king's, 13, 29-30, 33

Foreigners, as sheriffs, 159-161, 165, 183

Forisfactura, 30, 31, 32, 62, 66, 91, 97, 100

Fortieth, 148

Franchises, 146, 221, 233, 249, 252, 277

Frankpledge, 26, 69, 205; view of, 56-57, 74, 89, 119, 146, 156, 182, n. 103, 203, 245

Fugitives, 123, 197

Galleys, 152, 266

Gallows, 234

Gemot, of five boroughs, 10; of hundred, 24, 25-26, 56; of king's reeve, 5-6, 7, 11; of shire, 5-6, 8, 14-15, 19, 24-25, 57, 61

Geoffrey de Clinton, 81, 85, 86, n. 100, 104, n. 252; the younger, 108

Geoffrey de Mandeville, 48, n. 50, 51, 59, n. 148, 107

Gerard de Atheé, 160, 165, n. 200

Gersoma, 83, 84, 88, 99

Gilbert, sheriff of Henry I., 78, 80, 82, 100

Glamorgan, early sheriffs in, 108

Godric, sheriff of Berkshire, 27, 36, 42

Grith, 10, 21

Grooms, of sheriff, 189

Haimo, the *dapifer*, 46, n. 47, 51, 76
Hidage, 260
High-reeve, 5, 8
Hostages, 151
Household officials, as sheriffs, 35, 38, 47-48, 113
Hue and cry, 222, 224, 225, 228
Hugh of Buckland, 48, n. 49, 52, 56, 77, 78-79, 89, 91, n. 152, 104, n. 251
Hugh of Leicester, 78, 81
Hundred, bailiffs, 169, 188, 189, 191, 192, 200, 210, 244, 258; court of, see Court; exempt hundreds, 55; functions of, 20; reeve of, 22, 25, 26, 146
Hundredman, 33, 54

Imprisonment, false, 230
Indenture, between sheriffs, 187-188
Inquests, before sheriff, 58, 90, 121, 203, 204, 212, 216, 219, 261, 278
Inquest of Sheriffs, 113, 128, 132
Inward, 33, 59, 67

Jail delivery, 209, 210, 279, 281
Jails, 115-116, 151, 186-187, 196, 220, 230
Jews, 154, 214-215, 216, 218, 262; massacre of, 136
John, sheriffs of, 161-165, 168, 170
Judicial functions of sheriff, 24-25, 54-58, 89-90, 93-94, 119-120, 192-199
Jurors, 122, 139, 189, 210, 279-280, 281
Jury, 208, 210; of recognition, 58, 121; of presentment, 118, 122, 139, 200
Justice, the, 71-72, 97, 101-102, 104, 139, 150, 208, 261; of assize, 210; of the forest, 209; of jail delivery, 210, 233; local, 56, 57, 72, 102-103, 104, 107; sheriff as, 55, 58, 93-94, 114, 120-121, 137, 138, 141, 159; subordination of sheriff to, 71, 72, 100-102, 103, 126, 209

Knights, 154, 210; assigned, 116, 238; in county court, 193; of the shire, 172, 201-202, 235-236, 237; as sheriffs, 167, 183
Knight service, 117-118, 151-152, 153, 235

Lancaster, hereditary sheriff of, 182; Edmund, Earl of, 182; Thomas, earl of, 182

Landholding by sheriffs, 35-36, 47-49, note, 49, 167
London, sheriffs of, 48, n. 50, 77, n. 17, 79, n. 34, 81-82, 100, 107, 111, 113-114, 130, 155, 182
Longchamp, William, 136

Mainprise, 150, 232, 233
Mantell, Matthew, 145
Mark, Philip, 160, 163, 168
Markets, 153, 216
Manucaptor, 255, 257
Manumission, 120
Merchants, 154, 216, 217
Mangonels, 153
Miles of Gloucester, 83, 104, 107, 112
Military duties of sheriff, 27, 58-60, 117, 151-152, 234-237
Mill, the king's, 264
Ministerial functions of sheriff, 91, 121-122, 210-213
Murage, 219, 220, n. 113
Murdrum, 56, n. 110, 66, 69, 97, 132

Norfolk and Suffolk, rival sheriffs in, 161
Northumberland, tourn in, 203
Nuisances, 221

Oath, of fealty, 223; the sheriff's, 134, 171, 172, 176-177, 181, 184, 222, 255; tenants in chief's to sheriff, 135
Officials of state, as sheriffs, 163. See Household officials
Oppression, by king's reeve, 11, 14, 15; by sheriff, 36, 68, 69, 70, 98-100, 104, 107, 113, 139, 154, 155, 158, 160, 277-281, 284-285
Osbert, sheriff of Henry I., 77, 79, 89
Osward, sheriff of Kent, 36, 42
Outlawry, 57, 116, 119, 150, 196

Palatinates, sheriffs in, 44-45, 55, 109, n. 292, 137, 179, 181
Pasturage, 212
Pavage, 220
Peace, the king's, 28, 60, 90, n. 136; breach of, 97; the sheriff's, 119, 149; sheriff's guardianship of, 115-116, 222-229; his functions in relation to, 122-123, 149-150, 221-229
Perquisites, the sheriff's, 24, 53, 68-69, 99, 114, 204, 276-283
Peter of Rievaux, 169
Petrariae, 153

Pleas, of the crown, 30, 55, 101, 150, 159

Plevin, 156. See Mainprise

Prisage, 262, 266, 278

Prises, 157

→ Police powers, of hundred, 20; of reeve, 9-10; of sheriff, 28, 60, 115-116, 149-150; 221-222

Pontage, 219, 278

Posse, 222, 226-227, 273, n. 271

Priests, as sheriffs, 22, n. 39, 113

Prisoners, 100, 116-117, 150, 151, 156, 196, 227, 231, 233, 266, 282

Proclamation, 153, 199, 209, 210, 217-219, 225, 236

Proficuum, of the county, 147, 155, 192, 198, 243-244, 246, 276, 283

Proffer, the sheriff's, 147, 247-249, 250

Provisions of Oxford, 169, 170

Public works, 219-220

Purveyance, by sheriff, 149, 265, 266-269, 270, 278

Rapes, sheriffs in, 108 .

Record, in county court, 121, 147, 195, 198, 202; in hundred court, 202

Reeve, borough, 153; and four of vill, 154; as subordinate of sheriff, 26, 33, 53, 54, 56, 146

Relief, collection of, 67

Removal, of king's reeves, 21; of sheriff, 46, 177, 178, 183

Repairs, through sheriff, 149, 269-270

Resistance, to sheriff, 149, 221

Retrocomitatus, 192-193

Richard I., suspicion of sheriffs, 139

Richard, the Fleming, 145

Robert fitz Walter, sheriff, 80, n. 37, 81, 86, 87

Robert fitz Wymarc, 35, 36, n. 155, 38, n. 166, 43.

Roll, coroner's, 150, 195, 238; sheriff's, 150, 187, 188

Safe conduct, 223

St. Andrews, bishop of, 232

Schirreue, twelfth-century title of sheriff, 113

Scirman, 3, 19, 22, 25

Scotale, 157, 158, 277

Scutage, 133, 260-261

Sea coal, 217

Seal, the sheriff's, 115, 195, 268

Sergeants, 117, 151, 152, 161, n. 172, 190, 196, 223, 236

Services, rendered to sheriff, 33-34, 67

Sheriffscot, 246

Shire, origin, 1-2; *scir* of *Judicia Curtatis Lundonie*, 7-8, 9, 18

Shropshire and Staffordshire, election of sheriff, 185

Steward, lord's, 135, 147

Students, 224, 225

Summa, at exchequer, 253-254, 255

Summoners, 120, 121, 208

Superplusagium, 86, 126, 255

Tallage, 114, 132, 148, 261-262

Tallies, exchequer, 124, 254, 258-259; sheriff's, 258-259, 267, 268, 269-270, 278

Taxation, sheriff and, 260

Tenserie, 139, 152

Tenure of sheriffs, 23, 46, 144; hereditary sheriffs, 50-51, 52, 76-77, 84, 145, 169, 179-182; proposal for annual term, 170

Thirteenth, 148

Third penny, of borough, 25, n. 64, 45, 63; of county, 31, 65, n. 204, 125, 265

Tourn, the sheriff's, 119, 122, 150, 156, 175, 182, n. 103, 192, 195, 203, 221, 246, 279

Tournaments, 225, 226

Transport, 152. See Carriage

Treasurer, 273; and appointment of sheriffs, 178; the sheriff's, 189

Tun, the king's, 3, 4, 5, 6, 7, 17, 19

Tungerefa, 4, 5, 8

Ulcote, Philip de, 161, 168

Under-sheriff, 53, 88, 115, 144, 146, 147, 160, 178, 180, 189, 202, 207, 210, 231, 249

Urse d'Abetot, 47, n. 47, 48, n. 49, 50, 51, n. 63, 52, n. 80, 53, 58, 59, 65, n. 212, 69, n. 248, 70, 71, n. 260, 76, 79

Vicomte, Norman, 41, 42, 49, 53, 57, 59

View of account, 250

Vills, four neighbouring, 203, 204

Vipont, Robert de, 179; the younger, 180; his daughters, 180

Wales, sheriffs in, 175-176

Walter of Coutances, 137

Wapentake, see Hundred

Watch and ward, 150, 228, 278
Westmoreland, bailiffs in, 190, 196; hereditary sheriff of, 179-180; tourn in, 203
William of Eynesford, sheriff of Henry I., 79, 81, 86
William of Pont de l'Arche, 78, 80, 85, 86, n. 100

Wines, 216
Wool, 216, 262, 268, 281
Worcestershire, hereditary sheriff of, 180-181
Writs, of course, 121; judicial, 121, 146, 207, 210, 212-213; return of, 211-212; *replegiare*, 232; as controlling sheriff's office, 207, 239.

Date Due